WORLD HEALTH ORGANIZATION

INTERNATIONAL AGENCY FOR RESEARCH ON CANCER

IARC MONOGRAPHS
ON THE
EVALUATION OF THE CARCINOGENIC RISK OF CHEMICALS TO HUMANS

Some Aromatic Amines, Anthraquinones and Nitroso Compounds, and Inorganic Fluorides Used in Drinking-water and Dental Preparations

VOLUME 27

This publication represents the views and expert opinions
of an IARC Working Group on the
Evaluation of the Carcinogenic Risk of Chemicals to Humans
which met in Lyon,
10-17 February 1981

April 1982

INTERNATIONAL AGENCY FOR RESEARCH ON CANCER

IARC MONOGRAPHS

In 1971, the International Agency for Research on Cancer (IARC) initiated a programme on the evaluation of the carcinogenic risk of chemicals to humans, involving the production of critically evaluated monographs on individual chemicals. In 1980, the programme was expanded to include the evaluation of the carcinogenic risk associated with employment in specific occupations.

The objective of the programme is to elaborate and publish in the form of monographs critical reviews of data on carcinogenicity for chemicals and complex mixtures to which humans are known to be exposed, and on specific occupational exposures, to evaluate those data in terms of human risk with the help of international working groups of experts in chemical carcinogenesis and related fields, and to indicate where additional research efforts are needed.

International Agency for Research on Cancer 1982

ISBN 92 8 321227 4

PRINTED IN SWITZERLAND

CONTENTS

LIST OF PARTICIPANTS 5

NOTE TO THE READER 9

PREAMBLE 11
    Background 11
    Objective and Scope 11
    Selection of Chemicals for Monographs 12
    Working Procedures 13
    Data for Evaluations 13
    The Working Group 13
    General Principles for Evaluating the Carcinogenic Risk of Chemicals 14
    Explanatory Notes on the Monograph Contents 19

GENERAL REMARKS ON THE SUBSTANCES CONSIDERED 31

THE MONOGRAPHS 35

    Aromatic amines
    Aniline and aniline hydrochloride 39
    ortho- and *para*-Anisidine and their hydrochlorides 63
    4-Chloro-*ortho*- and 4-chloro-*meta*-phenylenediamine 81
    meta- and *para*-Cresidine 91
    2,4-Diaminoanisole and 2,4-diaminoanisole sulphate 103
    4,4'-Methylenebis(N,N-dimethyl)benzenamine 119
    1,5-Naphthalenediamine 127
    5-Nitro-*ortho*-anisidine 133
    2,2',5,5'-Tetrachlorobenzidine 141
    4,4'-Thiodianiline 147
    ortho-Toluidine and *ortho*-toluidine hydrochloride 155
    2,4,5- and 2,4,6-Trimethylaniline and their hydrochlorides 177

    Anthraquinones
    2-Aminoanthraquinone 191
    1-Amino-2-methylanthraquinone 199
    2-Methyl-1-nitroanthraquinone 205

    Nitroso compounds
    N-Nitrosodiphenylamine 213
    para-Nitrosodiphenylamine 227

    Inorganic fluorides used in drinking-water and dental preparations . . . 237

APPENDIX 1: EPIDEMIOLOGICAL EVIDENCE RELATING TO THE POSSIBLE CARCINOGENIC EFFECTS OF HAIR DYES IN HAIRDRESSERS AND USERS OF HAIR DYES . . 307

SUPPLEMENTARY CORRIGENDA TO VOLUMES 1-26 321

CUMULATIVE INDEX TO MONOGRAPHS 323

IARC WORKING GROUP ON THE EVALUATION OF THE CARCINOGENIC RISK OF CHEMICALS TO HUMANS:

SOME AROMATIC AMINES, ANTHRAQUINONES AND NITROSO COMPOUNDS, AND INORGANIC FLUORIDES USED IN DRINKING-WATER AND DENTAL PREPARATIONS

Lyon, 10-17 February 1981

Members[1]

B.K. Armstrong, NHRMRC Research Unit in Epidemiology and Preventive Medicine, Department of Medicine, Medical School Building, The Queen Elizabeth II Medical Centre, Nedlands, Western Australia 6009, Australia (*Chairman; co-rapporteur section 3.3*)

R. Bass, Institute of Pharmaceutics of the Federal Ministry of Health, Postal Box 330013, 1000 Berlin 33, Federal Republic of Germany

N. Breslow, Department of Biostatistics, SC-32, University of Washington, Seattle, WA 98195, USA

I.N. Chernozemsky, Head, Laboratory of Chemical Carcinogenesis and Testing, Institute of Oncology, Medical Academy, Sofia 1156, Bulgaria (*Vice-chairman; co-rapporteur section 3.1*)

G. Della Porta, Director, Division of Experimental Oncology, National Institute for the Study and Treatment of Tumours, Via Venezian 1, 20133 Milan, Italy

H.K. Kang, Health Scientist, Office of Carcinogen Identification and Classification, Health Standards Programs, Occupational Safety & Health Administration, US Department of Labor, Washington DC 20210, USA

B.J. Kilbey, University of Edinburgh, Department of Genetics and ARC Unit of Animal Genetics, Institute of Animal Genetics, West Mains Road, Edinburgh EH9 3JN, UK

C.M. King, Chairman, Department of Chemical Carcinogenesis, Michigan Cancer Foundation, 110 East Warren Avenue, Detroit, MI 48201, USA (*Co-rapporteur section 3.2*)

E. Kriek, Chief, Chemical Carcinogenesis Division, The Netherlands Cancer Institute, Antoni van Leeuwenhoekhuis Plesmanlaan 121, 1066 CX Amsterdam, The Netherlands

N.P. Napalkov, Director, N.N. Petrov Research Institute of Oncology, Leningradskaya Street 68, Pesochny-2, Leningrad 188646, USSR

H.S. Rosenkranz[2], Department of Microbiology, New York Medical College, Basic Science Building, Valhalla, NY 10595, USA (*Co-rapporteur section 3.2*)

---

[1] Unable to attend: E. Arrhenius, Vindkallsvaeden 3, 18261 Djursholm, Sweden; S.M. Brown, PO Box 3254, Berkeley, CA 94703, USA; P. Cole, University of Alabama, Department of Epidemiology, Tidwell Hall 203, Birmingham, AL 35295, USA; M.R. Parkhie, Visiting Scientist, MRC Toxicology Unit, Medical Research Council Laboratories, Woodmansterne Road, Carshalton, Surrey SM5 4EF, UK

[2] Present address: Director, Center for the Environmental Health Sciences, School of Medicine, Case Western Reserve University, Cleveland, OH 44106, USA

S.S. Thorgeirsson, Laboratory of Chemical Pharmacology, Developmental Therapeutics Program, Division of Cancer Treatment, National Cancer Institute, Building 37, Room 5A13, Bethesda, MD 20205, USA

J.M. Ward, Veterinary Pathologist, National Cancer Institute, National Toxicology Program, Bethesda, MD 20205, USA

M. Webb, Medical Research Council, Toxicology Unit, MRC Laboratories, Woodmansterne Road, Carshalton, Surrey SM5 4EF, UK

F. Zadjela, Director, U22 INSERM, Curie Institute, Bldg 110, 91405 Orsay, France

Representative from the National Cancer Institute

H. Kraybill, Scientific Coordinator for Environmental Cancer, Division of Cancer Cause and Prevention, Landow Building, Room 3C37, National Cancer Institute, Bethesda, MD 20205, USA

Representative from SRI International

M. McCaleb, Director, Chemical-Environmental Program, SRI International, 333 Ravenswood Avenue, Menlo Park, CA 94025, USA (*Rapporteur sections 2.1 & 2.2*)

Representative from the Chemical Manufacturers' Association

A.M. Kaplan, E.I. du Pont de Nemours Co., Haskell Laboratory, Elkton Road, Newark, DE 19711, USA

Representative from the Commission of the European Communities

M.-T. van der Venne, Directorate General for Employment and Social Affairs, Health & Safety Directorate, Commission of the European Communities, Jean Monnet Building, Plateau du Kirchberg, BP 1907, Luxembourg, Grand Duchy of Luxembourg

Representative from the European Chemical Industry Ecology and Toxicology Centre

J. Ishmael, ICI Central Toxicology Laboratory, Alderley Park, Macclesfield, Cheshire SK10 4TJ, UK

PARTICIPANTS

Observers

M. Gilbert, Scientific Affairs Officer, International Register of Potentially Toxic Chemicals, United Nations Environment Programme, 16 Avenue Jean Tremblay, Petit Saconnex, 1209 Geneva, Switzerland

J. Matthews, Association of Scientific, Technical and Managerial Staffs, Health & Safety Research Officer, Whitehall Office, Dane O'Coys Road, Bishops Stortford, Herts, UK

Secretariat[3]

C. Agthe, Division of Epidemiology and Biostatistics
A. Aitio, Division of Environmental Carcinogenesis
H. Bartsch, Division of Environmental Carcinogenesis (*Co-rapporteur section 3.2*)
J.R.P. Cabral, Division of Environmental Carcinogenesis
W. Davis, Programme of Research Training and Liaison (*Technical editor IARC*)
M. Friesen, Division of Environmental Carcinogenesis
L. Haroun, Division of Environmental Carcinogenesis (*Co-secretary*)
E. Heseltine, Charost, France (*Editor*)
A. Likhachev, Division of Environmental Carcinogenesis
D. Mietton, Division of Environmental Carcinogenesis (*Library assistant*)
R. Montesano, Division of Environmental Carcinogenesis (*Co-rapporteur section 3.1*)
I. O'Neill, Division of Environmental Carcinogenesis (*Rapporteur sections 1 & 2.3*)
C. Partensky, Division of Environmental Carcinogenesis (*Technical officer*)
I. Peterschmitt, Division of Environmental Carcinogenesis, Geneva (*Bibliographic researcher*)
R. Saracci, Division of Epidemiology and Biostatistics (*Co-rapporteur section 3.3*)
L. Simonato, Division of Epidemiology and Biostatistics
J. Wilbourn, Division of Environmental Carcinogenesis (*Co-secretary*)

Secretarial assistance

    M.-J. Ghess
    K. Masters
    S. Reynaud
    J. Smith

---

[3] Unable to attend: L. Tomatis, Director, Division of Environmental Carcinogenesis (*Head of the Programme*)

NOTE TO THE READER

The term 'carcinogenic risk' in the *IARC Monograph* series is taken to mean the probability that exposure to a chemical or complex mixture or employment in a particular occupation will lead to cancer in humans.

The fact that a monograph has been prepared on a chemical, complex mixture or occupation does not imply that a carcinogenic hazard is associated with the exposure, only that the published data have been examined. Equally, the fact that a chemical, complex mixture or occupation has not yet been evaluated in a monograph does not mean that it does not represent a carcinogenic hazard.

Anyone who is aware of published data that may alter an evaluation of the carcinogenic risk of a chemical, complex mixture or employment in an occupation is encouraged to make this information available to the Division of Environmental Carcinogenesis, International Agency for Research on Cancer, Lyon, France, in order that the chemical, complex mixture or occupation may be considered for re-evaluation by a future Working Group.

Although every effort is made to prepare the monographs as accurately as possible, mistakes may occur. Readers are requested to communicate any errors to the Division of Environmental Carcinogenesis, so that corrections can be reported in future volumes.

IARC MONOGRAPH PROGRAMME ON THE EVALUATION OF THE CARCINOGENIC RISK OF CHEMICALS TO HUMANS

PREAMBLE

BACKGROUND

In 1971, the International Agency for Research on Cancer (IARC) initiated a programme on the evaluation of the carcinogenic risk of chemicals to humans with the object of producing monographs on individual chemicals.[1] The criteria established at that time to evaluate carcinogenic risk to humans were adopted by all the working groups whose deliberations resulted in the first 16 volumes of the *IARC Monograph* series. In October 1977, a joint IARC/WHO *ad hoc* Working Group met to re-evaluate these guiding criteria; this preamble reflects the results of their deliberations(1) and those of a subsequent IARC *ad hoc* Working Group which met in April 1978(2).

A further *ad hoc* Working Group, which met in Lyon in April 1979 to prepare criteria to select chemicals for *IARC Monographs*(3), recommended that the *Monograph* programme be expanded to include consideration of human exposures in selected occupations. The Working Group which met in June 1980 therefore considered occupational exposures in wood, leather and some associated industries; their deliberations resulted in Volume 25 of the *Monograph* series.

OBJECTIVE AND SCOPE

The objective of the programme is to elaborate and publish in the form of monographs critical reviews of data on carcinogenicity for groups of chemicals to which humans are known to be exposed, to evaluate those data in terms of human risk with the help of international working groups of experts in chemical carcinogenesis and related fields, and to indicate where additional research efforts are needed.

The critical analyses of the data are intended to assist national and international authorities in formulating decisions concerning preventive measures. No recommendations are given concerning legislation, since this depends on risk-benefit evaluations, which seem best made by individual governments and/or other international agencies. In this connection, WHO recommendations on food additives(4), drugs(5), pesticides and contaminants(6) and occupational carcinogens(7) are particularly informative.

---

[1]Since 1972, the programme has undergone considerable expansion, primarily with the scientific collaboration and financial support of the US National Cancer Institute, Bethesda, MD.

Up to September 1981, 27 volumes of the *Monographs* had been published or were in press(8). A total of 564 compounds, industrial processes or occupational exposures had been evaluated or re-evaluated. For 42 chemicals, groups of chemicals, industrial processes or industrial exposures, a positive association or a strong suspicion of an association with human cancer has been found. For 22 of the individual chemicals, exposures are predominantly in occupational settings, although the general population may be exposed through environmental contamination. For 13 chemicals, human exposure was related to therapeutic uses; for one compound, exposure occurs *via* the diet. The preponderance of experimental data over epidemiological data on the 564 compounds or processes is striking: data on humans were available for only 70 of the chemicals, and 141 of them were evaluated for human carcinogenicity solely on the basis of *sufficient evidence* of carcinogenicity in experimental animals. The remainder could not be evaluated for carcinogenicity to humans.

The *IARC Monographs* are recognized as an authoritative source of information on the carcinogenicity of environmental chemicals. The first users' survey, made in 1976, indicates that the monographs are consulted routinely by various agencies in 24 countries. Each volume is printed in 4000 copies and distributed *via* the WHO publications service. (See last page for a listing of IARC publications and back outside cover for distribution and sales services.)

SELECTION OF CHEMICALS FOR MONOGRAPHS

The chemicals (natural and synthetic, including those which occur as mixtures and in manufacturing processes) are selected for evaluation on the basis of two main criteria: (a) there is evidence of human exposure, and (b) there is some experimental evidence of carcinogenicity and/or there is some evidence or suspicion of a risk to humans. In certain instances, chemical analogues are also considered. The scientific literature is surveyed for published data relevant to the monograph programme. In addition, the IARC *Survey of Chemicals Being Tested for Carcinogenicity*(9) often indicates those chemicals that may be scheduled for future meetings.

Inclusion of a chemical in a volume does not imply that it is carcinogenic, only that the published data have been examined. The evaluations must be consulted to ascertain the conclusions of the Working Group. Equally, the fact that a chemical has not appeared in a monograph does not mean that it is without carcinogenic hazard.

As new data on chemicals for which monographs have already been prepared and new principles for evaluating carcinogenic risk receive acceptance, re-evaluations will be made at subsequent meetings, and revised monographs will be published as necessary.

WORKING PROCEDURES

Approximately one year in advance of a meeting of a working group, a list of the substances to be considered is prepared by IARC staff in consultation with other experts. Subsequently, all relevant biological data are collected by IARC; in addition to the published literature, US Public Health Service Publication No. 149(10) has been particularly valuable and has been used in conjunction with other recognized sources of information on chemical carcinogenesis and systems such as CANCERLINE, MEDLINE and TOXLINE. The major collection of data and the preparation of first drafts for the sections on chemical and physical properties, on production, use, occurrence and on analysis are carried out by SRI International, Stanford, CA, USA under a separate contract with the US National Cancer Institute. Most of the data so obtained on production, use and occurrence refer to the United States and Japan; SRI International and IARC supplement this information with that from other sources in Europe. Bibliographical sources for data on mutagenicity and teratogenicity are the Environmental Mutagen Information Center and the Environmental Teratology Information Center, both located at the Oak Ridge National Laboratory, TN, USA.

Six to nine months before the meeting, reprints of articles containing relevant biological data are sent to an expert(s), or are used by the IARC staff, for the preparation of first draft monographs. These drafts are edited by IARC staff and are sent prior to the meeting to all participants of the Working Group for their comments. The Working Group then meets in Lyon for seven to eight days to discuss and finalize the texts of the monographs and to formulate the evaluations. After the meeting, the master copy of each monograph is verified by consulting the original literature, then edited and prepared for reproduction. The monographs are usually published within six months after the Working Group meeting.

DATA FOR EVALUATIONS

With regard to biological data, only reports that have been published or accepted for publication are reviewed by the working groups, although a few exceptions have been made. The monographs do not cite all of the literature on a particular chemical: only those data considered by the Working Group to be relevant to the evaluation of the carcinogenic risk of the chemical to humans are included.

Anyone who is aware of data that have been published or are in press which are relevant to the evaluations of the carcinogenic risk to humans of chemicals for which monographs have appeared is urged to make them available to the Division of Environmental Carcinogenesis, International Agency for Research on Cancer, Lyon, France.

THE WORKING GROUP

The tasks of the Working Group are five-fold: (a) to ascertain that all data have been collected; (b) to select the data relevant for the evaluation; (c) to ensure that the summaries of the data enable the reader to follow the reasoning of the committee; (d) to judge the significance of the results of experimental and epidemiological studies; and (e) to make an evaluation of the carcinogenic risk of the chemical.

Working Group participants who contributed to the consideration and evaluation of chemicals within a particular volume are listed, with their addresses, at the beginning of each publication (see p. 5). Each member serves as an individual scientist and not as a representative of any organization or government. In addition, observers are often invited from national and international agencies, organizations and industries.

GENERAL PRINCIPLES FOR EVALUATING THE CARCINOGENIC RISK OF CHEMICALS

The widely accepted meaning of the term 'chemical carcinogenesis', and that used in these monographs, is the induction by chemicals of neoplasms that are not usually observed, the earlier induction by chemicals of neoplasms that are usually observed, and/or the induction by chemicals of more neoplasms that are usually found - although fundamentally different mechanisms may be involved in these three situations. Etymologically, the term 'carcinogenesis' means the induction of cancer, that is, of malignant neoplasms; however, the commonly accepted meaning is the induction of various types of neoplasms or of a combination of malignant and benign tumours. In the monographs, the words 'tumour' and 'neoplasm' are used interchangeably. (In scientific literature the terms 'tumourigen', 'oncogen' and 'blastomogen' have all been used synonymously with 'carcinogen', although occasionally 'tumourigen' has been used specifically to denote a substance that induces benign tumours.)

Experimental Evidence

Qualitative aspects

Both the interpretation and evaluation of a particular study as well as the overall assessment of the carcinogenic activity of a chemical involve several qualitatively important considerations, including: (a) the experimental parameters under which the chemical was tested, including route of administration and exposure, species, strain, sex, age, etc.; (b) the consistency with which the chemical has been shown to be carcinogenic, e.g., in how many species and at which target organ(s); (c) the spectrum of neoplastic response, from benign neoplasm to multiple malignant tumours; (d) the stage of tumour formation in which a chemical may be involved: some chemicals act as complete carcinogens and have initiating and promoting activity, while others are promoters only; and (e) the possible role of modifying factors.

There are problems not only of differential survival but of differential toxicity, which may be manifested by unequal growth and weight gain in treated and control animals. These complexities are also considered in the interpretation of data.

Many chemicals induce both benign and malignant tumours. Few instances are recorded in which only benign neoplasms are induced by chemicals that have been studied extensively. Benign tumours may represent a stage in the evolution of a malignant neoplasm or they may be 'end-points' that do not readily undergo transition to malignancy. If a substance is found to induce only benign tumours in experimental animals, it should be suspected of being a carcinogen and requires further investigation.

Hormonal carcinogenesis

Hormonal carcinogenesis presents certain distinctive features: the chemicals involved occur both endogenously and exogenously; in many instances, long exposure is required; tumours occur in the target tissue in association with a stimulation of non-neoplastic growth, but in some cases hormones promote the proliferation of tumour cells in a target organ. Hormones that occur in excessive amounts, hormone-mimetic agents and agents that cause hyperactivity or imbalance in the endocrine system may require evaluative methods comparable with those used to identify chemical carcinogens; particular emphasis must be laid on quantitative aspects and duration of exposure. Some chemical carcinogens have significant side effects on the endocrine system, which may also result in hormonal carcinogenesis. Synthetic hormones and anti-hormones can be expected to possess other pharmacological and toxicological actions in addition to those on the endocrine system, and in this respect they must be treated like any other chemical with regard to intrinsic carcinogenic potential.

Quantitative aspects

Dose-response studies are important in the evaluation of carcinogenesis: the confidence with which a carcinogenic effect can be established is strengthened by the observation of an increasing incidence of neoplasms with increasing exposure.

The assessment of carcinogenicity in animals is frequently complicated by recognized differences among the test animals (species, strain, sex, age), route(s) of administration and in dose/duration of exposure; often, target organs at which a cancer occurs and its histological type may vary with these parameters. Nevertheless, indices of carcinogenic potency in particular experimental systems [for instance, the dose-rate required under continuous exposure to halve the probability of the animals remaining tumourless(11)] have been formulated in the hope that, at least among categories of fairly similar agents, such indices may be of some predictive value in other systems, including humans.

Chemical carcinogens differ widely in the dose required to produce a given level of tumour induction, although many of them share common biological properties, which include metabolism to reactive [electrophilic(12-14)] intermediates capable of interacting with DNA. The reason for this variation in dose-response is not understood, but it may be due either to differences within a common metabolic process or to the operation of qualitatively distinct mechanisms.

Statistical analysis of animal studies

Tumours which would have arisen had an animal lived longer may not be observed because of the death of the animal from unrelated causes, and this possibility must be allowed for. Various analytical techniques have been developed which use the assumption of independence of competing risks to allow for the effects of intercurrent mortality on the final numbers of tumour-bearing animals in particular treatment groups.

For externally visible tumours and for neoplasms that cause death, methods such as Kaplan-Meier (i.e., 'life-table', 'product-limit' or 'actuarial') estimates(11), with associated significance tests(15,16), have been recommended.

For internal neoplasms which are discovered 'incidentally'(15) at autopsy but which did not cause the death of the host, different estimates(17) and significance tests(15,16) may be necessary for the unbiased study of the numbers of tumour-bearing animals.

All of these methods(11,15-17) can be used to analyse the numbers of animals bearing particular tumour types, but they do not distinguish between animals with one or many such tumours. In experiments which end at a particular fixed time, with the simultaneous sacrifice of many animals, analysis of the total numbers of internal neoplasms per animal found at autopsy at the end of the experiment is straightforward. However, there are no adequate statistical methods for analysing the numbers of particular neoplasms that kill an animal. The design and statistical analysis of long-term carcinogenicity experiments were recently reviewed, in Supplement 2 to the *Monograph* series(18).

Evidence of Carcinogenicity in Humans

Evidence of carcinogenicity in humans can be derived from three types of study, the first two of which usually provide only suggestive evidence: (1) reports concerning individual cancer patients (case reports), including a history of exposure to the supposed carcinogenic agent; (2) descriptive epidemiological studies in which the incidence of cancer in human populations is found to vary (spatially or temporally) with exposure to the agent; and (3) analytical epidemiological studies (e.g., case-control or cohort studies) in which individual exposure to the agent is found to be associated with an increased risk of cancer.

An analytical study that shows a positive association between an agent and a cancer may be interpreted as implying causality to a greater or lesser extent, on the basis of the following criteria: (a) There is no identifiable positive bias. (By 'positive bias' is meant the operation of factors in study design or execution which lead erroneously to a more strongly positive association between an agent and disease than in fact exists. Examples of positive bias include, in case-control studies, better documentation of exposure to the agent for cases than for controls, and, in cohort studies, the use of better means of detecting cancer in individuals exposed to the agent than in individuals not exposed.) (b) The possibility of positive confounding has been considered. (By 'positive confounding' is meant a situation in which the relationship between an agent and a disease is rendered more strongly positive than it truly is as a result of an association between that agent and another agent which either causes or prevents the disease. An example of positive confounding is the association between coffee consumption and lung cancer, which results from their joint association with cigarette smoking.) (c) The association is unlikely to be due to chance alone. (d) The association is strong. (e) There is a dose-response relationship.

In some instances, a single epidemiological study may be strongly indicative of a cause-effect relationship; however, the most convincing evidence of causality comes when several independent studies done under different circumstances result in 'positive' findings.

Analytical epidemiological studies that show no association between an agent and a cancer ('negative' studies) should be interpreted according to criteria analogous to those listed above: (a) there is no identifiable negative bias; (b) the possibility of negative confounding has been considered; and (c) the possible effects of misclassification of exposure or outcome have been weighed. In addition, it must be recognized that in any study there are confidence limits around the estimate of association or relative risk. In a study regarded as 'negative', the upper confidence limit may indicate a relative risk substantially greater than unity; in that case, the study excludes only relative risks that are above the upper limit. This usually means that a 'negative' study must be large to be convincing. Confidence in a 'negative' result is increased when several independent studies carried out under different circumstances are in agreement. Finally, a 'negative' study may be considered to be relevant only to dose levels within or below the range of those observed in the study and is pertinent only if sufficient time has elapsed since first human exposure to the agent. Experience with human cancers of known etiology suggests that the period from first exposure to a chemical carcinogen to development of clinically observed cancer is usually measured in decades and may be in excess of 30 years.

The Working Group whose deliberations resulted in Supplement 1 to the *Monographs* (IARC, 1979) defined *sufficient evidence* of the carcinogenicity of a chemical to humans as that which provides a causal association between exposure and cancer; *limited evidence* was defined as that which indicates a possible carcinogenic effect in humans.

Relevance of Experimental Data to the Evaluation of Carcinogenic Risk to Humans

No adequate criteria are presently available to interpret experimental carcinogenicity data directly in terms of carcinogenic potential for humans. Nonetheless, utilizing data collected from appropriate tests in animals, positive extrapolations to possible human risk can be approximated.

Information compiled from the first 26 volumes of the *IARC Monographs*(19-21) shows that of the 41 chemicals, groups of chemicals, manufacturing processes or occupational exposures now generally accepted to cause or probably to cause cancer in humans, all but possibly two (arsenic and benzene) of those which have been tested appropriately produce cancer in at least one animal species. For several of the chemicals that are carcinogenic for humans (aflatoxins, 4-aminobiphenyl, diethylstilboestrol, melphalan, mustard gas and vinyl chloride), evidence of carcinogenicity in experimental animals preceded evidence obtained from epidemiological studies or case reports.

In general, the evidence that a chemical produces tumours in experimental animals is of two degrees: (a) *sufficient evidence* of carcinogenicity is provided by the production of malignant tumours; and (b) *limited evidence* of carcinogenicity reflects qualitative and/or quantitative limitations of the experimental results.

Sufficient evidence of carcinogenicity is provided by experimental studies that show an increased incidence of malignant tumours: (i) in multiple species or strains, and/or (ii) in multiple experiments (routes and/or doses), and/or (iii) to an unusual degree (with regard to incidence, site, type and/or precocity of onset). Additional evidence may be provided by data concerning dose-response, mutagenicity or structure.

For many of the chemicals evaluated in the first 27 volumes of the *IARC Monographs* for which there is *sufficient evidence* of carcinogenicity in animals, data relating to carcinogenicity for humans are either insufficient or nonexistent. In the absence of adequate data on humans, it is reasonable, for practical purposes, to regard such chemicals as if they presented a carcinogenic risk to humans.

In the present state of knowledge, it would be difficult to define a predictable relationship between the dose (mg/kg bw/day) of a particular chemical required to produce cancer in test animals and the dose which would produce a similar incidence of cancer in humans. The available data suggest, however, that such a relationship may exist(22,23), at least for certain classes of carcinogenic chemicals. Data that provide sufficient evidence of carcinogenicity in test animals may therefore be used in an approximate quantitative evaluation of the human risk at some given exposure level, provided that the nature of the chemical concerned and the physiological, pharmacological and toxicological differences between the test animals and humans are taken into account. However, no acceptable methods are currently available for quantifying the possible errors in such a procedure, whether it is used to generalize among species or to extrapolate from high to low doses. The methodology for such quantitative extrapolation to humans requires further development.

Evidence for the carcinogenicity of some chemicals in experimental animals may be *limited* for two reasons. Firstly, experimental data may be restricted to such a point that it is not possible to determine a causal relationship between administration of a chemical and the development of a particular lesion in the animals. Secondly, there are certain neoplasms, including lung tumours and hepatomas in mice, which have been considered of lesser significance than neoplasms occurring at other sites for the purpose of evaluating the carcinogenicity of chemicals. Such tumours occur spontaneously in high incidence in these animals, and their malignancy is often difficult to establish. An evaluation of the significance of these tumours following administration of a chemical is the responsibility of particular working groups preparing individual monographs, and it has not been possible to set down rigid guidelines; the relevance of these tumours must be determined by considerations which include experimental design and completeness of reporting.

Some chemicals for which there is *limited evidence* of carcinogenicity in animals have also been studied in humans with, in general, inconclusive results. While such chemicals may indeed be carcinogenic to humans, more experimental and epidemiological investigation is required.

Hence, *'sufficient evidence'* of carcinogenicity and *'limited evidence'* of carcinogenicity do not indicate categories of chemicals: the inherent definitions of those terms indicate varying degrees of experimental evidence, which may change if and when new data on the chemicals become available. The main drawback to any rigid classification of chemicals with regard to their carcinogenic capacity is the as yet incomplete knowledge of the mechanism(s) of carcinogenesis.

In recent years, several short-term tests for the detection of potential carcinogens have been developed. When only inadequate experimental data are available, positive results in validated short-term tests (see p. 23) are an indication that the compound is a potential carcinogen and that it should be tested in animals for an assessment of its carcinogenicity. Negative results from short-term tests cannot be considered sufficient evidence to rule out carcinogenicity. Whether short-term tests will eventually be as reliable as long-term tests in predicting carcinogenicity in humans will depend on further demonstrations of consistency with long-term experiments and with data from humans. Available screening assays are evaluated in Supplement 2 to the *Monographs*(18).

EXPLANATORY NOTES ON THE MONOGRAPH CONTENTS

Chemical and Physical Data (Section 1)

The Chemical Abstracts Services Registry Number, the latest Chemical Abstracts Primary Name (9th Collective Index)(24) and the IUPAC Systematic Name(25) are recorded in section 1. Other synonyms and trade names are given, but no comprehensive list is provided. Further, some of the trade names are those of mixtures in which the compound being evaluated is only one of the ingredients.

The structural and molecular formulae, molecular weight and chemical and physical properties are given. The properties listed refer to the pure substance, unless otherwise specified, and include, in particular, data that might be relevant to carcinogenicity (e.g., lipid solubility) and those that concern identification.

A separate description of the composition of technical products includes available information on impurities and formulated products.

Production, Use, Occurrence and Analysis (Section 2)

The purpose of section 2 is to provide indications of the extent of past and present human exposure to the chemical.

Synthesis

Since cancer is a delayed toxic effect, the dates of first synthesis and of first commercial production of the chemical are provided. In addition, methods of synthesis used in past and present commercial production are described. This information allows a reasonable estimate to be made of the date before which no human exposure could have occurred.

Production

Since Europe, Japan and the United States are reasonably representative industrialized areas of the world, most data on production, foreign trade and uses are obtained from those countries. It should not, however, be inferred that those nations are the sole or even the major sources or users of any individual chemical.

Production and foreign trade data are obtained from both governmental and trade publications by chemical economists in the three geographical areas. In some cases, separate production data on organic chemicals manufactured in the United States are not available because their publication could disclose confidential information. In such cases, an indication of the minimum quantity produced can be inferred from the number of companies reporting commercial production. Each company is required to report on individual chemicals if the sales value or the weight of the annual production exceeds a specified minimum level. These levels vary for chemicals classified for different uses, e.g., medicinals and plastics; in fact, the minimal annual sales value is between $1000 and $50000, and the minimal annual weight of production is between 450 and 22700 kg. Data on production in some European countries are obtained by means of general questionnaires sent to companies thought to produce the compounds being evaluated. Information from the completed questionnaires is compiled by country, and the resulting estimates of production are included in the individual monographs.

Volume 16 (1978) Some Aromatic Amines and Related Nitro Compounds - Hair Dyes, Colouring Agents, and Miscellaneous Industrial Chemicals (32 monographs), 400 pages

Volume 17 (1978) Some N-Nitroso Compounds (17 monographs), 365 pages

Volume 18 (1978) Polychlorinated Biphenyls and Polybrominated Biphenyls (2 monographs), 140 pages

Volume 19 (1979) Some Monomers, Plastics and Synthetic Elastomers, and Acrolein (17 monographs), 513 pages

Volume 20 (1979) Some Halogenated Hydrocarbons (25 monographs), 609 pages

Volume 21 (1979) Sex Hormones (II) (22 monographs), 583 pages

Volume 22 (1980) Some Non-Nutritive Sweetening Agents (2 monographs), 208 pages

Volume 23 (1980) Some Metals and Metallic Compounds (4 monographs), 438 pages

Volume 24 (1980) Some Pharmaceutical Drugs (16 monographs), 337 pages

Volume 25 (1981) Wood, Leather and Some Associated Industries (7 monographs), 412 pages

Volume 26 (1981) Some Antineoplastic and Immunosuppressive Agents (18 monographs), 411 pages

Volume 27 (1981) Some Aromatic Amines, Anthraquinones and Nitroso Compounds, and Inorganic Fluorides Used in Drinking-Water and Dental Preparations (18 monographs), 341 pages

9. IARC (1973-1979) *Information Bulletin on the Survey of Chemicals Being Tested for Carcinogenicity*, Numbers 1-8, Lyon, France

    Number 1 (1973) 52 pages
    Number 2 (1973) 77 pages
    Number 3 (1974) 67 pages
    Number 4 (1974) 97 pages
    Number 5 (1975) 88 pages
    Number 6 (1976) 360 pages
    Number 7 (1978) 460 pages
    Number 8 (1979) 604 pages

10. PHS 149 (1951-1976) Public Health Service Publication No. 149, *Survey of Compounds which have been Tested for Carcinogenic Activity*, Washington DC, US Government Printing Office

    1951 Hartwell, J.L., 2nd ed., Literature up to 1947 on 1329 compounds, 583 pages

    1957 Shubik, P. & Hartwell, J.L. Supplement 1, Literature for the years 1948-1953 on 981 compounds, 388 pages

    1969 Shubik, P. & Hartwell, J.L., edited by Peters, J.A., Supplement 2, Literature for the years 1954-1960 on 1048 compounds, 655 pages

    1971 National Cancer Institute, Literature for the years 1968-1969 on 882 compounds, 653 pages

    1973 National Cancer Institute, Literature for the years 1961-1967 on 1632 compounds, 2343 pages

    1974 National Cancer Institute, Literature for the years 1970-1971 on 750 compounds, 1667 pages

    1976 National Cancer Institute, Literature for the years 1972-1973 on 966 compounds, 1638 pages

11. Pike, M.C. & Roe, F.J.C. (1963) An actuarial method of analysis of an experiment in two-stage carcinogenesis. *Br. J. Cancer, 17,* 605-610
12. Miller, E.C. & Miller, J.A. (1966) Mechanisms of chemical carcinogenesis: nature of proximate carcinogens and interactions with macromolecules. *Pharmacol. Rev., 18,* 805-838
13. Miller, J.A. (1970) Carcinogenesis by chemicals: an overview - G.H.A. Clowes Memorial Lecture. *Cancer Res., 30,* 559-576
14. Miller, J.A. & Miller, E.C. (1976) *The metabolic activation of chemical carcinogens to reactive electrophiles.* In: Yuhas, J.M., Tennant, R.W. & Reagon, J.D., eds, *Biology of Radiation Carcinogenesis,* New York, Raven Press
15. Peto, R. (1974) Guidelines on the analysis of tumour rates and death rates in experimental animals. *Br. J. Cancer, 29,* 101-105
16. Peto, R. (1975) Letter to the editor. *Br. J. Cancer, 31,* 697-699
17. Hoel, D.G. & Walburg, H.E., Jr (1972) Statistical analysis of survival experiments. *J. natl Cancer Inst., 49,* 361-372
18. IARC (1980) *IARC Monographs on the Evaluation of the Carcinogenic Risk of Chemicals to Humans,* Supplement 2, *Long-term and Short-term Screening Assays for Carcinogens: A Critical Appraisal,* Lyon
19. IARC Working Group (1980) An evaluation of chemicals and industrial processes associated with cancer in humans based on human and animal data: *IARC Monographs* Volumes 1 to 20. *Cancer Res., 40,* 1-12
20. IARC (1979) *IARC Monographs on the Evaluation of the Carcinogenic Risk of Chemicals to Humans,* Supplement 1, *Chemicals and Industrial Processes Associated with Cancer in Humans,* Lyon
21. IARC (1979) *Annual Report, 1979,* Lyon, pp. 89-99
22. Rall, D.P. (1977) *Species differences in carcinogenesis testing.* In: Hiatt, H.H., Watson, J.D. & Winsten, J.A., eds, *Origins of Human Cancer,* Book C, Cold Spring Harbor, NY, Cold Spring Harbor Laboratory, pp. 1383-1390
23. National Academy of Sciences (NAS) (1975) *Contemporary Pest Control Practices and Prospects: the Report of the Executive Committee,* Washington DC
24. Chemical Abstracts Services (1978) *Chemical Abstracts Ninth Collective Index (9CI), 1972-1976,* Vols 76-85, Columbus, OH
25. International Union of Pure & Applied Chemistry (1965) *Nomenclature of Organic Chemistry,* Section C, London, Butterworths
26. WHO (1958) Second Report of the Joint FAO/WHO Expert Committee on Food Additives. Procedures for the testing of intentional food additives to establish their safety and use. *WHO tech. Rep. Ser., No. 144*
27. WHO (1967) Scientific Group. Procedures for investigating intentional and unintentional food additives. *WHO tech. Rep. Ser., No. 348*
28. Berenblum, I., ed. (1969) Carcinogenicity testing. *UICC tech. Rep. Ser., 2*
29. Sontag, J.M., Page, N.P. & Saffiotti, U. (1976) Guidelines for carcinogen bioassay in small rodents. *Natl Cancer Inst. Carcinog. tech. Rep. Ser., No. 1*
30. Committee on Standardized Genetic Nomenclature for Mice (1972) Standardized nomenclature for inbred strains of mice. Fifth listing. *Cancer Res., 32,* 1609-1646

31. Wilson, J.G. & Fraser, F.C. (1977) *Handbook of Teratology*, New York, Plenum Press
32. Ames, B.N., Durston, W.E., Yamasaki, E. & Lee, F.D. (1973) Carcinogens are mutagens: a simple test system combining liver homogenates for activation and bacteria for detection. *Proc. natl Acad. Sci. (USA), 70*, 2281-2285
33. McCann, J., Choi, E., Yamasaki, E. & Ames, B.N. (1975) Detection of carcinogens as mutagens in the *Salmonella*/microsome test: assay of 300 chemicals. *Proc. natl Acad. Sci. (USA), 72*, 5135-5139
34. McCann, J. & Ames, B.N. (1976) Detection of carcinogens as mutagens in the *Salmonella*/microsome test: assay of 300 chemicals: discussion. *Proc. natl Acad. Sci. (USA), 73*, 950-954
35. Sugimura, T., Sato, S., Nagao, M., Yahagi, T., Matsushima, T., Seino, Y., Takeuchi, M. & Kawachi, T. (1977) *Overlapping of carcinogens and mutagens*. In: Magee, P.N., Takayama, S., Sugimura, T. & Matsushima, T., eds, *Fundamentals in Cancer Prevention*, Baltimore, University Park Press, pp. 191-215
36. Purchase, I.F.M., Longstaff, E., Ashby, J., Styles, J.A., Anderson, D., Lefevre, P.A. & Westwood, F.R. (1976) Evaluation of six short-term tests for detecting organic chemical carcinogens and recommendations for their use. *Nature, 264*, 624-627
37. Vogel, E. & Sobels, F.H. (1976) *The function of Drosophila in genetic toxicology testing*. In: Hollaender, A., ed., *Chemical Mutagens: Principles and Methods for Their Detection*, Vol. 4, New York, Plenum Press, pp. 93-142
38. San, R.H.C. & Stich, H.F. (1975) DNA repair synthesis of cultured human cells as a rapid bioassay for chemical carcinogens. *Int. J. Cancer, 16*, 284-291
39. Pienta, R.J., Poiley, J.A. & Lebherz, W.B. (1977) Morphological transformation of early passage golden Syrian hamster embryo cells derived from cryopreserved primary cultures as a reliable *in vitro* bioassay for identifying diverse carcinogens. *Int. J. Cancer, 19*, 642-655
40. Zimmermann, F.K. (1975) Procedures used in the induction of mitotic recombination and mutation in the yeast *Saccharomyces cerevisiae. Mutat. Res., 31*, 71-86
41. Ong, T.-M. & de Serres, F.J. (1972) Mutagenicity of chemical carcinogens in *Neurospora crassa. Cancer Res., 32*, 1890-1893
42. Huberman, E. & Sachs, L. (1976) Mutability of different genetic loci in mammalian cells by metabolically activated carcinogenic polycyclic hydrocarbons. *Proc. natl Acad. Sci. (USA), 73*, 188-192
43. Krahn, D.F. & Heidelburger, C. (1977) Liver homogenate-mediated mutagenesis in Chinese hamster V79 cells by polycyclic aromatic hydrocarbons and aflatoxins. *Mutat. Res., 46*, 27-44
44. Kuroki, T., Drevon, C. & Montesano, R. (1977) Microsome-mediated mutagenesis in V79 Chinese hamster cells by various nitrosamines. *Cancer Res., 37*, 1044-1050
45. Searle, A.G. (1975) The specific locus test in the mouse. *Mutat. Res., 31*, 277-290
46. Evans, H.J. & O'Riordan, M.L. (1975) Human peripheral blood lymphocytes for the analysis of chromosome aberrations in mutagen tests. *Mutat. Res., 31*, 135-148
47. Epstein, S.S., Arnold, E., Andrea, J., Bass, W. & Bishop, Y. (1972) Detection of chemical mutagens by the dominant lethal assay in the mouse. *Toxicol. appl. Pharmacol., 23*, 288-325

48. Perry, P. & Evans, H.J. (1975) Cytological detection of mutagen-carcinogen exposure by sister chromatid exchanges. *Nature, 258*, 121-125
49. Stetka, D.G. & Wolff, S. (1976) Sister chromatid exchanges as an assay for genetic damage induced by mutagen-carcinogens. I. *In vivo* test for compounds requiring metabolic activation. *Mutat. Res., 41*, 333-342
50. Bartsch, H. & Grover, P.L. (1976) *Chemical carcinogenesis and mutagenesis.* In: Symington, T. & Carter, R.L., eds, *Scientific Foundations of Oncology,* Vol. IX, *Chemical Carcinogenesis,* London, Heinemann Medical Books Ltd, pp. 334-342
51. Hollaender, A., ed. (1971a,b, 1973, 1976) *Chemical Mutagens: Principles and Methods for Their Detection,* Vols 1-4, New York, Plenum Press
52. Montesano, R. & Tomatis, L., eds (1974) *Chemical Carcinogenesis Essays (IARC Scientific Publications No. 10),* Lyon, International Agency for Research on Cancer
53. Ramel, C., ed. (1973) Evaluation of genetic risk of environmental chemicals: report of a symposium held at Skokloster, Sweden, 1972. *Ambio Spec. Rep., No. 3*
54. Stoltz, D.R., Poirier, L.A., Irving, C.C., Stich, H.F., Weisburger, J.H. & Grice, H.C. (1974) Evaluation of short-term tests for carcinogenicity. *Toxicol. appl. Pharmacol., 29*, 157-180
55. Montesano, R., Bartsch, H. & Tomatis, L., eds (1976) *Screening Tests in Chemical Carcinogenesis (IARC Scientific Publications No. 12),* Lyon, International Agency for Research on Cancer
56. Committee 17 (1976) Environmental mutagenic hazards. *Science, 187*, 503-514

GENERAL REMARKS ON THE SUBSTANCES CONSIDERED

Volume 27 of the *IARC Monographs on the Evaluation of the Carcinogenic Risk of Chemicals to Humans* comprises 18 monographs, in which the carcinogenicity of some 29 chemicals is evaluated. The compounds include some aromatic amines and anthraquinones, which are used mainly as dyestuff intermediates, two nitroso compounds, and inorganic fluorides used in the fluoridation of drinking-water or in toothpastes or other tooth-care products. Occupational exposure to fluoride compounds and exposure of the general population as a result of gross contamination of the environment with fluorides may be considered at a future date.

A previous monograph on 2,4-diaminoanisole (IARC, 1978a) was updated, together with a section on the epidemiological evidence relating to possible carcinogenic risks associated with use of hair dyes by hairdressers and consumers (IARC, 1978b).

The Working Group also prepared a monograph on formaldehyde; but since the results of some experimental studies on carcinogenicity were available only in preliminary form, it was left to the discretion of the Secretariat to re-submit the monograph to a subsequent IARC Working Group, which would be able to evaluate the carcinogenicity of formaldehyde on the basis of the results of the completed studies. Since these results will soon be available, the IARC Working Group that convened in Lyon from 13-20 October 1981 reconsidered the monograph on formaldehyde as an extra item on its agenda.

The Working Group drew attention to the importance of data on purity for chemicals used in animal carcinogenicity experiments and in mutagenicity tests. Many of the chemicals used in the carcinogenicity studies were technical grades, varying in purity from relatively high to indeterminate. Chemical analysis was generally limited to identifying the major component(s), although sometimes the several data given in an analysis were not correlated to provide a specific positive identification. Minor contaminants were usually not identified. Although the Working Group was aware that occupational exposure is often to crude compounds and mixtures, the consequences arising from the use in carcinogenicity tests of substantially impure materials may limit the applicability of the results and make more difficult an evaluation of the pure chemical (see Murrill *et al.*, 1980).

In only a few of the mutagenicity studies and other related assays reported in this volume had the chemicals been analysed and their purity stated. Most of the reports considered included no information on the purity of the compounds tested. In two of the studies (Dunkel & Simmon, 1980; Rosenkranz & McCoy, 1981), the chemicals used were from the same batch as that used in the corresponding animal bioassays conducted under the auspices of the US National Cancer Institute (NCI).

The carcinogenicity studies evaluated in this volume include 20 long-term studies undertaken in testing laboratories under contract to the NCI. In these studies, inbred Fischer (F344) rats and hybrid (C57BL x C3H) F_1 (B6C3F_1) mice obtained from commercial suppliers were used (Goodman *et al.*, 1979; Ward *et al.*, 1979). Both F344 rats and B6C3F_1 mice develop tumours of types similar to those found in other strains of rats and mice, but the incidence of the tumours is characteristic for each strain. The tumours found most commonly in F344 rats are those of the testis, pituitary, adrenals and haematopoietic system (Goodman et al., 1979); and those in B6C3F$_1$ mice are found in the liver, lung and haemotopoietic system (Ward et al., 1979). The variation in tumour incidence in the historical controls must be considered in evaluating these bioassay data; for example, the mean incidence of liver tumours in approximately 1800 control male B6C3F$_1$ mice was 30%, and in groups of 20-50 controls the range was 5-55% (Tarone et al., 1981).

The test doses used in the NCI studies were established in range-finding studies of 7-13 weeks' duration, carried out to estimate a maximally tolerated dose (MTD). Generally, MTDs were established as those doses that produced no more than a 10% depression of mean body weight gain compared with controls, and no mortality due to toxicity. In the carcinogenicity studies, the MTD and half the MTD were usually used as test doses.

The pathology protocol used was standard in most NCI experiments. The preputial, clitoral and Zymbal glands were examined histologically only when grossly abnormal. All animals were necropsied, whether they died or were sacrificed. The diagnoses were incorporated on computer input forms by a laboratory pathologist, coded by a Systematized Nomenclature of Pathology (SNOP) code and entered into the Carcinogenesis Bioassay Data System (CBDS). Computer tables thus generated were used in the report of the bioassay. Tumours were diagnosed and classified according to general criteria used in human and veterinary pathology (Turusov, 1973, 1976, 1979; Squire et al., 1978; Ward et al., 1980). For liver tumours in the mouse, the classifications of Frith & Ward (1980) and of Turusov & Takayama (1979) were used; and for the rat, the classifications of Squire & Levitt (1975) and of the Institute of Laboratory Animal Resources (1980).

The statistical analyses of the test data followed the procedure described by Gart et al. (1979). Probabilities of survival were estimated for each treatment group by the product-limit procedure of Kaplan and Meier. For a typical experiment, in which the compound was tested in three groups (control, low-dose and high-dose), the statistical significance of the differences among treatment groups and controls in tumour incidence at each site was evaluated by three statistical tests. The Cochran-Armitage statistic for trend in proportions, with continuity correction, was used to test the null hypothesis of no difference among the three dose groups against an alternative of increasing incidence with increasing dose. Two additional tests, based on the Fisher exact statistic for comparison of two proportions, were applied separately to compare incidence in the low-dose *versus* control, and high-dose *versus* control groups. When there was early mortality in one or more treatment groups, the denominators of the proportions were adjusted to reflect the numbers of animals still alive either at 52 weeks or at the age at which the first tumour was observed, if earlier. No life-table comparisons of tumour incidence were reported.

Because these three statistical tests examine somewhat different hypotheses, situations arise in which only one of them yields significant results at conventional ($P < 0.05$) levels. This is true especially when the statistical results are of borderline significance and do not themselves provide convincing evidence for a carcinogenic effect or lack thereof. In such situations other sources of information were considered in arriving at a judgement as to whether the tumour incidence had in fact been increased. These included data from previous experiments in the same series (Goodman et al., 1979; Tarone et al., 1981; Ward et al., 1979) and whether the agent caused precancerous lesions in the putative target organ.

While the assessments made in the NCI reports were noted by the Working Group, the data were considered on their own merits, and an evaluation was reached on the basis of all information available at the time of the meeting.

The data from the NCI carcinogenicity studies were presented in much greater detail than is commonly found in such reports, and it was therefore possible for the Working Group to distinguish certain shortcomings in the conduct of the experiments - shortcomings which could also apply to studies carried out elsewhere and not reported in the same detail.

A note on the aromatic amines

Although specific information on the absorption, distribution and metabolism of many of the aromatic amines considered in this volume was not available, experience with this class of compounds suggests that in humans absorption may occur *via* the respiratory and gastrointestinal tracts and through the skin. Aromatic amines and amides are readily metabolized by oxidation of their carbon or nitrogen atoms in both experimental animals and man. The amines and hydroxyl groups can be conjugated with glucuronate, sulphate or acetate. The formation of nucleic acid adducts and mutagenic and carcinogenic metabolites appears to involve *N*-oxidation (Kriek & Westra, 1979). Toxic effects of aromatic amines include methaemoglobinaemia and haemolysis (Kiese, 1966).

References

Dunkel. V.C. & Simmon, V.F. (1980) *Mutagenic activity of chemicals previously tested for carcinogenicity in the National Cancer Institute Bioassay Program*. In: Montesano, R., Bartsch, H. & Tomatis, L., eds, *Molecular and Cellular Aspects of Carcinogen Screening Tests (IARC Scientific Publications No. 27)*. Lyon, International Agency for Research on Cancer, pp. 283-302.

Frith, C.H. & Ward, J.M. (1980) A morphologic classification of proliferative and neoplastic hepatic lesions in mice. *J. environ. Pathol. Toxicol., 3*, 329-351

Gart, J.J., Chu, K.C. & Tarone, R.E. (1979) Statistical issues in interpretation of chronic bioassay tests for carcinogenicity. *J. natl Cancer Inst., 62*, 957-974

Goodman, D.G., Ward, J.M., Squire, R.A., Chu, K.C. & Linhart, M.S. (1979) Neoplastic and nonneoplastic lesions in aging F344 rats. *Toxicol. appl. Pharmacol., 48*, 237-248

IARC (1978a) *IARC Monographs on the Evaluation of the Carcinogenic Risk of Chemicals to Man*, Vol. 16, *Some Aromatic Amines and Related Nitro Compounds - Hair Dyes, Colouring Agents and Miscellaneous Industrial Chemicals*, Lyon, pp. 51-62

IARC (1978b) *IARC Monographs on the Evaluation of the Carcinogenic Risk of Chemicals to Man*, Vol. 16, *Some Aromatic Amines and Related Nitro Compounds - Hair Dyes, Colouring Agents and Miscellaneous Industrial Chemicals*, Lyon, pp. 27-37

Institute of Laboratory Animal Resources (1980) Histologic typing of liver tumors of the rat. *J. natl Cancer Inst.*, *64*, 177-206

Kiese, M. (1966) The biochemical production of ferrihemoglobin-forming derivatives from aromatic amines, and mechanisms of ferrihemoglobin formation. *Pharmacol. Rev.*, *18*, 1091-1161

Kriek, E. & Westra, J.G. (1979) *Metabolic activation of aromatic amines and amides and interaction with nucleic acids*. In: Grover, P.L., ed., *Chemical Carcinogens and DNA*, Vol. 2, Boca Raton, FL, CRC Press, Inc., pp. 1-28

Murrill, E.A., Woodhouse, E.J., Olin, S.S. & Jameson, C.W. (1980) Carcinogenesis testing and analytical chemistry. *Anal. Chem.*, *52*, 1188A- 1195A

Rosenkranz, H.S. & McCoy, E.C. (1981) *Report to the National Cancer Institute* (Contract No. 1-CP-65855), Bethesda, MD, National Cancer Institute, Division of Cancer Cause and Prevention

Squire, R.A. & Levitt, M.H. (1975) Report of a workshop on classification of specific hepatocellular lesions of rats. *Cancer Res.*, *35*, 3214-3223

Squire, R.A., Goodman, D.G., Valerio, M.G., Fredrickson, T.N., Strandberg, J.D., Levitt, M.H., Lingeman, C.H., Harshbarger, J.C. & Dawe, C.J. (1978) *Tumors. Liver and biliary tract*. In: Benirschke, K., Garner, F.M. & Jones, T.C., eds, *Pathology of Laboratory Animals*, Vol. 2, New York, Springer, pp. 1151-1161

Tarone, R.E., Chu, K.C. & Ward, J.M. (1981) Variation in incidence of selected tumours in F344 rats and B6C3F$_1$ mice used as controls in carcinogenesis bioassays. *J. natl Cancer Inst.*, *66*, 1175-1181

Turusov, V.S., ed. (1973) *Pathology of Tumours in Laboratory Animals*, Vol. 1, *Tumours of the Rat*, Part 1 (*IARC Scientific Publications No. 5*), Lyon, International Agency for Research on Cancer

Turusov, V.S., ed. (1976) *Pathology of Tumours in Laboratory Animals*, Vol. 1, *Tumours of the Rat*, Part 2, (*IARC Scientific Publications No. 6*), Lyon, International Agency for Research on Cancer

Turusov, V.S., ed. (1979) *Pathology of Tumours in Laboratory Animals*, Vol. 2, *Tumours of the Mouse* (*IARC Scientific Publications No. 23*), Lyon, International Agency for Research on Cancer

Turusov, V.S. & Takayama, S. (1979) *Tumours of the liver*. In: Turusov, V.S., ed., *Pathology of Tumours in Laboratory Animals*, Vol. 2, *Tumours of the Mouse* (*IARC Scientific Publications No. 23*), Lyon, International Agency for Research on Cancer, pp. 193-233

Ward, J.M., Goodman, D.G., Squire, R.A., Chu, K.C. & Linhart, M.S. (1979) Neoplastic and nonneoplastic lesions in aging (C57BL/6N x C3H/HeN)F$_1$ (B6C3F$_1$) mice. *J. natl cancer Inst.*, *63*, 849-854

Ward, J.M., Sagartz, J.W. & Casey, H.W. (1980) *Pathology of the aging F344 rat (Syllabus)*. In: *Registry of Veterinary Pathology*, Washington DC, Armed Forces Institute of Pathology

THE MONOGRAPHS

AROMATIC AMINES

ANILINE and ANILINE HYDROCHLORIDE

Aniline was considered by a previous Working Group, in June 1973 (IARC, 1974). Since that time, new data have become available, and these have been incorporated into the monograph and taken into consideration in the present evaluation.

1. Chemical and Physical Data

ANILINE

1.1 Synonyms and trade names

Chem. Abstr. Services Reg. No.: 62-53-3
Chem. Abstr. Name: Benzenamine
IUPAC Systematic Name: Aniline
Synonyms: Aminobenzene; aminophen; aniline oil; C.I. Oxidation Base 1; C.I. 76000; phenylamine; appears in early literature as 'benzidam', 'cyanol', 'krystallin' and 'kyanol'
Trade names: Anyvim; Blue Oil

1.2 Structural and molecular formulae and molecular weight

C_6H_7N Mol. wt: 93.1

1.3 Chemical and physical properties of the pure substance

From Windholz (1976) or Northcott (1978), unless otherwise specified
(a) *Description*: Colourless, oily liquid (when freshly distilled) with a characteristic odour and burning taste
(b) *Boiling-point*: 184.4°
(c) *Melting point*: -6.2°
(d) *Density*: d_{20}^{20} 1.022
(e) *Refractive index*: n_D^{20} 1.5863
(f) *Spectroscopic data*: Infra-red, nuclear magnetic resonance, ultra-violet (Sadtler Research Laboratories, Inc., undated) and mass spectral data (Mass Spectrometry Data Centre, 1974) have been reported.

(g) Solubility: Miscible with water (3.5 wt % at 25° and 6.4 wt % at 90°) and with most organic solvents

(h) Viscosity: 4423-4435 Nsm^{-2} at 20°

(i) Volatility: Volatile with steam; vapour pressure, 50 mm at 102°

(j) Stability: Darkens on exposure to air and light; combustible (flash-point, 76°)

(k) Reactivity: Reacts with acids to form salts and with alkali or alkaline earth metals (with evolution of hydrogen) to form metal anilides

1.4 Technical products and impurities

Aniline is available in the US as a chemically pure grade (99.9% minimal purity) and as a technical grade (99.5% minimal purity), with the following specifications: appearance, colourless to pale-yellow liquid; boiling-point, 184.2-184.6°; maximal distillation range (start dry), 1.0°; a maximum of 0.04% moisture; a maximum of 2 mg/kg nitrobenzene; maximal freezing-point, -6.2°; and specific gravity at 25/4°, 1.017-1.019 (Northcott, 1978). Samples taken before 1900 are alleged also to have contained 4-aminobiphenyl as an impurity (Walpole *et al.*, 1952).

Aniline is available in Japan as a technical grade (99% minimal purity), with the following specifications: distillation range, 95% volume in the range 183-186°; a maximum of 50 mg/kg nitrobenzene; minimal freezing-point, -6.5°; and specific gravity at 15/4°, 1.025-1.028.

ANILINE HYDROCHLORIDE

1.1 Synonyms and trade names

Chem. Abstr. Services Reg. No.: 142-04-1

Chem. Abstr. Name: Benzenamine, hydrochloride

IUPAC Systematic Name: Aniline hydrochloride

Synonyms: Aniline chloride; aniline, hydrochloride; aniline salt; anilinium chloride; C.I. 76001; phenylamine hydrochloride

1.2 Structural and molecular formulae and molecular weight

$C_6H_7N \cdot HCl$ Mol. wt: 129.6

1.3 Chemical and physical properties of the pure substance

From Windholz (1976) or Hawley (1977)
(a) *Description*: White crystals
(b) *Boiling-point*: 245°
(c) *Melting-point*: 198°
(d) *Solubility*: Soluble in water (1:1), ethanol and diethyl ether
(e) *Stability*: Darkens on exposure to air and light

1.4 Technical products and impurities

No data were available to the Working Group.

2. Production, Use, Occurrence and Analysis

2.1 Production and use

(a) Production

Aniline was first produced by Unverdorben in 1826 by the dry distillation of indigo. It was isolated from coal-tar in 1834, first synthesized in 1842 and first manufactured commercially in 1847 (Kouris & Northcott, 1963).

Aniline and its hydrochloride have been produced commercially in the US for at least 60 years (US Tariff Commission, 1922). For many years aniline was made by the Bechamp reduction of nitrobenzene with iron and hydrochloric acid; and, until 1966, amination of chlorobenzene with ammonia was used. Presently, aniline is produced in the US and in several western European countries by the catalytic hydrogenation of nitrobenzene. A US plant using the Bechamp process to produce aniline and pigment-grade iron oxide is expected to start up in 1981 (Northcott, 1978).

Aniline has been produced commercially in Japan since before 1945. Although ammonolysis of phenol was reported to be the method employed (Northcott, 1978), all production is now based on catalytic hydrogenation of nitrobenzene.

Total US production of aniline by eight companies in 1979 amounted to 313 million kg (US International Trade Commission, 1980a), compared with 275 million kg produced by eight companies in 1978 (US International Trade Commission, 1979). One US company reported the production of an undisclosed amount (see preamble, p. 20) of aniline hydrochloride in 1979 (US International Trade Commission, 1980a). Imports of aniline through the principal US customs districts in 1979 totaled 19 657 kg; no imports of the hydrochloride were reported (US International Trade Commission, 1980b).

The annual capacity for aniline production in western Europe was approximately 500 million kg in 1979, the Federal Republic of Germany and the UK being the largest producing countries. Aniline is believed to be produced by three companies in Spain, two companies each in Belgium, the Federal Republic of Germany and the UK, and by one company in each of the following countries: France, Italy, Portugal and Switzerland. Aniline hydrochloride is believed to be produced by two companies in the UK and by one company in the Federal Republic of Germany.

Total production of aniline in Japan in 1979 by four companies is estimated to have been 79.1 million kg; approximately 7.3 million kg were exported. Total production of aniline hydrochloride is estimated to have been 45 thousand kg, all of which was used in Japan.

(b) Use

Aniline was used as follows in the US in 1979: 50% for production of 4,4'-methylenediphenyl diisocyanate (MDI) and polymethylene polyphenyl isocyanate (so-called 'polymeric MDI'); 27% for production of rubber chemicals; 6% for synthesis of dyes and intermediates; 5% for hydroquinone production; 3% for production of drugs; and 9% for miscellaneous applications, including production of herbicides and fibres (Anon., 1979).

MDI and polymeric MDI, important intermediates for a variety of polyurethane products, were the subject of an earlier IARC Monograph (IARC, 1979). US production of polymeric MDI by three companies in 1979 was 221 million kg (US International Trade Commission, 1980a).

Among the rubber-processing agents made from aniline are several important vulcanization accelerators and antioxidants, including aldehyde-aniline condensates (e.g., the n-butyraldehyde condensate), guanidines (e.g., I,3-diphenylguanidine), N-phenyl-2-naphthylamine (see IARC, 1978), thiazoles (e.g., 2,2'-dithiobisbenzothiazole and other derivatives of 2-mercaptobenzothiazole), and a variety of derivatives of diphenyl amine (which is made from aniline).

According to The Society of Dyers & Colourists (1971, 1975, 1976, 1978, 1980), 179 dyes can be prepared from aniline. Another 700 dyes listed in the *Colour Index* can be made from aniline; however, very few dyes based on aniline or aniline derivatives are produced in commercially significant quantities (Northcott, 1978); of 33 typical dyes derived from aniline (Kouris & Northcott, 1963), only sixteen were reported to be produced commercially in the US in 1979. Production data were published for six of these; those produced in the largest volumes were Basic Orange 2 (275 thousand kg), Solvent Yellow 14 (159 thousand kg), Acid Red 1 (128 thousand kg) and Acid Black 1 (169 thousand kg). Another dye made from aniline in significant quantities is Direct Orange 102: US production by five companies in 1979 amounted to 192 thousand kg (US International Trade Commission, 1980a).

Hydroquinone was evaluated in an earlier monograph (IARC, 1977). Data on US production of technical-grade hydroquinone in recent years were not available; however, total sales of the three US producers in 1977 amounted to 5.4 million kg (US International Trade Commission, 1978).

Aniline is used as an intermediate for the production of a variety of pharmaceutical products. A major derivative is believed to be acetanilide, which is used in the production of several sulfa drugs, including sulfafurazole and sulfamethoxazole, which were evaluated by the IARC (1980a). Phenazopyridine, another drug based on aniline, was evaluated by the same Working Group (IARC, 1980b).

Of the miscellaneous uses of aniline, production of 2,6-diethylaniline is believed to be significant, since this chemical is a key intermediate in the production of the herbicide 2-chloro-2',6'-diethyl-*N*-(methoxymethyl)- acetanilide; 40-45 million kg were used in the US in 1978. The intermediate produced in the conversion of aniline to MDI, 4,4'-methylenedianiline, is also used to make *trans,trans*-bis-*para*-aminocyclohexyl methane, which is condensed with dodecanedioic acid to produce an alicyclic nylon fibre with silk-like properties.

Use of aniline in western Europe in 1979 is estimated to have been as follows: 60-65% for production of isocyanates (MDI and polymeric MDI); 20-22% for production of rubber chemicals; 9-10% for synthesis of dyes and intermediates; and 4-10% for miscellaneous uses.

Approximately 50% of the aniline consumed in Japan in 1979 went into the production of isocyanates (principally MDI); 30% was used to produce rubber chemicals; 10% for dye manufacture; and 10% for miscellaneous applications, including the production of pharmaceuticals.

Permissible levels of aniline in the working environment have been established by regulation or recommended guidelines in at least 17 countries. These standards are listed in Table 1.

The American Conference of Governmental Industrial Hygienists (1980) has adopted: (1) a threshold limit value-time-weighted average for dermal exposure to aniline and its homologues of 10 mg/m^3 air (2 ppm) for any eight-hour working day or 40-hour work week; and (2) a threshold limit value-short-term exposure limit of 20 mg/m^3 air (5 ppm) for a period of up to 15 min, not to occur more than four times per day.

As of 28 September 1979, notification must be given to the US Environmental Protection Agency (EPA) of any discharge into waterways of mixtures containing 454 kg or more of aniline (US Environmental Protection Agency, 1979). Since the EPA has identified aniline as a toxic waste, as of 19 November 1980, persons who generate, transport, treat, store or dispose of aniline must comply with the regulations of the Federal hazardous waste management programme. Included in the list of hazardous wastes are: 'distillation bottoms from aniline production', 'process residues from aniline extraction from the production of aniline' and 'combined wastewater streams generated from nitrobenzene/aniline production' (US Environmental Protection Agency, 1980, 1981).

Additionally, as of 20 November 1980, shipments of aniline are subject to a variety of labelling, packaging, quantity and shipping restrictions consistent with the designation of aniline as a hazardous material (US Department of Transportation, 1980).

Table 1. National occupational exposure limits for aniline[a]

Country	Year	Concentration (mg/m^3)	(ppm)	Interpretation[b]	Status
Australia	1973	19	5	TWA	Guideline
Belgium	1974	19	5	TWA	Guideline
Bulgaria	1971	–	–	Ceiling	Regulation
Czechoslovakia	1976	5	–	TWA	Regulation
		20	–	Ceiling	
Finland	1975	19	5	TWA	Regulation
German Democratic Republic	1973	30	–	Max. average ceiling (30 min)	Regulation
		10	–	TWA	
Federal Republic of Germany	1979	19	5	TWA	Guideline
Hungary	1974	5	–	TWA[c]	Regulation
Italy	1975	19	–	TWA	Guideline
Japan	1975	19	5	TWA	Guideline
The Netherlands	1973	19	5	TWA[d]	Guideline
Poland	1976	5	–	Ceiling	Regulation
Romania	1975	4	–	TWA	Regulation
		6	–	Ceiling	Regulation
Sweden	1974	19	5	TWA	Guideline
Switzerland	1976	19	5	TWA	Regulation
US[e]	1980	19	5	TWA	Regulation
USSR	1977	0.1	–	Ceiling	Regulation
Yugoslavia	1971	5	0.25	Ceiling	Regulation

[a] From International Labour Office (1977)
[b] TWA - time-weighted average
[c] May be exceeded five times per shift as long as average does not exceed value
[d] Skin irritant notation added
[e] From US Occupational Safety & Health Administration (1980)

2.2 Occurrence

(a) Water and sediments

Aniline has been reported in surface water samples taken from some rivers in the Federal Republic of Germany in concentrations of 0.5-3.7 µg/l (Neurath et al., 1977). It has also been detected in rivers in The Netherlands, both in the Waal River (Meijers & van der Leer, 1976) and in the Rhine delta (Greve & Wegman, 1975). Microgram levels of aniline in river water and soil samples in the Federal Republic of Germany were attributed to industrial chemical waste (Herzel & Schmidt, 1977). Aniline has been identified in the US at unspecified levels in surface water, finished drinking-water and river water; it has also been found in industrial effluents from oil shale recovery and oil refineries, and from chemical and coal conversion plants (Shackelford & Keith, 1976; Hushon et al., 1980).

(b) Food, beverages and animal feeds

In one study carried out in the Federal Republic of Germany, aniline was found to be present in samples of several fruits and vegetables at levels ranging from <0.1-30.9 mg/kg, and in rapeseed cake, an animal feed, at a concentration of 120 mg/kg (Neurath et al., 1977). It has also been found as a volatile component of black tea (Vitzthum et al., 1975).

(c) Tobacco and tobacco smoke

Aniline has been found in tobacco smoke at concentrations of 50.6-577 ng/cigarette (Patrianakos & Hoffmann, 1979). It has also been detected in the steam volatiles from the distillation of the leaves of one type of tobacco (Latakia) (Irvine & Saxby, 1969).

(d) Other

Aniline has been detected in swine manure: 8.2 µg/l in the raw liquid, 16 µg/l after anaerobic digestion and l5 µg/l after aerobic digestion (Yasuhara & Fuwa, 1979).

2.3 Analysis

An IARC manual (Egan et al., 1981) gives selected methods for the analysis of aromatic amines, including aniline. Typical methods for the analysis of aniline are summarized in Table 2.

Table 2. Methods for the analysis of aniline

Sample matrix	Sample preparation	Assay procedure[a]	Limit of detection	Reference
Bulk chemical	Spot on chromatographic paper; spray with sodium nitroprusside solution and ammonium hydroxide	PC[b]	not given	Sugumaran & Vaidyanathan (1978)
	Dissolve in ethanol; add hydrochloric acid and *para*-dimethylaminocinnamaldehyde; add water	S[b]	0.1 mg/l	Qureshi & Khan (1976)
Air	Draw air through silica gel adsorbent; elute with 95% ethanol containing 0.1% heptanol	GC-FID	not given	Wood & Anderson (1975)
Water	Extract with benzene; derivatize with perfluoromethoxypropionic anhydride	GC-ECD	80 mg/l	Kulikova *et al.* (1979)
Cigarette smoke	Trap smoke in hydrochloric acid solution; extract with diethyl ether; add sodium hydroxide; extract with diethyl ether; dry; add pentafluoropropionic anhydride, sodium bicarbonate and ethyl acetate; wash reaction mixture with hydrochloric acid and sodium bicarbonate; elute on chromatotraphic column with hexane and benzene	GC-ECD	50 pg/cigarette	Patrianakos & Hoffmann (1979)

[a] Abbreviations: PC, paper chromatography; S, spectrophotometry; GC-FID, gas chromatography-flame ionization detection; GC-ECD, gas chromatography-electron capture detection

[b] Not specific for aniline

3. Biological Data Relevant to the Evaluation of Carcinogenic Risk to Humans

3.1 Carcinogenicity studies in animals[1]

In many of the early animal experiments, the purity of the aniline used is unknown, and the experimental techniques employed were often inadequate. Bonser (1947) critically reviewed some of these early studies, paying special attention to the pathological findings reported; in particular, the positive results reported by Yamazaki & Sato (1937) were considered unreliable. In other studies, the period of exposure was often insufficient, possibly due to the toxicity of the doses given. The Working Group that convened in 1973 (IARC, 1974) agreed with these assessments and considered some of the more adequate studies available. Since that time new data have become available.

(a) Oral administration

Mouse: Groups of 50 male and 50 female B6C3F$_1$ mice, six weeks of age, were fed diets containing 6000 or 12 000 mg/kg aniline hydrochloride (of indeterminate purity, with at least two impurities detected by thin-layer chromatography) for 103 weeks; surviving mice were sacrificed at 107 weeks. These doses were selected on the basis of a range-finding study [see section 3.2*(a)*]. A group of 50 mice of each sex served as matched controls. Mean body weight gain of treated male mice was depressed throughout the study; that of female mice was similar to that of controls. By the end of the study, 60-86% of the animals were still alive. No tumour was seen in statistically significantly increased incidence in treated mice of either sex when compared with controls (National Cancer Institute, 1978).

Rat: Forty young Osborne-Mendel rats of both sexes were fed diets containing 330 mg/kg aniline hydrochloride for 420-1032 days (average, 654). Hepatomas were found in four rats and sarcomas of the spleen in three; fibrosis and splenic haemorrhage were found in 14 and 23 rats, respectively. The sexes of these animals were not given. No control group was used, but the authors noted the rarity of liver and splenic tumours in the animals they used (White *et al.*, 1948). [The Working Group noted the lack of controls and the inadequate reporting of the data.]

Aniline hydrochloride was administered in the drinking-water to 50 100-day-old random-bred rats (sex unspecified), to provide an intake of 22 mg/day. Half of the rats lived for more than 425 days, and the last rat to day 750; the total doses administered were 14-16.5 g/rat. Only the bladder, liver, spleen and kidneys were examined; no tumours were observed (Druckrey, 1950). [The Working Group noted the low dose administered and that there were no controls.]

---

[1] The Working Group was aware of a completed but as yet unpublished study by oral administration of aniline hydrochloride to rats (IARC, 1979b).

Groups of 50 male and 50 female Fischer 344 rats, six weeks of age, were fed diets containing 3000 or 6000 mg/kg aniline hydrochloride (of indeterminate purity, with at least two impurities detected by thin-layer chromatography) for 103 weeks; surviving rats were sacrificed at 107-108 weeks. These doses were selected on the basis of a range-finding study [see section 3.2 (a)]. A group of 25 rats of each sex served as matched controls. Adequate numbers of rats survived to the end of the experiment: 68, 68 and 54% of males and 64, 88 and 82% of females were still alive in the control, low-dose and high-dose groups, respectively. Fibrosarcomas or sarcomas were seen in a dose-related trend ($P < 0.001$) in the spleen and/or abdominal cavity of 0/25 control males, 5/50 low-dose males and 18/48 high-dose males ($P < 0.001$). Haemangiosarcomas were seen in a dose-related trend ($P < 0.001$) in the spleen and body cavities of 0/25 controls, 19/50 low-dose ($P < 0.001$) and 21/48 high-dose males ($P < 0.001$). In female rats, fibrosarcomas or sarcomas arising in the spleen or body cavities were seen in a dose-related trend ($P < 0.009$) in 0/24 controls, 1/50 low-dose and 7/50 high-dose animals (National Cancer Institute, 1978).

Groups of 28 male Wistar rats received aniline in drinking-water at concentrations of 0, 300, 600 or 1200 mg/l for 80 weeks. No marked adverse effects on body weight were observed. The effective numbers of rats at 80 weeks were 15, 10, 18 and 16 in the control, low-, mid- and high-dose groups, respectively. Papillomas of the forestomach were observed in 0/15, 1/10, 1/18 and 0/16 rats in the four groups, respectively; pituitary adenomas were seen only in 1/10 animals given the low dose (Hagiwara et al., 1980). [The Working Group noted the poor survival and the inadequate duration of treatment.]

(b) Subcutaneous and/or intramuscular injection

Mouse: Repeated s.c. injections of 1.25 mg aniline in lard produced no tumours in 15/30 stock and D mice that survived two years (Shear & Stewart, 1941). No tumours were observed after 15 months in 20 C mice given eight s.c. injections of aniline (5 mg in olive oil), or after 12 months in 11 mice given 13 s.c. injections of aniline hydrochloride (4 mg in aqueous solution) (Hartwell & Andervont, 1951). [No controls were reported.]

(c) Carcinogenicity of metabolites

para-Aminophenol and *ortho*-aminophenol hydrochloride each gave negative results when fed to groups of 12 rats at concentrations of 0.09-0.2% in the diet for periods of 270 to 331 days (Ekman & Strombeck, 1949a,b; Miller & Miller, 1948). *ortho*-Aminophenol was ineffective as a carcinogen in mice that received an implant in a cholesterol pellet in the lumen of the urinary bladder (Clayson et al., 1958). Phenylhydroxylamine did not increase the incidence of tumours in groups of 12-15 rats given repeated i.p. injections two to three times per week for three months and observed for a period of one to two years (Miller et al., 1966; Belman et al., 1968). Repeated s.c. injections of 2.5 mg nitrosobenzene twice weekly for eight weeks produced no tumours in 20 rats observed up to eight months after the start of the experiment. No sarcomas occurred at the injection site in 20 control animals, whereas with

2-nitrofluorene local sarcomas occurred in 10 out of 20 rats (Miller et al., 1965). [The Working Group noted the inadequate conduct and/or reporting of these studies.]

3.2 Other relevant biological data

(a) Experimental systems

Toxic effects

The oral LD_{50} of aniline in male rats was 440 mg/kg bw (Jacobson, 1972; Matrka et al., 1978). The acute dermal LD_{50} on the abraded skin was 840 mg/kg bw in rabbits and 2200 mg/kg bw in guinea-pigs; the dermal LD_{50} on intact skin was 1320 mg/kg bw in guinea-pigs (Roudabush et al., 1965). Exposure of rats to 950 mg/m^3 of air (250 ppm) aniline vapour for 4 h caused the death of approximately 50% of animals (Carpenter et al., 1949).

In mice, rats, rabbit, cats and dogs, the major acute toxic effects of aniline included methaemoglobinaemia and haemolysis (Spicer 1950; Kiese, 1966).

Repeated daily s.c. administration of aniline to rats caused a decrease in steroidogenesis and lipid accumulation in corpora lutea and adrenal cortex (Hatakeyama et al., 1971; Horvath et al., 1971; Kovacs et al., 1971).

In an eight-week subchronic feeding study, Fischer rats and B6C3F$_1$ mice were given diets containing up to 1% aniline hydrochloride. Male and female rats fed the highest concentration had a 25% reduction in mean body weight gain; no effect was seen in mice. In carcinogenicity studies, male mice showed a depression of mean body weight gain and chronic inflammatory biliary lesions. Rats given 0.3 and 0.6% aniline in the diet showed fatty metamorphosis, fibrosis and papillary hyperplasia of the spleen, as well as haemosiderosis of the liver (only at the high dose) and kidney and endometrial stromal polyps. No such dose-related lesions were found in mice (National Cancer Institute, 1978).

Effects on reproduction and prenatal toxicity

No data were available to the Working Group.

Absorption, distribution, excretion and metabolism

In experimental animals, aniline is rapidly absorbed after oral administration, application to the skin or inhalation (Carpenter et al., 1949; Roudabush et al., 1965; Kiese, 1966). After i.v. administration to rats, radiolabelled aniline is distributed throughout the body; highest concentrations of radioactivity were found in blood, liver, kidney, bladder and gut; after 0.5 and 6 h of dosing, radioactivity was concentrated in the stomach and small intestine (Irons et al., 1980).

Plasma clearance of aniline in rats was increased after pretreatment with phenobarbital or benzo[a]pyrene (Wisniewska-Knypl & Jabonska, 1975).

All species tested (rabbit, rat, mouse, guinea-pig, gerbil, hamster, cat and dog) oxidize aniline to *ortho-* and *para-*aminophenol, which are excreted in the urine as various conjugates (Williams, 1959; Parke, 1960). The ratios of these isomers in the urine differ widely in the various species. Small amounts of free aniline, phenylsulphamic acid, and aniline-*N*-glucuronide have been found in the urine of some species after administration of aniline (Boyland et al., 1957; Parke, 1960). *meta-*Aminophenol has been detected in trace quantities in the urine of dogs and rabbits (Parke, 1960).

ortho- and *para-*Aminophenyl- and *para-*acetylaminophenylmercapturic acids are also excreted in rats, and *para-*acetylaminophenol and *para-*acetylaminophenylmercapturic acid in rabbits (Boyland et al., 1963). Acetanilide has been detected in the urine of rabbits but not in the urine of dogs (Williams, 1959).

No phenylhydroxylamine (*N*-hydroxyaniline) has been detected in the urine of animals given aniline; however, phenylhydroxylamine and nitrosobenzene appear in the blood of treated dogs and cats. The formation of phenylhydroxylamine from aniline appears to be the major cause of the methaemoglobinaemia that follows its administration. The *N*-hydroxylation of aniline by hepatic microsomal preparations from several species has been observed *in vitro* (Kiese, 1966).

Mutagenicity and other short-term tests

Aniline did not exhibit preferential toxicity for DNA repair-deficient strains of *Escherichia coli* in the presence or absence of uninduced rat liver microsomes (Rosenkranz & Poirier, 1979); the hydrochloride gave negative results with *Bacillus subtilis* (Kawachi et al., 1980).

Aniline and its hydrochloride have been studied extensively in the *Salmonella typhimurium* mutagenicity assay system, both in the presence and the absence of liver microsomes derived from uninduced or phenobarbital- or Aroclor (PCB)-induced mice, rats and hamsters. The results consistently indicate that aniline and its hydrochloride, tested at doses of up to 2500 µg/plate, are not mutagenic in the standard *Salmonella* mutagenicity assay system (McCann et al., 1975; Sugimura et al., 1976; Garner & Nutman, 1977; Purchase et al., 1978; Rosenkranz & Poirier, 1979; Simmon, 1979a; Kawachi et al., 1980; Mori et al., 1980; Rosenkranz & McCoy, 1981). However, in the presence of norharman (200 µg per plate) and hepatic microsomes from PCB-induced rats, aniline was mutagenic for *Salmonella* strain TA98 (Nagao et al,. 1977).

Aniline and its hydrochloride (same sample as used in the NCI carcinogenicity tests) were not mutagenic for *E. coli* WP2 *uvrA* in the presence of liver microsomes from Aroclor-induced or uninduced rats, hamsters and mice (Rosenkranz & McCoy, 1981).

Aniline did not induce mitotic recombination in *Saccharomyces cerevisiae* D3, even in the presence of liver microsomes derived from Aroclor-induced rats (Simmon, 1979b).

Aniline did not induce DNA damage in primary rat hepatocyte cultures (Williams, 1980). The hydrochloride did, however, induce an increase in sister chromatid exchanges in cultured Chinese hamster pseudodiploid (Don) (Abe & Sasaki, 1977) and lung fibroblasts (Kawachi *et al.*, 1980). Aniline did not induce DNA damage in Chinese hamster lung fibroblasts (V79), as assessed by the alkaline elution technique, even when exposure was for 4 h to 10 mM in the presence of rat microsomes (Swenberg *et al.*, 1976).

The hydrochloride did not induce chromosomal aberrations in cultured Chinese hamster pseudodiploid (Don) (Abe & Sasaki, 1977) or lung fibroblasts (Kawachi *et al.*, 1980).

Aniline has been reported to be slightly, but significantly, mutagenic for the thymidine kinase locus of cultured L5178Y mouse lymphoma cells in the presence of liver microsomes from Aroclor-treated rats (Amacher *et al.*, 1980).

No transformation was observed with up to 10 μg/ml in Syrian hamster embryo cells (Pienta, 1980), or with up to 250 μg/ml in baby hamster kidney cells (BHK21) (Purchase *et al.*, 1978).

Aniline hydrochloride was not mutagenic for the silkworm (*Bombyx mori*) when up to 25 μg were injected into oocytes in the pupal stage (Kawachi *et al.*, 1980; Tazima, 1980).

Injection of rats with aniline hydrochloride did not produce chromosomal aberrations in bone-marrow cells (Kawachi *et al.*, 1980); however, evidence of DNA binding has been reported in rat liver after i.p. injection (Lutz, 1979).

Although aniline *per se* is not mutagenic in the *S. typhimurium* assay system, the urines of male, Sprague-Dawley rats that received 300 mg/kg bw aniline orally were mutagenic for *S. typhimurium* when the assay was performed in the presence of liver microsomes from PCB-induced rats (Tanaka *et al.*, 1980).

(b) Humans

Toxic effects

Single oral doses of 25-65 mg/person of aniline caused a dose-dependent increase in methaemoglobin formation. Doses of 45-65 mg/person also produced a slight increase in serum bilirubin in two subjects (Jenkins *et al.*, 1972).

Effects on reproduction and prenatal toxicity

No data were available to the Working Group.

Absorption, distribution, excretion and metabolism

Metabolic conversion of aniline to urinary conjugates of *para*-aminophenol has been observed in man (Williams, 1959). The urinary excretion of these metabolites has been found to reflect the absorption of aniline vapour through the skin and respiratory tract (Piotrowski, 1957; Dutkiewicz, 1961; Dutkiewicz & Piotriwski, 1961; Piotrowski, 1972). The methaemoglobinaemia produced in humans by aniline probably results from its *N*-hydroxylation.

Mutagenicity and chromosomal effects

No data were available to the Working Group.

3.3 Case reports and epidemiological studies of carcinogenicity in humans

Following the original observations of Rehn (1895), many cases of bladder cancer were reported in men involved in the manufacture or use of aniline (International Labour Office, 1921; Berenblum, 1932; Gehrmann, 1936; Hueper, 1938; Aboulker & Smagghe, 1953). In most of these reports, contact with aniline was not distinguished from contact with other possibly carcinogenic compounds with which it was commonly associated. In addition, the aniline in common use during the early part of the twentieth century contained, as impurities, other compounds now known to be carcinogenic for the bladder (Walpole *et al.*, 1954). Several studies have been undertaken since these problems were recognized which allow an evaluation of the carcinogenicity of aniline *per se*.

Goldblatt (1949) reported details of 99 cases of bladder tumour arising between 1934 and 1947 in some 4000 men employed in two chemical factories in the UK. Three papillomas arose in men involved only in the manufacture of aniline. Larger numbers of tumours, consisting of more carcinomas than papillomas, arose in men exposed only to 1-naphthylamine, to 2-naphthylamine or to benzidine. [The number of men exposed only to aniline was not given, and the possibility that aniline was contaminated with other amines was not addressed.]

Gehrmann *et al.* (1949) observed no bladder tumours, in spite of regular cystoscopic examinations, in the work force of a factory in which only aniline was made for a period of 25 years. The factory was described as having a 'substantial' number of employees, many of whom had been employed there for the full 25 years of operation. Exposure to aniline in the early days of operation of the factory had been sufficient to produce acute toxic symptoms. [The size of the population at risk is not stated, nor was there any indication of follow-up of men who had left the factory.]

Case & Pearson (1954) and Case et al. (1954) reported the results of follow-up of 4622 men employed for more than six months in the UK chemical industry between 1910 and 1952. In 1233 men exposed to aniline but not to benzidine, 1- or 2-naphthylamine, magenta or auramine, one death from bladder cancer was observed compared with 0.83 expected from data on the mortality of men in England and Wales. Among all 4622 men, 127 deaths from bladder cancer were observed, whereas 4.1 were expected.

Vigliani & Barsotti (1961) reported that one case of bladder tumour (a papilloma) occurred between 1931 and 1960 in workers exposed to aniline in six Italian dyestuffs factories. [The size of the population at risk is not given, nor was there any indication that workers who had left the factory were followed up.]

Temkin (1963) observed no bladder tumours in cystoscopic surveys of people involved in the manufacture of aniline in the USSR. [The number of individuals exposed to aniline is not given, but it was said to be 'large'; nor was there any indication that people who had left the factory were followed up.]

4. Summary of Data Reported and Evaluation

4.1 Experimental data

Aniline hydrochloride was tested in mice by dietary administration, producing no carcinogenic effects. It was also tested by oral administration in rats; in one experiment by dietary administration it produced fibrosarcomas, sarcomas and haemangiosarcomas of the spleen or the peritoneal cavity.

Aniline was inactive in bacterial and mammalian DNA repair assays, in tests for mitotic recombination with yeast and in cell transformation assays. It did not induce chromosomal aberrations in mammalian cells or in animals. It was not mutagenic for the silkworm; nor was it mutagenic to *Salmonella typhimurium* unless both norharman and hepatic microsomes were present. Urines from treated rats were mutagenic for *S. typhimurium* with metabolic activation. Aniline induced sister chromatid exchanges in cultured mammalian cells.

4.2 Human data

Aniline has been produced commercially since 1847. Its numerous applications as a chemical intermediate could result in occupational exposure. Contamination of the general environment has been reported to occur.

The high risk of bladder cancer observed originally in workers in the aniline dye industry was probably due to exposure to chemicals other than aniline. Studies of individuals exposed to aniline but to no other known blader carcinogens have shown little evidence of increased risk. The best of these reported one death from bladder cancer in 1223 men producing or using aniline, with 0.83 deaths expected from population rates. The degree of confidence which can be placed in the negative results obtained in the other studies is difficult to assess because of the absence of estimates of expected numbers of bladder cancers and the presumed lack of follow-up of workers who had left the industry.

4.3 Evaluation

There is *limited evidence* [1] for the carcinogenicity of aniline hydrochloride in experimental animals. The available epidemiological data are insufficient to allow a conclusion as to the carcinogenicity of aniline. On the basis of all the available data, no evaluation could be made of the carcinogenicity of aniline to humans.

---

[1] See preamble, pp. 18-19.

5. References

Abe, S. & Sasaki, M. (1977) Chromosome aberrations and sister chromatid exchanges in Chinese hamster cells exposed to various chemicals. *J. natl Cancer Inst.*, *58*, 1635-1641

Aboulker, P. & Smagghe, G. (1953) Bladder tumours in dye workers (21 cases). Usefulness of systemic screening and choice of methods (Fr). *Arch. Mal. prof.*, *14*, 380-386

Amacher, D.E., Paillet, S.C., Turner, G.N., Ray, V.A. & Salsburg, D.S. (1980) Point mutations at the thymidine kinase locus in L5178Y mouse lymphoma cells. II. Test validation and interpretation. *Mutat. Res.*, *72*, 447-474

American Conference of Governmental Industrial Hygienists (1980) *Threshold Limit Values for Chemical Substances and Physical Agents in the Workroom Environment with Intended Changes for 1980*, Cincinnati, OH, p. 9

Anon. (1979) *Chemical Profiles*, 1 October, New York, Schnell Publishing Co., Inc.

Belman, S., Troll, W., Teebor, G. & Mukai, F. (1968) The carcinogenic and mutagenic properties of N-hydroxy-aminonaphthalenes. *Cancer Res.*, *28*, 535-542

Berenblum, I. (1932) Aniline cancer. *Cancer Rev.*, *7*, 337-355

Bonser, G.M. (1947) Experimental cancer of the bladder. *Br. med. Bull.*, *4*, 379-382

Boyland, E., Manson, D. & Orr, S.F.D. (1957) The conversion of arylamines into arylsulphamic acids and arylamine-N-glucosiduronic acids. *Biochem. J.*, *65*, 417-423

Boyland, E., Manson, D. & Nery, R. (1963) Mercapturic acids as metabolites of aniline and 2-naphthylamine. *Biochem. J.*, *86*, 263-271

Carpenter, C.P., Smyth, H.F., Jr & Pozzani, U.C. (1949) The assay of acute vapor toxicity, and the grading and interpretation of results on 96 chemical compounds. *J. ind. Hyg. Toxicol.*, *31*, 343-346

Case, R.A.M. & Pearson, J.T. (1954) Tumours of the urinary bladder in workmen engaged in the manufacture and use of certain dyestuff intermediates in the British chemical industry. II. Further consideration of the role of aniline and of the manufacture of auramine and magenta (Fuchsine) as possible causative agents. *Br. J. ind. Med.*, *11*, 213-216

Case, R.A.M., Hosker, M.E., McDonald, D.B. & Pearson, J.T. (1954) Tumours of the urinary bladder in workmen engaged in the manufacture and use of certain dyestuff intermediates in the British chemical industry. I. The role of aniline, benzidine, *alpha*-naphthylamine, and *beta*-naphthylamine. *Br. J. ind. Med.*, *11*, 75-104

Clayson, D.B., Jull, J.W. & Bonser, G.M. (1958) The testing of *ortho* hydroxy-amines and related compounds by bladder implantation and a discussion of their structural requirements for carcinogenic activity. *Br. J. Cancer*, *12*, 222-230

Druckrey, H. (1950) Contribution to the pharmacology of carcinogenic compounds. Study with aniline (Ger.). *Arch. exp. Pathol. Pharmakol.*, *210*, 137-158

Dutkiewicz, T. (1961) Aniline vapour absorption in men (Pol.). *Med. Pracy*, *12*, 1-14

Dutkiewicz, T. & Piotrowski, J.K. (1961) Experimental investigations on the quantitative estimation of aniline absorption in man. *Pure appl. Chem.*, *3*, 319-323

Egan, H., Fishbein, L., Castegnaro, M., O'Neill, I.K. & Bartsch, H., eds (1981) *Environmental Carcinogens - Selected Methods of Analysis*, Vol. 4, *Some Aromatic Amines and Azodyes in the General and Industrial Environment (IARC Scientific Publications No. 40)*, Lyon, International Agency for Research on Cancer

Ekman, B. & Strombeck, J.P. (1949a) The effect of feeding aniline on the urinary bladder of rats. *Acta pathol. microbiol. scand.*, *26*, 472-479

Ekman, B. & Strombeck, J.P. (1949b) The effect of some splitproducts of 2,3'-azotoluene on the urinary bladder in the rat and their excretion on various diets. *Acta pathol. microbiol. scand.*, *26*, 447-467

Garner, R.C. & Nutman, C.A. (1977) Testing of some azo dyes and their reduction products for mutagenicity using *Salmonella typhimurium* TA 1538. *Mutat. Res*, *44*, 9-19

Gehrmann, G.H. (1936) Papilloma and carcinoma of the bladder in dye workers. *J. Am. med. Assoc.*, *107*, 1436-1439

Gehrmann, G.H., Foulger, J.H. & Fleming, A.J. (1949) *Occupational carcinoma of the bladder.* In: *Proceedings of the 9th International Congress of Industrial Medicine, London, 1948*, Bristol, Wright, p. 472-475

Goldblatt, M.W. (1949) Vesical tumours induced by chemical compounds. *Br. J. ind. Med.*, *6*, 65-81

Greve, P.A. & Wegman, R.C.C. (1975) Determination and occurrence of aromatic amines and their derivatives in Dutch surface waters (Ger.). *Schriftenr. Ver. Wasser-, Boden-, Lufthyg. (Berlin) Dahlem*, *46*, pp. 59-80 [*Chem. Abstr.*, *85*, 130181a]

Hagiwara, A., Masayuki, A., Hirose, M., Nakanowatari, J., Tsuda, H. & Ito, N. (1980) Chronic effects of norharman in rats treated with aniline. *Toxicol. Lett.*, *6*, 71-75

Hartwell, J.L. & Andervont, H.B. (1951) In: Hartwell, J.L., ed., *Survey of Compounds which have been Tested for Carcinogenic Activity (US Public Health Publication No. 149)*, 2nd ed., Washington DC, US Government Printing Office, p. 50

Hatakeyama, S., Kovacs, K., Yeghiayan, E. & Blascheck, J.A. (1971) Aniline-induced changes in the corpora lutea of rats. *Am. J. Obstet. Gynecol.*, *109*, 469-476

Hawley, G.G., ed. (1977) *The Condensed Chemical Dictionary*, 9th ed., New York, Van Nostrand-Reinhold, pp. 60-61

Herzel, F. & Schmidt, G. (1977) Aniline and urea derivatives in water and soil samples (Ger.). *Gesund.-Ing.*, *98*, 221-226 [*Chem. Abstr.*, *89*, 79885m]

Horvath, E., Kovacs, K. & Yeghiayan, E. (1971) Histochemical study of the 'adrenocortical lipid hyperplasia' induced in rats by aniline. *Acta histochem.*, *39*, 154-161

Hueper, W.C. (1938) 'Aniline tumors' of the bladder. *Arch. Pathol.*, *25*, 856-899

Hushon, J., Clerman, R., Small, R., Sood, S., Taylor, A. & Thoman, D. (1980) *An Assessment of Potentially Carcinogenic, Energy-Related Contaminants in Water.* Prepared for US Department of Energy and National Cancer Institute, McLean, VA, MITRE Corporation, p. 102

IARC (1974) *IARC Monographs on the Evaluation of Carcinogenic Risk of Chemicals to Man*, Vol. 4, *Some aromatic amines, hydrazine and related substances, N-nitroso compounds and miscellaneous alkylating agents*, Lyon, pp. 27-39

IARC (1977) *IARC Monographs on the Evaluation of the Carcinogenic Risk of Chemicals to Man*, Vol. 15, *Some Fumigants, the Herbicides 2,4-D and 2,4,5-T, Chlorinated Dibenzodioxins and Miscellaneous Industrial Chemicals*, Lyon, pp. 155-175

IARC (1978) *IARC Monographs on the Evaluation of the Carcinogenic Risk of Chemicals to Man*, Vol. 16, *Some Aromatic Amines and Related Nitro Compounds - Hair Dyes, Colouring Agents and Miscellaneous Industrial Chemicals*, Lyon, pp. 325-341

IARC (1979) *IARC Monographs on the Evaluation of the Carcinogenic Risk of Chemicals to Humans*, Vol. 19, *Some Monomers, Plastics and Synthetic Elastomers, and Acrolein*, Lyon, pp. 303-341

IARC (1980a) *IARC Monographs on the Evaluation of the Carcinogenic Risk of Chemicals to Humans*, Vol. 24, *Some Pharmaceutical Drugs*, Lyon, pp. 275-295

IARC (1980b) *IARC Monographs on the Evaluation of the Carcinogenic Risk of Chemicals to Humans*, Vol. 24, *Some Pharmaceutical Drugs*, Lyon, pp. 163-173

IARC (1981) *Information Bulletin on the Survey of Chemicals Being Tested for Carcinogenicity*, No. 9, Lyon, p. 206

International Labour Office (1921) Cancer of the bladder among workers in aniline factories. *Studies and Reports*, Series F, No. 1, Geneva, pp. 2-26

International Labour Office (1977) *Occupational Exposure Limits for Airborne Toxic Substances (Occupational Safety & Health Series No. 37)*, Geneva, pp. 44-45

Irons, R.D., Gross, E.A. & White, E.L. (1980) Aniline: evidence for an enterogastric cycle in the rat. *Food Cosmet. Toxicol.*, *18*, 393-397

Irvine, W.J. & Saxby, M.J. (1969) Steam volatile amines of Latakia tobacco leaf. *Phytochemistry*, *9*, 473-476

Jacobson, K.H. (1972) Acute oral toxicity of mono- and di-alkyl ring-substituted derivatives of aniline. *Toxicol. appl. Pharmacol.*, *22*, 153-154

Jenkins, F.P., Robinson, J.A., Gellatly, J.B.M. & Salmond, G.W.A. (1972) The no-effect dose of aniline in human subjects and a comparison of aniline toxicity in man and the rat. *Food Cosmet. Toxicol.*, *10*, 671-679

Kawachi, T., Yahagi, T., Kada, T., Tazima, Y., Ishidate, M., Sasaki, M. & Sugiyama, T. (1980) *Cooperative programme on short-term assays for carcinogenicity in Japan*. In: Montesano, R., Bartsch, H. & Tomatis, L., eds, *Molecular and Cellular Aspects of Carcinogen Screening Tests (IARC Scientific Publications No. 27)*, Lyon, International Agency for Research on Cancer, pp. 323-330

Kiese, M. (1966) The biochemical production of ferrihemoglobin-forming derivatives from aromatic amines, and mechanisms of ferrihemoglobin formation. *Pharmacol. Rev.*, *18*, 1091-1161

Kouris, C.S. & Northcott, J. (1963) *Aniline and its derivatives*. In: Kirk, R.E. & Othmer, D.F., eds, *Encyclopedia of Chemical Technology*, 2nd ed., Vol. 2, New York, John Wiley & Sons, pp. 411-419

Kovacs, K., Blascheck, J.A., Yeghiayan, E., Hatakeyama, S. & Gardell, C. (1971) Adrenocortical lipid hyperplasia induced in rats by aniline. A histologic and electron microscopic study. *Am. J. Pathol.*, *62*, 17-34

Kulikova, G.S., Kirichenko, V.E. & Pashkevich, K.I. (1979) Gas-liquid chromatographic determination of aniline derivatives in water. *Zh. anal. Khim.*, *34*, 790-793

Lutz, W.K. (1979) *In vivo* covalent binding of organic chemicals to DNA as a quantitative indicator in the process of chemical carcinogenesis. *Mutat. Res.*, *65*, 289-356

Mass Spectrometry Data Centre (1974) *Eight Peak Index of Mass Spectra*, 2nd ed., Reading, UK, Atomic Weapons Research Establishment

Matrka, M., Rambousek, V. & Zverina, Z. (1978) Toxicity of *p*-substituted derivatives of aniline in experimental rats (Slov.). *Cs. Hyg.*, *23*, 168-172

McCann, J., Choi, E., Yamasaki, E. & Ames, B.N. (1975) Detection of carcinogens as mutagens in the *Salmonella*/microsome test: assay of 300 chemicals. *Proc. natl Acad. Sci. (USA)*, *72*, 5135-5139

Meijers, A.P. & van der Leer, R. C. (1976) The occurrence of organic micropollutants in the river Rhine and the river Maas in 1974. *Water Res.*, *10*, 597-604

Miller, J.A. & Miller, E.C. (1948) The carcinogenicity of certain derivatives of *p*-dimethylaminoazobenzene in the rat. *J. exp. Med.*, *87*, 139-156

Miller, E.C., McKechnie, D., Poirier, M.M. & Miller, J.A. (1965) Inhibition of amino acid incorporation *in vitro* by metabolites of 2-acetylaminofluorene and by certain nitroso compounds. *Proc. Soc. exp. Biol. Med.*, *120*, 538-541

Miller, E.C., Lotlikar, P.D., Pitot, H.C., Fletcher, T.L. & Miller, J.A. (1966) *N*-Hydroxy metabolites of 2-acetylaminophenanthrene and 7-fluoro-2-acetylaminofluorene as proximate carcinogens in the rat. *Cancer Res.*, *26*, 2239-2247

Mori, Y., Niwa, T., Hori, T. & Yoyoshi, K. (1980) Mutagenicity of 3'-methyl-*N,N*-dimethyl-4-amino azobenzene metabolites and related compounds. *Carcinogenesis*, *1*, 121-127

Nagao, M., Yahagi, T., Honda, M., Seino, Y., Matsushima, T. & Sugimura, T. (1977) Demonstration of mutagenicity of aniline and *o*-toluidine by norharman. *Proc. Jpn Acad., Ser. B*, *53*, 34-37

National Cancer Institute (1978) *Bioassay of Aniline Hydrochloride for Possible Carcinogenicity* (Tech. Rep. Ser. No. 130; *DHEW Publ. No. (NIH) 78-1385*), Washington DC, US Government Printing Office

Neurath, G.B., Duenger, M., Pein, F.G., Ambrosius, D. & Schreiber, O. (1977) Primary and secondary amines in the human environment. *Food Cosmet. Toxicol.*, *15*, 275-282

Northcott, J. (1978) *Amines, aromatic (aniline)*. In: Kirk, R.E. & Othmer, D.F., eds, *Encyclopedia of Chemical Technology*, 3rd ed., Vol. 2, New York, John Wiley & Sons, pp. 309-321

Parke, D.V. (1960) The metabolism of [^{14}C]aniline in the rabbit and other animals. *Biochem J.*, *77*, 493-503

Patrianakos, C. & Hoffmann, D. (1979) Chemical studies on tobacco smoke. LXIV. On the analysis of aromatic amines in cigarette smoke. *J. anal. Toxicol.*, *3*, 150-154

Pienta, R.J. (1980) *Transformation of Syrian hamster embryo cells by diverse chemicals and correlation with their reported carcinogenic and mutagenic activities*. In: de Serres, F.J. & Hollaender, A., eds, *Chemical Mutagens*, Vol. 6, New York, Plenum Press, pp. 175-202

Piotrowski, J.K. (1957) Quantitative estimation of aniline absorption through the skin in man. *J. Hyg. Epidemiol. Microbiol. Immunol. (Praha)*, *1*, 23-32

Piotrowski, J.K. (1972) Certain problems of exposure tests for aromatic compounds. *Prac. Lek.*, *24*, 94-97

Purchase, I.F.H., Longstaff, E., Ashby, J., Styles, J.A., Anderson, D., Lefevre, P.A. & Westwood, F.R. (1978) An evaluation of 6 short-term tests for detecting organic chemical carcinogens. *Br. J. Cancer*, *37*, 873-959

Qureshi, M. & Khan, I.A. (1976) Detection and spectrophotometric determination of some aromatic nitrogen compounds with *p*-dimethylaminocinnamaldehyde. *Anal. chim. Acta*, *86*, 309-311

Rehn, L. (1895) Bladder tumours in Fuchsine workers (Ger.) *Arch. klin. Chir.*, *50*, 588-600

Rosenkranz, H.S. & McCoy, E.C. (1981) *Aniline and Aniline Hydrochloride*. (Contract No 1-CP-65855), Bethesda, MD, National Cancer Institute, Division of Cancer Cause and Prevention, pp. 739-747, 837-844

Rosenkranz, H.S. & Poirier, L.A. (1979) Evaluation of the mutagenicity and DNA-modifying activity of carcinogens and noncarcinogens in microbial systems. *J. natl Cancer Inst.*, 62, 873-892

Roudabush, R.L., Terhaar, C.J., Fassett, D.W. & Dziuba, S.P. (1965) Comparative acute effects of some chemicals on the skin of rabbits and guinea pigs. *Toxicol. appl. Pharmacol.*, 7, 559-565

Sadtler Research Laboratories, Inc. (undated) *Sadtler Standard Spectra*, Philadelphia, PA

Shackelford, W.M. & Keith, L.H. (1976) *Frequency of Organic Compounds Identified in Water* (EPA-600/4-76-062), Athens, GA, Environmental Research Laboratory, pp. 9, 58, 59

Shear, M.J. & Stewart, H.L. (1941) In: Hartwell, J.L., ed. (1951) *Survey of Compounds which Have Been Tested for Carcinogenic Activity* (*US Public Health Service Publication No. 149*), 2nd ed., Washington DC, US Government Printing Office, p. 50

Simmon, V.F. (1979a) *In vitro* mutagenicity assays of chemical carcinogens and related compounds with *Salmonella typhimurium*. *J. natl Cancer Inst.*, 62, 893-899

Simmon, V.F. (1979b) *In vitro* assays for recombinogenic activity of chemical carcinogens and related compounds with *Saccharomyces cerevisiae* D3. *J. natl Cancer Inst.*, 62, 901-909

The Society of Dyers & Colourists (1971) *Colour Index*, 3rd ed., Vol. 4, Bradford, Yorkshire, Perkin House, p. 4699

The Society of Dyers & Colourists (1975) *Colour Index*, revised 3rd ed., Vol. 6, Bradford, Yorkshire, Lund Humphries, p. 6405

The Society of Dyers & Colourists (1976) *Colour Index*, revised 3rd ed., *Additions and Amendments*, No. 20, Bradford, Yorkshire, Perkin House, p. 584

The Society of Dyers & Colourists (1978) *Colour Index*, revised 3rd ed., *Additions and Amendments*, No. 28, Bradford, Yorkshire, Perkin House, p. 742

The Society of Dyers & Colourists (1980) *Colour Index*, revised 3rd ed., *Additions and Amendments*, No. 34, Bradford, Yorkshire, Perkin House, p. 861

Spicer, S.S. (1950) Species differences in susceptibility to methemoglobin formation. *J. Pharmacol. exp. Ther.*, 99, 185-194

Sugimura, T., Sato, S., Nagao, M., Yahagi, T., Matsushima, T., Seino, T., Takeuchi, M. & Kawachi, T. (1976) *Overlapping of carcinogens and mutagens*. In: Magee, P.N., Takayama, S., Sugimura, T. & Matsushima, T., eds, *Fundamentals in Cancer Prevention*, Tokyo, University of Tokyo Press/Baltimore, University Park Press, pp. 191-215

Sugumaran, M. & Vaidyanathan, C.S. (1978) A method for detecting free phenol, aniline, indole and quinone derivatives in paper chromatography. *J. Indian Inst. Sci.*, 60, 51-56

Swenberg, J.A., Petzold, G.L. & Harbach, P.R. (1976) *In vitro* DNA damage/alkaline elution assay for predicting carcinogenic potential. *Biochem. biophys. Res. Commun.*, 72, 732-738

Tanaka, K.-I., Marui, S. & Mii, T. (1980) Mutagenicity of extracts of urine from rats treated with aromatic amines. *Mutat. Res.*, 79, 173-176

Tazima, Y. (1980) *Chemical mutagenesis in the silkworm.* In: de Serres, F.J. & Alexander, A., eds, *Chemical Mutagens,* Vol. 6, New York, Plenum, pp. 203-238

Temkin, I.S. (1963) *Industrial Bladder Carcinogenesis,* London, Pergamon, pp. 34-38

US Department of Transportation (1980) Transport of hazardous wastes and hazardous substances. *US Code Fed. Regul., Title 49,* Parts 171-174, 176, 177; *Fed. Regist., 45, No. 101,* 34560, 34594

US Environmental Protection Agency (1979) Water programs; determination of reportable quantities for hazardous substances. *US Code Fed. Regul., Title 40,* Part 117; *Fed. Regist., 44, No. 169,* 50766-50779

US Environmental Protection Agency (1980) Hazardous waste management system: identification and listing of hazardous wastes. *US Code Fed. Regul., Title 40,* Part 261; *Fed. Regist., 45, No. 98,* 33084, 33124-33127

US Environmental Protection Agency (1981) Hazardous waste management system; identification and listing of hazardous wastes. *US Code Fed. Regul., Title 40,* Part 261; *Fed. Regist., 46, No. 11,* 4614-4620

US International Trade Commission (1978) *Synthetic Organic Chemicals, US Production and Sales, 1977 (USITC Publication 920),* Washington DC, US Government Printing Office, pp. 47, 70

US International Trade Commission (1979) *Synthetic Organic Chemicals, US Production and Sales, 1978 (USITC Publication 1001),* Washington DC, US Government Printing Office, pp. 43, 49

US International Trade Commission (1980a) *US Production and Sales 1979 (USITC Publication 1099),* Washington DC, US Government Printing Office, pp. 25, 31, 48, 65-66, 68, 71, 74, 77, 80, 94

US International Trade Commission (1980b) *Imports of Benzenoid Chemicals and Products, 1979 (USITC Publication 1083),* Washington DC, US Government Printing Office, p. 11

US Occupational Safety & Health Administration (1980) Air contaminants. *US Code Fed. Regul., Title 29,* Part 1910.1000, p. 26

US Tariff Commission (1922) *Census of Dyes and Other Synthetic Organic Chemicals, 1921 (Tariff Information Series No. 26),* Washington DC, US Government Printing Office, p. 21

Vigliani, E.C. & Barsotti, M. (1961) Environmental tumors of the bladder in some Italian dye-stuff factories. *Med. Lav., 52,* 241-250

Vitzthum, O.G., Werkhoff, P. & Hubert, P. (1975) New volatile constituents of black tea aroma. *J. agric. Food Chem., 23,* 999-1003

Walpole, A.L., Williams, M.H.C. & Roberts, D.C. (1952) The carcinogenic action of 4-aminodiphenyl and 3:2'-dimethyl-4-aminodiphenyl. *Br. J. ind. Med., 9,* 255-263

Walpole, A.L., Williams, M.H.C. & Roberts, D.C. (1954) Tumours of the urinary bladder in dogs after ingestion of 4-aminodiphenyl. *Br. J. ind. Med., 11,* 105-109

White, F.R., Eschenbrenner, A.B. & White, J. (1948) Oral administration of *p*-aminodimethylaniline, aniline and *p*-aminoazobenzene and the development of tumors in rats. *Unio int. contra cancrum Acta, 6,* 75-78

Williams, R.T. (1959) *Detoxification Mechanisms,* 2nd ed., London, Chapman & Hall Ltd, pp. 430-432, 464-465

Williams, G.M. (1980) *The detection of chemical mutagens/carcinogens by DNA repair and mutagenesis in liver cultures.* In: de Serres, F.J. & Hollaender, A., eds, *Chemical Mutagens*, Vol. 6, New York, Plenum, pp. 61-79

Windholz, M., ed. (1976) *The Merck Index*, 9th ed., Rahway, NJ, Merck & Co., pp. 89-90

Wisniewska-Knypl, J.M. & Jabonska, J.K. (1975) The rate of aniline metabolism *in vivo* in rats exposed to aniline and drugs. *Xenobiotica, 5*, 511-519

Wood, G.O. & Anderson, R.G. (1975) Personal air sampling for vapors of aniline compounds. *Am. ind. Hyg. Assoc. J., 36*, 538-548

Yamazaki, J. & Sato, S. (1937) Experimental production of bladder tumours with aniline and *o*-aminoazotoluene in rabbits (Ger.). *Jpn. J. Dermatol. Urol., 42*, 332-342

Yasuhara, A. & Fuwa, K. (1979) Odor and volatile compounds in liquid swine manure. III. Volatile and odorous components in anaerobically or aerobically digested liquid swine manure. *Bull. chem. Soc. Jpn, 52*, 114-117

ortho- and para-ANISIDINE and their HYDROCHLORIDES

1. Chemical and Physical Data

ortho-ANISIDINE

1.1 Synonyms and trade names

Chem. Abstr. Services Reg. No.: 90-04-0
Chem. Abstr. Name: Benzenamine, 2-methoxy-
IUPAC Systematic Name: o-Anisidine
Synonyms: 2-Aminoanisole; ortho-aminoanisole; 1-amino-2-methoxybenzene; 2-anisidine; ortho-anisylamine; 2-methoxy-1-aminobenzene; 2-methoxyaniline; ortho-methoxyaniline; 2-methoxybenzenamine; ortho-methoxyphenylamine

1.2 Structural and molecular formulae and molecular weight

C_7H_9NO Mol. wt: 123.2

1.3 Chemical and physical properties of the pure substance

From Weast (1979), unless otherwise specified
(a) Description: Yellowish liquid (Windholz, 1976)
(b) Boiling-point: 224°
(c) Melting-point: 6.2°
(d) Density: d_{20}^{20} 1.0923
(e) Refractive index: n_D^{20} 1.5713
(f) Spectroscopic data: Infra-red, nuclear magnetic resonance, ultra-violet (Sadtler Research Laboratories, Inc., undated) and mass spectral data (Mass Spectrometry Data Centre, 1974) have been reported.
(g) Solubility: Slightly miscible with water; miscible with acetone, benzene, diethyl ether and ethanol
(h) Volatility: Volatile in steam (Windholz, 1976); vapour pressure is 40 mm at 132°
(i) Stability: Turns brown in air (Windholz, 1976)

1.4 Technical products and impurities

ortho-Anisidine is available in western Europe as a liquid containing a minimum of 99% active ingredient and in Japan as a pale-yellow to light red-brown liquid containing a minimum of 99% active ingredient and a maximum of 0.2% water, with a minimal freezing-point of 5°.

ortho-ANISIDINE HYDROCHLORIDE

1.1 Synonyms and trade names
Chem. Abstr. Services Reg. No.: 134-29-2
Chem. Abstr. Name: Benzenamine, 2-methoxy-, hydrochloride
IUPAC Systematic Name: *o*-Anisidine hydrochoride
Synonyms: 2-Aminoanisole hydrochloride; *ortho*-aminoanisole hydrochloride; 1-amino-2-methylbenzene hydrochloride; 2-anisidine hydrochloride; *ortho*-anisylamine hydrochloride; C.I. 37115; 2-methoxy-1- aminobenzene hydrochloride; 2-methoxyaniline hydrochloride; *ortho*- methoxyaniline hydrochloride; 2-methoxybenzeneamine hydrochloride; *ortho*- methoxyphenylamine hydrochloride

Trade name: Fast Red BB Base

1.2 Structural and molecular formulae and molecular weight

$C_7H_9NO \cdot HCl$ Mol. wt: 159.6

1.3 Chemical and physical properties of the pure substance

Spectroscopic data: Infra-red, nuclear magnetic resonance and ultra-violet spectral data have been reported (Sadtler Research Laboratories, Inc., undated).

1.4 Technical products and impurities

No data were available to the Working Group

para-ANISIDINE

1.1 Synonyms and trade names

Chem. Abstr. Services Reg. No.: 104-94-9
Chem. Abstr. Name: Benzenamine, 4-methoxy-
IUPAC Systematic Name: *p*-Anisidine
Synonyms: 4-Aminoanisole; *para*-aminoanisole; 1-amino-4-methoxybenzene; 4-anisidine; *para*-anisylamine; 4-methoxy-1-aminobenzene; 4-methoxyaniline; *para*-methoxyaniline; 4-methoxybenzeneamine; *para*-methoxyphenylamine

1.2 Structural and molecular formulae and molecular weight

1.2 Structural and mole C_7H_9NO Mol. wt: 123.2

1.3 Chemical and physical properties of the pure substance

From Weast (1979), unless otherwise specified
(a) Description: Crystals (Windholz, 1976)
(b) Boiling-point: 243°
(c) Melting-point: 57.2°
(d) Density: d_4^{57} 1.071
(e) Refractive index: n_D^{67} 1.5559
(f) Spectroscopic data: Infra-red, nuclear magnetic resonance, ultra-violet (Sadtler Research Laboratories, Inc., undated) and mass spectral data (Mass Spectrometry Data Centre, 1974) have been reported.
(g) Solubility: Soluble in water, acetone and benzene; very soluble in diethyl ether and ethanol

1.4 Technical products and impurities

para-Anisidine is available in western Europe as flakes containing a minimum of 99% active ingredient and in Japan as off-white to light yellowish-brown flakes containing a minimum of 99% active ingredient and a maximum of 0.3% water, with a minimal melting-point of 57°.

para-ANISIDINE HYDROCHLORIDE

1.1 Synonyms and trade names
Chem. Abstr. Services Reg. No.: 20265-97-8
Chem. Abstr. Name: Benzenamine, 4-methoxy-, hydrochloride
IUPAC Systematic Name: *p*-Anisidine hydrochloride
Synonyms: 4-Aminoanisole hydrochloride; *para*-aminoanisole hydrochloride; 1-amino-4-methoxybenzene hydrochloride; 4-anisidine hydrochloride; *para*-anisidine monohydrochloride; *para*-anisylamine hydrochloride; 4-methoxy-1-aminobenzene hydrochloride; 4-methoxyaniline hydrochloride; *para*-methoxyaniline hydrochloride; 4-methoxybenzeneamine hydrochloride; *para*-methoxyphenylamine hydrochloride

1.2 Structural and molecular formulae and molecular weight

$C_7H_9NO \cdot HCl$ Mol. wt: 159.6

1.3 Chemical and physical properties of the pure substance

From Weast (1979), unless otherwise specified
(a) Description: Leaflets or needles
(b) Melting-point: 236°
(c) Solubility: Soluble in water and ethanol
(d) Spectroscopic data: Infra-red, nuclear magnetic resonance and ultra-violet spectral data have been reported (Sadtler Research Laboratories, Inc., undated).

1.4 Technical products and impurities

No data were available to the Working Group

2. Production, Use, Occurrence and Analysis

2.1 Production and use

ortho-ANISIDINE and *ortho*-ANISIDINE HYDROCHLORIDE

(a) Production

ortho-Anisidine was first produced by Muehlhauser in 1881 by the reduction of 2-nitroanisole with tin and hydrochloric acid (Windholz, 1976). It is believed to be made commercially from 2-nitroanisole by either catalytic hydrogenation or the Bechamp reduction (with iron filings and hydrochloric acid).

ortho-Anisidine is believed to be produced by two companies in the Federal Republic of Germany and by one company each in France and Italy. The hydrochloride is not believed to be produced in commercial quantities in western Europe.

Total production of *ortho*-anisidine in Japan by three companies is estimated to have been 669 thousand kg in 1979, as compared with nearly 1 million kg in 1974. Approximately 200 thousand kg of the hydrochloride were produced.

ortho-Anisidine was produced commercially in the US for over 50 years (US Tariff Commission, 1922), but it is not thought to be made there at present. One company reported commercial production of an undisclosed amount (see preamble, p. 20) in 1979 (US International Trade Commission, 1980a), and two reported production in 1978 (US International Trade Commission, 1979a). Three US companies reported production in 1977;

the production of one company was 4.5-45.4 thousand kg (US Environmental Protection Agency, 1980). In 1973, the last year in which total US production data were reported, three companies reported a total of 920 thousand kg (US International Trade Commission, 1975). Commercial production of *ortho*-anisidine hydrochloride was last reported in the US in 1957 (US Tariff Commission, 1958).

Imports of *ortho*-anisidine through the principal US customs districts in 1979 totaled 1.26 million kg; no imports of the hydrochloride were reported (US International Trade Commission, 1980b). Imports in 1978 were 1.12 million kg, principally from the Federal Republic of Germany, Italy and Poland (US International Trade Commission, 1979b).

(b) Use

The principal commercial use of *ortho*-anisidine is believed to be as an intermediate in the manufacture of dyes. It has also been reported to be an intermediate in the manufacture of synthetic guaiacol and its derivatives.

The Society of Dyers & Colourists (1971, 1975, 1976, 1978) reports that 45 dyes and pigments can be prepared from *ortho*-anisidine. Only nine of these were produced commercially in the US in 1979. Separate production data were published for five dyes: Direct Red 72 (five companies; 266 thousand kg); Disperse Orange 29 (five companies; 197 thousand kg); Direct Yellow 44 (three companies; 95 thousand kg sales); Direct Red 24 (five companies; 94 thousand kg); and Acid Red 4 (five companies; 17 thousand kg). Three derivatives of *ortho*-anisidine are intermediates for dyes and pigments produced commercially in the US: *ortho*-acetoacetanisidide (in four pigments); 4,4'-cyclohexylidenedi-*ortho*-anisidine (in one dye); and 3-hydroxy-2-naphth-*ortho*-anisidide (in one dye and two pigments). Separate production data for 1979 are available for three of the pigments made from *ortho*-acetoacetanisidide: Pigment Yellow 17 (489 thousand kg), Pigment Yellow 74 (731 thousand kg) and Pigment Yellow 73 (392 thousand kg) (US International Trade Commission, 1980a). In 1978, 22 thousand kg of another *ortho*- acetoacetanisidide-based pigment, Pigment Yellow 65, were sold. Total US sales of Pigment Red 9, derived from 3-hydroxy-2-naphth-*ortho*-anisidide, in that year were 8 thousand kg (US International Trade Commission, 1979a).

ortho-Anisidinomethanesulphonic acid, a dye intermediate believed to be produced from *ortho*-anisidine, is also manufactured commercially in the US, but no information was available on the quantity produced.

ortho-Anisidine is believed to be used in the commercial manufacture of synthetic guaiacol (*ortho*-methoxyphenol). This chemical, which may also be isolated from beechwood creosote, and several of its derivatives (the glyceryl ether, carbonate and salicylate, as well as potassium guaiacolsulphonate) are used as expectorants in human and veterinary medicine. Combined US sales of guaiacol, its glyceryl ether and potassium guaiacolsulphonate were last reported separately in 1972, when they amounted to 600 thousand kg (US Tariff Commission,

1974). Another derivative, guaiacol glyceryl ether carbamate (methocarbamol), is used as a skeletal muscle relaxant in human medicine.

In western Europe, *ortho*-anisidine is used for the production of dyes, pharmaceuticals and textile-processing chemicals. In Japan, it is used for the production of dyes.

Permissible levels of *ortho*-anisidine in the working environment have been established by regulation or recommended guidelines in at least 10 countries. These standards are listed in Table 1.

Table 1. National occupational exposure limits for ortho- and para-anisidine[a]

Country	Year	Concentration (mg/m^3)	(ppm)	Interpretation[b]	Status
Australia	1973	0.5	–	TWA	Guideline
Belgium	1974	0.5	0.1	TWA	Guideline
Finland	1975	0.5	–	TWA	Regulation
Federal Republic of Germany	1979	0.5	0.1	TWA	Guideline
The Netherlands	1973	0.5	0.1	TWA	Guideline
Romania	1975	0.3	–	TWA	Regulation
		0.5	–	Ceiling	Regulation
Switzerland	1976	0.5	0.1	TWA	Regulation
US[c]	1980	0.5	0.1	TWA	Regulation
USSR	1977	1.0[d]	–	Ceiling	Regulation
Yugoslavia	1971	0.5	–	Ceiling	Regulation

[a] From International Labour Office (1977)
[b] TWA - time-weighted average
[c] From US Occupational Safety & Health Administration (1980)
[d] *para*-Anisidine only

para-ANISIDINE and *para*-ANISIDINE HYDROCHLORIDE

(a) Production

para-Anisidine was first synthesized by Cahours in 1850 by the reduction of *para*-nitroanisole with ethanolic ammonium sulphide (Windholz, 1976). It is believed to be produced commercially from 4-nitroanisole by catalytic hydrogenation or the Bechamp reduction (with iron filings and hydrochloric acid). It can also be prepared by methylation of *para*-aminophenol (Hawley, 1977).

para-Anisidine is believed to be produced by two companies in the Federal Republic of Germany and by one company in France and one in Italy. The hydrochloride is not thought to be produced in commercial quantities in western Europe.

Commercial production of *para*-anisidine in Japan began before 1945; total production in 1979 by five companies was about 1.14 million kg. An estimated 110 thousand kg of the hydrochloride were also produced.

para-Anisidine was first produced in the US as early as 1937 (US Tariff Commission, 1938), but it is not believed to be made there at present. Commercial production in the US was last reported by one company in 1977 in an undisclosed amount (see preamble, p. 20) (US International Trade Commission, 1978). The hydrochloride is available in the US in laboratory quantities only.

In 1979, 183 thousand kg of *para*-anisidine were imported through the principal US customs districts; no imports of the hydrochloride were reported (US International Trade Commission, 1980b).

(b) Use

para-Anisidine is used almost exclusively as an intermediate in the manufacture of dyes. The Society of Dyers & Colourists (1971) reported that seven dyes and pigments can be prepared from it. Two of these were produced commercially in the US in 1979: Azoic Coupling Component 11 (3-hydroxy-2- napth-*para*-anisidide) and Pigment Red 190 (US International Trade Commission, 1980a), but separate production data were not available.

para-Anisidine has been reported to be used in small quantities as an analytical reagent and as an intermediate in the synthesis of compounds in the form of liquid crystals.

In western Europe, *para*-anisidine is used for the production of dyes and pharmaceuticals, and in Japan mainly for the production of dyes.

Permissible levels of *para*-anisidine in the working environment in various countries are listed in Table 1.

2.2 Occurrence

(a) Natural occurrence

ortho- and *para*-anisidine and their hydrochlorides are not known to occur as natural products.

(b) Water and sediments

Both *ortho-* and *para*-anisidine have been identified in the surface water of effluents from chemical plants (Shackelford & Keith, 1976) and from oil refineries (Hushon et al., 1980) in the US.

(c) Tobacco and tobacco smoke

ortho-Anisidine has been identified in cigarette smoke (Patrianakos & Hoffmann, 1979).

2.3 Analysis

An IARC manual (Egan et al., 1981) gives selected methods for the analysis of aromatic amines. Typical methods for the analysis of *ortho-* and *para*-anisidine in various matrices are summarized in Table 2.

Table 2. Methods for the analysis of ortho- and para-anisidine

Sample matrix	Sample preparation	Assay procedure[a]	Limit of detection	Reference
Amine mixtures	Dissolve in isopropanol and hydrochloric acid; elute with aqueous methanol; develop with N,N-dimethyl-*para*-aminobenzaldehyde in ethanol and glacial acetic acid	TLC	not given	Lepri et al. (1974)
	Apply to TLC plate; elute; spray plate with aqueous selenium dioxide or phenol in 20% sodium bicarbonate with an overspray of hydrochloric acid	TLC	1-2 mg	Mitchell & Waring (1978)

Sample matrix	Sample preparation	Assay procedure[a]	Limit of detection	Reference
	Inject into column packed with mixture of aminopropyltri-methoxysilane bonded to copper (II) ion and Partisil-10; elute with methanol and cyclohexane	HPLC-FL	not given	Chow & Grushka (1977)
Air	Draw air through silica gel adsorbant; elute with 95% ethanol containing 0.1% heptanol	GC-FID	not given	Wood & Anderson (1975)
Waste water	Adjust to pH 11 with potassium hydroxide; extract with Freon-TF	GC-FID[b]	0.6 ng	Austern et al. (1975)

[a] Abbreviations: TLC - thin-layer chromatography; HPLC-FL - high-performance liquid chromatography - fluorescence detection; GC-FID - gas chromatography - flame ionization detection
[b] *ortho*-Anisidine only

3. Biological Data Relevant to the Evaluation of Carcinogenic Risk to Humans

3.1 Carcinogenicity studies in animals

ortho-ANISIDINE HYDROCHLORIDE

Oral administration

Mouse: Groups of 55 male and 55 female B6C3F$_1$ mice, six weeks of age, were fed diets containing 2500 or 5000 mg/kg *ortho*-anisidine hydrochloride (purity >99% by titration, corroborated by a combination of spectroscopic and chromatographic techniques) for 103 weeks; surviving mice were sacrificed at 105 weeks. The doses were selected on the basis of a range-finding study [see section 3.2(a)]. A group of 55 mice of each sex served as matched controls. Mean body weight gain of treated male and female mice was depressed as compared with that of controls throughout the study. By the end of the study, 78% of treated male mice, 80% of control males, 69-76% of treated females and 80% of control females were still alive.

Transitional-cell carcinomas of the urinary bladder were found in 0/48 control males, 0/55 low-dose males, 15/53 high-dose males (P < 0.001), 0/50 control females, 0/51 low-dose females and 18/50 high-dose females (P < 0.001). Transitional-cell papillomas were found in 0/48 control males, 2/55 low-dose males, 7/53 high-dose males, 0/50 control females, 1/51 low-dose females and 4/50 high-dose females. Bladder hyperplasia was found in 1/48 control males, 2/55 low-dose males, 21/53 high-dose males, 0/50 control females, 1/51 low-dose females and 12/50 high-dose females (National Cancer Institute, 1978a).

Rat: Groups of 55 male and 55 female Fischer 344 rats, six weeks of age, were fed diets containing 5000 or 10000 mg/kg *ortho*-anisidine hydrochloride (same sample as used above) for 103 and 83-88 weeks, respectively; surviving rats were sacrificed at 106-107 weeks. The doses were selected on the basis of a range-finding study [see section 3.2(a)]. A group of 55 animals of each sex served as matched controls. Mean body weight gain of treated rats was depressed as compared with that of controls throughout the study. A positive dose-related trend in mortality (P < 0.001) was seen for animals of each sex; none of the rats that received the high dose survived to termination of the study, although 80-89% survived past week 52. Mortality was due to the development of tumours of the urinary bladder. Transitional-cell carcinomas were found in a dose-related trend (P < 0.001) in 0/51 control males, 50/54 low-dose males (P < 0.001) and 51/52 high-dose males (P < 0.001), and in 0/49 control females, 41/49 low-dose females (P < 0.001) and 50/51 high-dose females (P < 0.001). Transitional-cell papillomas of the bladder were found in a few treated rats. Transitional-cell carcinomas of the renal pelvis were found in a dose-related trend (P < 0.005) in 3/55 low-dose males (not significant) and in 7/53 high-dose males (P = 0.006); only 1/54 high-dose females had this tumour. Hydronephrosis, epithelial hyperplasia of the urinary tract and renal papillary necrosis were found in some treated rats. Thyroid follicular-cell tumours (adenomas, carcinomas, cystadenomas and cystadenocarcinomas) were found in a dose-related trend (P = 0.009) in 0/53 male controls, 7/40 low-dose males (P = 0.002) and 6/40 high-dose males (P = 0.005); there was no significant increase in the incidence of thyroid tumours in females (National Cancer Institute, 1978a).

para-ANISIDINE HYDROCHLORIDE

Oral administration

Mouse: Groups of 55 male and 55 female B6C3F$_1$ mice, six weeks of age, were fed diets containing 5000 or 10000 mg/kg *para*-anisidine hydrochloride for 103 weeks; surviving mice were sacrificed at 105 weeks. Two separate batches of this substance, with grossly different melting-points and ultra-violet spectra, were used for the bioassay; the ultra-violet spectra indicated that the first batch may not have been *para*-anisidine hydrochloride, while the second batch was substantially impure. The doses were selected on the basis of a range-finding study [see section 3.2(a)]. A group of 55 mice of each sex served as matched controls. Mean body weight gain of treated mice was depressed as compared with that of controls throughout the study. Survival (76-91%) was comparable among control and treated groups. No tumours were found in statistically significantly increased incidences in any treatment group when compared with controls (National Cancer Institute, 1978b).

Rat: Groups of 55 male and 55 female Fischer 344 rats, six weeks of age, were fed diets containing 3000 or 6000 mg/kg *para*-anisidine hydrochloride (same preparations as used above) for 103 weeks; surviving rats were sacrificed at 105-106 weeks. The doses were selected on the basis of a range-finding study [see section 3.2(a)]. A group of 55 animals of each sex served as matched controls. Mean body weight gain of treated rats was depressed as compared with that of controls throughout the study. Survival (65-91%) was comparable among control and treated groups. Preputial gland adenomas or carcinomas were found in 1/54 control males, 8/54 low-dose males (P = 0.016) and 3/55 high-dose males (not significant); the dose-related trend was, however, not significant. A transitional-cell papilloma of the bladder was found in only 1/55 males that received the higher dose; and transitional-cell carcinomas of the bladder were found in only 2/51 females that received the lower dose. None of these findings was statistically significant (National Cancer Institute, 1978b).

3.2 Other relevant biological data

(a) *Experimental systems*

Toxic effects

ortho-ANISIDINE and *ortho*-ANISIDINE HYDROCHLORIDE

The oral LD_{50} of *ortho*-anisidine has been reported to be 2000 mg/kg bw in rats, 1400 mg/kg bw in mice and 870 mg/kg bw in rabbits. Subacute effects include haematological changes, anaemia and nephrotoxicity (Prosolenko, 1975).

In seven-week subchronic feeding studies, male and female $B6C3F_1$ mice were fed diets containing up to 30000 mg/kg *ortho*-anisidine hydrochloride. A dose-dependent depression in mean body weight gain of up to 40% was observed. The spleens of animals given >10000 mg/kg of diet were black and enlarged. When chronic doses of 2500 or 5000 mg/kg of diet were given in long-term tumour induction studies, female mice developed more cystic hyperplasia of the uterus and endometrium than did control animals. At the higher dose level, animals of both sexes had an increased incidence of hyperplasia of the bladder (National Cancer Institute, 1978a).

Subchronic feeding of diets containing up to 30000 mg/kg of diet *ortho-* anisidine hydrochloride to Fischer 344 rats for seven weeks led to reductions in weight gain of up to 52% in males and 27% in females. Feeding of diets containing 1000 or 3000 mg/kg resulted in granular spleens in males but not in females; with >10000 mg/kg, the spleens of animals of both sexes were dark and granular. In carcinogenicity experiments, a dose of 10000 mg/kg of diet resulted in depressions of weight gain of 21% for males and 11% for females. Male and female rats fed diets containing 5000 or 10000 mg/kg *ortho*-anisidine hydrochloride developed non-neoplastic lesions of the thyroid gland and kidney more frequently than did control animals (National Cancer Institute, 1978a).

para-ANISIDINE and *para*-ANISIDINE HYDROCHLORIDE

The oral LD_{50} of *para*-anisidine has been reported to be 810 mg/kg bw (Fitzgerald *et al.*, 1974) and 1300 mg/kg bw in mice, 1400 mg/kg bw in rats and 2900 mg/kg bw in rabbits. Subacute effects include haematological changes, anaemia and nephrotoxicity (Prosolenko, 1975).

Male and female $B6C3F_1$ mice survived the administration of up to 30000 mg/kg *para*-anisidine hydrochloride in the diet for 8 weeks. This dose led to darkened spleens, and to weight depressions of 38 and 29% in male and female animals, respectively (National Cancer Institute, 1978b).

In eight-week subchronic studies, male and female Fischer 344 rats survived administration of up to 10000 mg/kg *para*-anisidine hydrochloride in the diet. The weights of animals given the higher level of compound were reduced by 21 and 13% in male and female animals, respectively, and their spleens were darkened. A daily dose of 30000 mg/kg of diet was lethal (National Cancer Institute, 1978b).

Effects on reproduction and prenatal toxicity

No data were available to the Working Group.

Absorption, distribution, excretion and metabolism

Experiments *in vitro* with microsomes from rat liver showed that the *ortho* isomer is dealkylated more readily than the *para* derivative: the capacity of microsomes to dealkylate *para*-anisidine was increased by 3-methylcholanthrene (Schmidt *et al.*, 1973).

Mutagenicity and other short-term tests

ortho-Anisidine hydrochloride (same sample as used in the carcinogenicity tests) induced reverse mutations in *Salmonella typhimurium* strains TA1538, TA98 and TA100 in the presence of liver microsomes from Aroclor-induced and uninduced mice, rats and hamsters at concentrations of ≥ 333 μg per plate (Dunkel & Simmon, 1980; Rosenkranz & McCoy, 1981).

para-Anisidine was not mutagenic in *S. typhimurium* when tested at up to 2500 μg per plate in the presence of hepatic microsomes from Aroclor-induced rats. Up to 250 μg/ml of this compound did not produce morphological transformations of cultured baby hamster kidney (BHK21) cells in the presence of liver microsomes from Aroclor-induced rats (Purchase et al., 1978).

(b) Humans

No data were available to the Working Group.

3.3 Case reports and epidemiological studies of carcinogenicity in humans

No data were available to the Working Group.

4. Summary of Data Reported and Evaluation

4.1 Experimental data

ortho-Anisidine hydrochloride was tested in mice and rats by dietary administration. It was carcinogenic in both species, producing transitional- cell carcinomas of the urinary bladder. *para*-Anisidine hydrochloride, tested in the same way, did not produce carcinogenic effects in mice, and the data in rats were inadequate for evaluation; there is, in addition, some doubt about the nature of the compound tested.

While *ortho*-anisidine hydrochloride was mutagenic for *Salmonella typhimurium*, *para*-anisidine was not. *para*-Anisidine did not transform morphologically BHK21 cells in culture.

4.2 Human data

ortho-Anisidine has been produced commercially for over 50 years. Its use as an intermediate in the manufacture of dyes, pigments and synthetic guaiacol could result in occupational exposure. *para*-Anisidine has been produced commercially since at least 1937. Its use as a dye intermediate could result in occupational exposure.

No case report or epidemiological study was available to the Working Group.

4.3 Evaluation

There is *sufficient evidence*[1] for the carcinogenicity of *ortho*-anisidine hydrochloride in experimental animals. In the absence of data on humans, *ortho*-anisidine should be regarded, for practical purposes, as if it presented a carcinogenic risk to humans.

The available data were inadequate to evaluate the carcinogenicity of *para*-anisidine hydrochloride in experimental animals. No evaluation of the carcinogenicity of *para*-anisidine to humans could be made.

---

[1] See preamble, p. 18.

5. References

Austern, B.M., Dobbs, R.A. & Cohen, J.M. (1975) Gas-chromatographic determination of selected organic compounds added to waste-water. *Environ. Sci. Technol.*, 9, 588-590

Chow, F.K. & Grushka, E. (1977) Separation of aromatic amine isomers by high pressure liquid chromatography with a copper (II)-bonded phase. *Anal. Chem.*, 49, 1756-1761

Dunkel, V.C. & Simmon, V.F. (1980) *Mutagenic activity of chemicals previously tested for carcinogenicity in the National Cancer Institute Bioassay Program.* In: Montesano, R., Bartsch, H. & Tomatis, L., eds, *Molecular and Cellular Aspects of Carcinogen Screening Tests (IARC Scientific Publications No. 27)*, Lyon, International Agency for Research on Cancer, pp. 283-302

Egan, H., Fishbein, L., Castegnaro, M., O'Neill, I.K. & Bartsch, H., eds (1981) *Environmental Carcinogens - Selected Methods of Analysis*, Vol. 4, *Some Aromatic Amines and Azodyes in the General and Industrial Environment (IARC Scientific Publications No. 40)*, Lyon, International Agency for Research on Cancer

Fitzgerald, T.J., Doull, J. & DeFeo, F.G. (1974) Radioprotective activity of *p*-aminopropiophenone. A structure-activity investigation. *J. med. Chem.*, 17, 900-902

Hawley, G.G., ed. (1977) *The Condensed Chemical Dictionary*, 9th ed., New York, Van Nostrand-Reinhold, p. 62

Hushon, J., Clerman, R., Small, R., Sood, S., Taylor, A. & Thoman, D. (1980) *An Assessment of Potentially Carcinogenic, Energy-Related Contaminants in Water.* Prepared for US Department of Energy and the National Cancer Institute, McLean, VA, MITRE Corporation, p. 103

International Labour Office (1977) *Occupational Exposure Limits for Airborne Toxic Substances (Occupational Safety & Health Series No. 37)*, Geneva, pp. 44, 45

Lepri, L., Desideri, P.G. & Coas, V. (1974) Chromatographic and electrophoretic behaviour of primary aromatic amines on anion-exchange thin layers. *J. Chromatogr.*, 90, 331-339

Mass Spectrometry Data Centre (1974) *Eight Peak Index of Mass Spectra*, 2nd ed., Reading, UK, Atomic Weapons Research Establishment

Mitchell, S.C. & Waring, R.H. (1978) Detection of aminophenols, aromatic amines and related compounds on thin-layer plates. *J. Chromatogr.*, 151, 249-251

National Cancer Institute (1978a) *Bioassay of o-Anisidine Hydrochloride for Possible Carcinogenicity (Tech. Rep. Ser. No. 89; DHEW Publ. No. (NIH) 78-1339)*, Washington DC, US Government Printing Office

National Cancer Institute (1978b) *Bioassay of p-Anisidine Hydrochloride for Possible Carcinogenicity (Tech. Rep. Ser. No. 116; DHEW Publ. No. (NIH) 78-1371)*, Washington DC, US Government Printing Office

Patrianakos, C. & Hoffmann, D. (1979) Chemical studies on tobacco smoke. LXIV. On the analysis of aromatic amines in cigarette smoke. *J. anal. Toxicol.*, 3, 150-154

Prosolenko, N.V. (1975) Comparative toxicological evaluation of methoxyanilines (*o*- and *p*-anisidines). *Tr. Khar'k. Gos. med. Inst.*, 124, 11-14.

Purchase, I.F.H., Longstaff, E., Ashby, J., Styles, J.A., Anderson, D., Lefevre, P.A. & Westwood, F.R. (1978) An evaluation of 6 short-term tests for detecting organic chemical carcinogens. *Br. J. Cancer*, 37, 873-959

Rosenkranz, H.S. & McCoy, E.C. (1981) *o-Anisidine Hydrochloride* (Contract No. 1-CP-65855), Bethesda, MD, National Cancer Institute, Division of Cancer Cause and Prevention, pp. 986-994

Sadtler Research Laboratories, Inc. (undated) *Sadtler Standard Spectra*, Philadelphia, PA

Schmidt, H.L., Moeller, M.R. & Weber, N. (1973) Influence of substrates on the microsomal dealkylation of aromatic *N*-, *O*- and *S*-alkyl compounds (Ger.) *Biochem. Pharmacol.*, 22, 2989-2996

Shackelford, W.M. & Keith, L.H. (1976) *Frequency of Organic Compounds Identified in Water (EPA-600/4-76-062)*, Athens, GA, Environmental Research Laboratory, pp. 9, 60

The Society of Dyers & Colourists (1971) *Colour Index*, 3rd ed., Vol. 4, Bradford, Yorkshire, Lund Humphries, pp. 4696, 4705-4706, 4791

The Society of Dyers & Colourists (1975) *Colour Index*, revised 3rd ed., Vol. 6, Bradford, Yorkshire, Perkin House, p. 6405

The Society of Dyers & Colourists (1976) *Colour Index*, revised 3rd ed., *Additions and Amendments*, No. 20, Bradford, Yorkshire, Perkin House, pp. 584-585

The Society of Dyers & Colourists (1978) *Colour Index*, revised 3rd ed., *Additions and Amendments*, No. 28, Bradford, Yorkshire, Perkin House, p. 742

US Environmental Protection Agency (1980) *Chemicals in Commerce Information System (CICIS)*, Washington DC, Office of Pesticides and Toxic Substances, Chemical Information Division

US International Trade Commission (1975) *Synthetic Organic Chemicals, US Production and Sales, 1973 (USITC Publication 728)*, Washington DC, US Government Printing Office, pp. 22, 28

US International Trade Commission (1978) *Synthetic Organic Chemicals, US Production and Sales, 1977 (USITC Publication 920)*, Washington DC, US Government Printing Office, p. 54

US International Trade Commission (1979a) *Synthetic Organic Chemicals, US Production and Sales, 1978 (USITC Publication 1001)*, Washington DC, US Government Printing Office, pp. 49, 83-84, 90, 98-100, 131, 135-136

US International Trade Commission (1979b) *Imports of Benzenoid Chemicals and Products, 1978 (USITC Publication 990)*, Washington DC, US Government Printing Office, pp. 7, 11

US International Trade Commission (1980a) *Synthetic Organic Chemicals, US Production and Sales, 1979 (USITC Publication 1099)*, Washington DC, US Government Printing Office, pp. 31, 65-67, 71, 76, 79, 81, 84-85, 103, 107, 110

US International Trade Commission (1980b) *Imports of Benzenoid Chemicals and Products, 1979 (USITC Publication 1083)*, Washington DC, US Government Printing Office, p. 12

US Occupational Safety & Health Administration (1980) Air contaminants. *US Code Fed. Regul.*, Title 29, Part 1910.1000

US Tariff Commission (1922) *Census of Dyes and Other Synthetic Organic Chemicals, 1921 (Tariff Information Series No. 26)*, Washington DC, US Government Printing Office, p. 21

US Tariff Commission (1938) *Dyes and Other Synthetic Organic Chemicals in the United States, 1937 (Report No. 132, Second Series)*, Washington DC, US Government Printing Office, p. 12

US Tariff Commission (1958) *Synthetic Organic Chemicals, US Production and Sales, 1957* (*Report No. 203, Second Series*), Washington DC, US Government Printing Office, pp. 63, 161

US Tariff Commission (1974) *Synthetic Organic Chemicals, US Production and Sales, 1972* (*TC Publication 681*), Washington DC, US Government Printing Office, pp. 104, 115

Weast, R.C., ed. (1979) *CRC Handbook of Chemistry and Physics*, 60th ed., Cleveland, OH, Chemical Rubber Co., pp. C-116 - C-117, D-209

Windholz, M., ed. (1976) *The Merck Index*, 9th ed., Rahway, NJ, Merck & Co., p. 91

Wood, G.O. & Anderson, R.G. (1975) Personal air sampling for vapors of aniline compounds. *Am. ind. Hyg. Assoc. J., 36*, 538-548

4-CHLORO-*ortho*-PHENYLENEDIAMINE and 4-CHLORO-*meta*-PHENYLENEDIAMINE

1. Chemical and Physical Data

4-CHLORO-*ortho*-PHENYLENEDIAMINE

1.1 Synonyms and trade names

Chem. Abstr. Services Reg. No.: 95-83-0
Chem. Abstr. Name: 1,2-Benzenediamine, 4-chloro-
IUPAC Systematic Name: 4-Chloro-*ortho*-phenylenediamine
Synonyms: 2-Amino-4-chloroaniline; 4-chloro-1,2-diaminobenzene; 4-chloro-1,2-phenylenediamine; *para*-chloro-*ortho*-phenylenediamine; C.I. 76015; 1,2-diamino-4-chlorobenzene; 3,4-diaminochlorobenzene; 3,4-diamino-1-chlorobenzene
Trade name: Ursol Olive 6G

1.2 Structural and molecular formulae and molecular weight

$C_6H_7ClN_2$ Mol. wt: 142.6

1.3 Chemical and physical properties of the pure substance

From Weast (1979), unless otherwise specified
(a) *Description*: Crystals
(b) *Melting-point*: 76°
(c) *Solubility*: Slightly soluble in water; soluble in benzene and petroleum ether; very soluble in ethanol and diethyl ether
(d) *Spectroscopic data*: Infra-red, nuclear magnetic resonance and ultra-violet spectral data have been reported (Sadtler Research Laboratories, Inc., undated).

1.4 Technical product and impurities

4-Chloro-*ortho*-phenylenediamine is available in the US as a crystalline powder containing maxima of 2% ash and 2% moisture and with a melting-range of 70-73° (Fairmount Chemical Co., Inc., 1972).

4-CHLORO-*meta*-PHENYLENEDIAMINE

1.1 Synonyms and trade names
Chem. Abstr. Services Reg. No.: 5131-60-2
Chem. Abstr. Name: 1,3-Benzenediamine, 4-chloro-
IUPAC Systematic Name: 4-Chloro-*meta*-phenylenediamine
Synonyms: 1-Chloro-2,4-diaminobenzene; 4-chloro-1,3-phenylenediamine; C.I. 76027

1.2 Structural and molecular formulae and molecular weight

$C_6H_7ClN_2$ Mol. wt: 142.6

1.3 Chemical and physical properties of the pure substance

From Weast (1979), unless otherwise specified
(a) Description: Crystals
(b) Melting-point: 91°
(c) Solubility: Soluble in ethanol; slightly soluble in water; insoluble in petroleum ether
(d) Spectroscopic data: Infra-red, nuclear magnetic resonance and ultra-violet spectral data have been reported (Sadtler Research Laboratories, Inc., undated).

1.4 Technical products and impurities

4-Chloro-*meta*-phenylenediamine is available in western Europe as crystals containing 95% active ingredient, or as a 10% solution. Traces of iron and chloronitroanilines may be present.

2. Production, Use, Occurrence and Analysis

2.1 Production and use

4-CHLORO-*ortho*-PHENYLENEDIAMINE

(a) Production

4-Chloro-*ortho*-phenylenediamine was first produced by Laubenheimer in 1876 by the reduction of 4-chloro-1,2-dinitrobenzene (Prager *et al.*, 1930).

It has been produced commercially in the US since 1941-1943 (US Tariff Commission, 1945). At present, one US company manufactures it; production in 1977 was 0.45-4.5 thousand kg (US Environmental Protection Agency, 1980).

It is believed to be produced by one company in the Federal Republic of Germany. It is not produced in commercial quantities in Japan.

(b) Use

4-Chloro-*ortho*-phenylenediamine has been used as a dye intermediate (Fairmount Chemical Co., Inc., 1972). The Society of Dyers & Colourists (1971) reported that this compound can be used as an oxidation base, and one dye can be prepared from it; its use as a hair-dye component has been patented (Tucker, 1967). However, no evidence was found that it is presently employed commercially in the US as a dye or as a dye intermediate.

4-Chloro-*ortho*-phenylenediamine is believed to be used to produce 5-chlorobenzotriazole. An unidentified isomer of chlorobenzotriazole serves as an intermediate and as a photographic chemical (Hawley, 1977).

ortho-Phenylenediamines (including 4-chloro-*ortho*-phenylenediamine) and their salts are permitted for use in cosmetics in the European Communities as oxidizing colouring agents in hair dyes, only if the concentration in the finished product is below 6% (calculated as the free base) (Commission of the European Communities, 1976).

4-CHLORO-*meta*-PHENYLENEDIAMINE

(a) Production

4-Chloro-*meta*-phenylenediamine was first produced by Beilstein and Kurbatow in 1879 by the reduction of 4-chloro-1,3-dinitrobenzene with zinc chloride and hydrochloric acid (Prager *et al.*, 1930). It can also be prepared by the reduction of 1-chloro-2,4-dinitrobenzene with iron.

One US company presently produces undisclosed amounts (see preamble, p. 20) of 4-chloro-*meta*-phenylenediamine. Imports through principal US customs districts were last reported in 1970, when they amounted to 4.5 thousand kg (US Tariff Commission, 1971).

This chemical is believed to be produced by one company in the Federal Republic of Germany. It is not made in commercial quantities in Japan.

(b) Use

4-Chloro-*meta*-phenylenediamine is believed to be used as a dye intermediate and as a rubber-processing agent.

According to The Society of Dyers & Colourists (1971), this compound can be used as an oxidation base, and five dyes can be prepared from it; its use as a hair-dye component has been patented (Tucker, 1967). However, no evidence was found that it is presently used commercially in the US as a dye or as a dye intermediate.

It is believed to be used as a dye intermediate and as a rubber- processing chemical in western Europe. *meta*-Phenylenediamines (including 4-chloro-*meta*-phenylenediamine) salts are permitted for use in cosmetics in the European Communities as oxidizing colouring agents in hair dyes only if the concentration in the finished product is less than 6% (calculated as the free base) (Commission of the European Communities, 1976).

2.2 Occurrence

4-Chloro-*ortho*- and 4-chloro-*meta*-phenylenediamine are not known to occur as natural products. No data on their occurrence in the environment were available to the Working Group.

2.3 Analysis

4-Chloro-*ortho*- and 4-chloro-*meta*-phenylenediamine can be identified in hair dyes by thin-layer chromatography (Kottemann, 1966). An IARC manual (Egan *et al.*, 1981) gives selected methods for the analysis of aromatic amines. No information on other methods of analysis for these compounds was available to the Working Group.

3. Biological Data Relevant to the Evaluation of Carcinogenic Risk to Humans

3.1 Carcinogenicity studies in animals

4-CHLORO-*ortho*-PHENYLENEDIAMINE

Oral administration

Mouse: Groups of 50 male and 50 female B6C3F$_1$ mice, six weeks of age, were fed 4-chloro-*ortho*-phenylenediamine (of indeterminate purity, with at least two impurities found by thin-layer chromatography) in the diet. The concentrations initially administered were 10000 and 20000 mg/kg of diet; at week 34, they were reduced by one-half because of mortality and excess depression of weight gain. Animals were treated for 78 weeks and observed for an additional 17-18 weeks. A group of 50 mice of each sex served as matched controls. By the end of the study, 84, 84 and 70% of males, and 72, 88 and 78% of females were still alive in the control, low-dose and high-dose groups, respectively. The incidences of liver tumours in both male and female treated mice were higher than those in control mice: hepatocellular adenomas plus carcinomas: control males, 15/50 (10 with carcinoma); low-dose males, 28/49 ($P = 0.006$) (18 with carcinoma); high-dose males, 34/47 ($P < 0.001$) (26 with carcinoma); control females, 0/46; low-dose females, 11/48 ($P < 0.001$) (4 with carcinoma); high-dose females, 10/47 ($P < 0.001$) (6 with carcinoma) (National Cancer Institute, 1978a; Weisburger *et al.*, 1980).

Rat: Groups of 50 or 49 male and 50 female Fischer 344 rats, six weeks of age, were fed diets containing 5000 or 10000 mg/kg 4-chloro-*ortho*-phenylenediamine (same sample as used above) for 78 weeks and observed for an additional 27-28 weeks. The doses were selected on the basis of a range-finding study [see section 3.2(*a*)]. A group of 50 rats of each sex served as matched controls. By the end of the study, 64, 80 and 56% of males and 72, 84 and 54% of females were still alive in the control, low-dose and high-dose groups, respectively. Carcinomas of the urinary bladder occurred in a dose-related trend ($P < 0.001$) in treated rats of both sexes: transitional-cell carcinomas, 0/48 male controls, 7/42 low-dose males, and 18/49 high-dose males; papillary or transitional-cell carcinomas, 0/47 female controls, 5/46 low-dose females, and 22/45 high-dose females. Squamous-cell papillomas and carcinomas of the forestomach developed in 4/48 high-dose male rats and squamous-cell papillomas in 3/46 females (National Cancer Institute, 1978a; Weisberger *et al.*, 1980).

4-CHLORO-*meta*-PHENYLENEDIAMINE

Oral administration

Mouse: Two groups of 50 male and 50 or 49 female B6C3F$_1$ mice, 6 weeks of age, were fed 4-chloro-*meta*-phenylenediamine (purity not established, with at least two impurities detected by thin-layer chromatography), at doses of 10000 and 20000 mg/kg of diet for 29 weeks, and then 5000 and 10000 mg/kg of diet for 49 weeks, and were observed for an additional 17 weeks. A group of 50 animals of each sex served as matched controls and were kept under observation for 97 weeks. A dose-related depression in body weight gain was apparent in animals of both sexes. At 78 weeks, five high-dose and five control animals of each sex were sacrificed. By the end of the study, 84, 82 and 80% of males, and 72, 74 and 78% of females were still alive in the control, low-dose and high-dose groups, respectively. A statistically significant (P = 0.01), dose-related increase in the incidence of hepatocellular adenomas plus carcinomas was observed among treated female mice: controls, 0/46; low-dose, 11/44 (P < 0.001) (8 with carcinomas); high-dose, 8/45 (P = 0.003) (5 with carcinomas) (National Cancer Institute, 1978b; Weisberger et al., 1980). [The Working Group noted that the incidence of liver tumours in the control group was lower than that observed in historical controls, and that the incidence in the treated groups was within the normal range seen in historical controls.]

Rat: Two groups of Fischer 344 rats, one of 50 males and 50 females and one of 49 males and 50 females, 6 weeks of age, were fed diets containing 2000 or 4000 mg/kg 4-chloro-*meta*-phenylenediamine (same sample as used above) for 78 weeks and were observed for an additional 26-27 weeks. A group of 50 rats of each sex served as matched controls and were kept under observation for 106 weeks. No depression of body weight gain was observed in treated animals of either sex. At week 78, five high-dose and five control animals were sacrificed. By the end of the study 64, 86 and 73% of males and 72, 78 and 66% of females were still alive in the control, low-dose and high-dose groups, respectively. Statistically significant (P < 0.05) increases in the incidences of adrenal pheochromocytomas and of interstitial-cell tumours of the testis were observed in treated males (National Cancer Institute, 1978b; Weisberger et al., 1980). [The Working Group noted that the observed differences do not appear to be significant when historical controls are taken into account.]

3.2 Other relevant biological data

(a) Experimental systems

Toxic effects

No data on LD_{50} values were available to the Working Group.

In eight-week subchronic feeding studies, male and female Fischer 344 rats and $B6C3F_1$ mice were fed diets containing up to 3% 4-chloro-*ortho*- or 4-chloro-*meta*-phenylenediamine. All rats fed 3% of either compound and all mice fed the *ortho* isomer died; whereas only one mouse fed 3% of the *meta* isomer died by the end of the study. A 25% depression of weight gain was observed in both rats and mice fed 1% of the *ortho* isomer; and reductions of 20% were seen in rats fed 1% of the *meta* isomer and of 10% in mice fed 3% of the *meta* isomer. No compound-related organ lesions were observed in chronic studies in rats with doses of up to 1.4% 4-chloro-*ortho*- and 0.4% 4-chloro-*meta*-phenylenediamine or in mice with doses of 1.4% and 1.4%, respectively (National Cancer Institute, 1978a,b; Weisburger *et al.*, 1980).

Effects on reproduction and prenatal toxicity

No data were available to the Working Group.

Absorption, distribution, excretion and metabolism

No data were available to the Working Group.

Mutagenicity and other short-term tests

4-Chloro-*ortho*-phenylenediamine induced reverse mutations in *Salmonella typhimurium* strain TA98 when assayed in the presence of a rat liver activation system (Yoshikawa *et al.*, 1976).

(b) Humans

No data were available to the Working Group.

3.3 Case reports and epidemiological studies of carcinogenicity in humans

No data were available to the Working Group, but see Appendix 1.

4. Summary of Data Reported and Evaluation

4.1 Experimental data

Both 4-chloro-*ortho*-phenylenediamine and 4-chloro-*meta*-phenylenediamine were tested for carcinogenicity in mice and rats by dietary administration. 4-Chloro-*ortho*-phenylenediamine was carcinogenic in mice, producing hepatocellular carcinomas in animals of both sexes, and in rats of both sexes, producing benign and malignant tumours of the urinary bladder. The results of the studies with 4-chloro-*meta*-phenylenediamine in mice were inconclusive; those in rats were not indicative of a carcinogenic effect.

4-Chloro-*ortho*-phenylenediamine was mutagenic to *Salmonella typhimurium* after metabolic activation. No data were available on the mutagenicity of 4-chloro-*meta*-phenylenediamine.

4.2 Human data

4-Chloro-*ortho*- and 4-chloro-*meta*-phenylenediamine have been produced commercially since the early 1940s. The extent of present human exposure is unknown. Both have been patented for use as hair-dye components.

No case report or epidemiological study was available to the Working Group, but see Appendix 1.

4.3 Evaluation

There is *sufficient evidence*[1] for the carcinogenicity of 4-chloro-*ortho*-phenylenediamine in experimental animals. In the absence of data on humans, 4-chloro-*ortho*- phenylenediamine should be regarded, for practical purposes, as if it presented a carcinogenic risk to humans.

The available data were inadequate to evaluate the carcinogenicity of 4-chloro-*meta*-phenylenediamine in experimental animals. No evaluation of the carcinogenicity of 4-chloro-*meta*-phenylenediamine to humans could be made.

For a consideration of the possible carcinogenic risk of hair dyes, see Appendix 1.

---

[1] See preamble, p. 18.

5. References

Commission of the European Communities (1976) Council Directive of 27 July 1976 on the approximation of the laws of the Member States relating to cosmetic products. *Off. J. Eur. Communities, L262*, 184-185

Egan, H., Fishbein, L., Castegnaro, M., O'Neill, I.K. & Bartsch, H., eds (1981) *Environmental Carcinogens - Selected Methods of Analysis*, Vol. 4, *Some Aromatic Amines and Azodyes in the General and Industrial Environment (IARC Scientific Publications No. 40)*, Lyon, International Agency for Research on Cancer

Fairmount Chemical Co., Inc. (1972) *Production Specification:* p-Chloro-o-Phenylenediamine, Newark, NJ

Hawley, G.G., ed. (1977) *The Condensed Chemical Dictionary*, 9th ed., New York, Van Nostrand-Reinhold, p. 195

Kotteman, C.M. (1966) Two-dimensional thin layer chromatographic procedure for the identification of dye intermediates in arylamine oxidation hair dyes. *J. Assoc. off. anal. Chem., 49*, 954-959

National Cancer Institute (1978a) *Bioassay of 4-Chloro-o-phenylenediamine for Possible Carcinogenicity (Tech. Rep. Ser., No. 63; DHEW Publ. No. (NIH) 78-1313)*, Washington DC, US Government Printing Office

National Cancer Institute (1978b) *Bioassay of 4-Chloro-m-phenylenediamine for Possible Carcinogenicity (Tech. Rep. Ser., No. 85; DHEW Publ. No. (NIH) 78-1335)*, Washington DC, US Government Printing Office

Prager, B., Jacobson, P., Schmidt, P. & Stern, D., eds (1930) *Beilstein's Handbuch der Organischen Chemie*, 4th ed., Vol. 13, Berlin, Springer, pp. 25, 53

Sadtler Research Laboratories, Inc. (undated) *Sadtler Standard Spectra*, Philadelphia, PA

The Society of Dyers & Colourists (1971) *Colour Index*, 3rd ed., Vol. 4, Bradford, Yorkshire, Lund Humphries, pp. 4820, 4822

Tucker, H.H. (1967) Hair coloring with oxidation dye intermediates. *J. Soc. Cosmet. Chem., 18*, 609-628

US Environmental Protection Agency (1980) *Chemicals in Commerce Information System (CICIS)*, Washington DC, Office of Pesticides and Toxic Substances, Chemical Information Division

US Tariff Commission (1945) *Synthetic Organic Chemicals, US Production and Sales, 1941-43 (Report No. 153, Second Series)*, Washington DC, US Government Printing Office, p. 73

US Tariff Commission (1971) *Imports of Benzenoid Chemicals and Products, 1970 (TC Publication 413)*, Washington DC, US Government Printing Office, p. 13

Weast, R.C., ed. (1979) *CRC Handbook of Chemistry and Physics*, 60th ed., Cleveland, OH, Chemical Rubber Co., p. C-153

Weisburger, E.K., Murthy, A.S.K., Fleischman, R.W. & Hagopian, M. (1980) Carcinogenicity of 4-chloro-*o*-phenylenediamine, 4-chloro-*m*-phenylanediamine and 2-chloro-*p*-phenylenediamine in Fischer 344 rats and B6C3F, mice. *Carcinogenesis, 1*, 495-499

Yoshikawa, K., Uchino, H. & Kurata, H. (1976) Studies on the mutagenicity of hair dye (Jpn.). *Eisei Shikensho Hokoku (Bull. natl Inst. Hyg. Sci., Tokyo), 94*, 28-32

meta- and *para*-CRESIDINE

1. Chemical and Physical Data

meta-CRESIDINE

1.1 Synonyms and trade names

Chem. Abstr. Services Reg. No.: 102-50-1
Chem. Abstr. Name: Benzenamine, 4-methoxy-2-methyl-
IUPAC Systematic Name: *meta*-Cresidine
Synonyms: 4-Methoxy-2-methylaniline; 2-methyl-*para*-anisidine; 2-methyl-4-methoxyaniline

1.2 Structural and molecular formulae and molecular weight

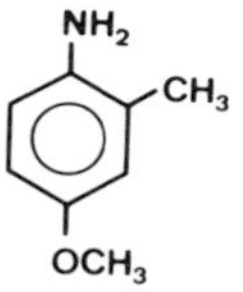

$C_8H_{11}NO$ Mol. wt: 137.2

1.3 Chemical and physical properties of the pure substance

From Weast (1979)
(a) Description: Crystals
(b) Boiling-point: 248-249°
(c) Melting-point: 29-30°
(d) Refractive index: n_D^{20} 1.5647
(e) Solubility: Soluble in ethanol and hot petroleum ether

1.4 Technical products and impurities

No data were available to the Working Group.

para-CRESIDINE

1.1 Synonyms and trade names

Chem. Abstr. Services Reg. No.: 120-71-8
Chem. Abstr. Name: Benzenamine, 2-methoxy-5-methyl-
IUPAC Systematic Name: *para*-Cresidine
Synonyms: *meta*-Amino-*para*-cresol, methyl ether; 1-amino-2- methoxy-5-methylbenzene; 3-amino-4-methoxytoluene; Azoic Red 36; C.I. Azoic Red 83 (Component); cresidine; 2-methoxy-5-methylaniline; 4-methyl-2-aminoanisole; 5-methyl-*ortho*-anisidine
Trade name: Kresidin

1.2 Structural and molecular formulae and molecular weight

$C_8H_{11}NO$ Mol. wt: 137.2

1.3 Chemical and physical properties of the pure substance

From Weast (1979), unless otherwise specified
(a) *Description*: White crystals (Hawley, 1977)
(b) *Boiling-point*: 235°
(c) *Melting-point*: 93-94°
(d) *Solubility*: Soluble in ethanol, diethyl ether, benzene and hot petroleum ether; slightly soluble in hot water
(e) *Volatility*: Volatile with steam (Hawley, 1977)

1.4 Technical products and impurities

para-Cresidine is available in the US as a technical grade containing a minimum of 99% active ingredient, a maximum of 0.5% moisture, with a minimal solidification-point of 51° and producing a clear solution in 10% hydrochloric acid (The Sherwin-Williams Co., 1980). It is available in western Europe as crystals containing a minimum of 99% active ingredient. Preparations available in Japan contain a minimum of 99% active ingredient, with a minimal freezing-point of 51°.

2. Production, Use, Occurrence and Analysis

2.1 Production and use

meta-CRESIDINE

(a) Production

meta-Cresidine is believed to be produced by one company in the Federal Republic of Germany. Is not produced commercially in the US or Japan; and no imports of *meta*-cresidine through the principal US customs districts have been reported since 1976, when 600 kg were imported (US International Trade Commission, 1977a).

(b) Use

The only evidence that *meta*-cresidine has been used commercially is the statement by The Society of Dyers & Colourists (1971) that an azoic dye component, Azoic Coupling Component 24, can be made from *meta*-cresidine; however, the same source indicates elsewhere that this dye component is made from an isomer of *meta*-cresidine, 3-methyl-*para*-anisidine. Commercial production of Azoic Coupling Component 24 in the US was last reported by one company in 1974 (US International Trade Commission, 1976).

para-CRESIDINE

(a) Production

para-Cresidine can be prepared by the methylation of 2-nitro-*para*-cresol to the corresponding methyl ether, followed by reduction of the nitro group to an amine (Hawley, 1977). This is believed to be the method used for commercial production in the US. In Japan, production is believed to be based on the methylation of *para*-cresol, followed by nitration with nitric and sulphuric acids and reduction of the resulting 3-nitro-4-methoxytoluene with iron powder.

para-Cresidine has been produced in the US since 1926 (US Tariff Commission, 1927). Production in 1977 by the one US company believed to produce it at present was 45.4-454 thousand kg (US Environmental Protection Agency, 1980). Imports through the principal US customs districts in 1979 totalled 267 thousand kg (US International Trade Commission, 1980a).

para-Cresidine is believed to be produced by two companies in the Federal Republic of Germany and by one company in Italy. Commercial production in Japan was started prior to 1945; total production in 1979 by two companies is estimated to have been 505 thousand kg.

(b) Use

para-Cresidine is believed to be used almost exclusively as a dye intermediate in the US, western Europe and Japan. The Society of Dyers & Colourists (1971, 1975, 1976, 1977) reported that it can be used as an azoic dye component, as Azoic Red 83 (Component) (subsequently renamed Azoic Red 36) and that 56 dyes can be prepared from it. However, Azoic Red 36 is no longer produced commercially in the US, and only eight of the dyes were reported to be made there in 1979. Separate production data are available for only one of these dyes, Food Red 17 (FD&C Red No. 40): total US production by four companies in 1979 was 1.21 million kg (US International Trade Commission, 1980b). This was a marked increase over the 866 thousand kg reported by five companies in 1978 (US International Trade Commission, 1979); the increase is believed to be related to concern about the possible carcinogenic effect of another red food dye, Food Red 9 (Amaranth; FD&C Red No. 2) (see IARC, 1975). Separate data were published in 1977 for one other of the dyes, Direct Orange 72, when five companies produced 107 thousand kg (US International Trade Commission, 1978). Total US sales of another of the dyes, Direct Orange 34, in 1976 by four companies amounted to 28 thousand kg (US International Trade Commission, 1977b). In 1974, separate data were published on the following dyes that are also based on *para*-cresidine: Direct Violet 9 (five companies; 72 thousand kg produced); Direct Orange 34 (seven companies, 28 thousand kg produced); and Direct Blue 126 (four companies; 22 thousand kg sold) (US International Trade Commission, 1976).

Commercial production of the following *para*-cresidine derivatives, all of which are believed to be dye intermediates, was reported in the US in 1979: 5-methyl-*ortho*-anisidinesulphonic acid (also reported as 4-amino-5-methoxy-2-methylbenzenesulphonic acid); 4-[(4-amino-5- methoxy-*ortho*-tolyl)azo]-4-hydroxy-2,7-naphthalenedisulphonic acid, benzene-sulphonate; 3-[(4-amino-5-methoxy-*ortho*-tolyl)azo]-1,5 naphthalenedisulphonic acid; and 7-[(4-amino-5-methoxy-*ortho*-tolyl)azo]-1,3-naphthalenedisulphonic acid (US International Trade Commission, 1980b).

2.2 Occurrence

meta-Cresidine and *para*-cresidine are not known to occur as natural products. No data on their occurrence in the environment were available to the Working Group.

2.3 Analysis

An IARC manual (Egan *et al.*, 1981) gives selected methods for the analysis of aromatic amines, including *meta*- and *para*-cresidine. No other data on methods of analysis for these compounds were available to Working Group.

3. Biological Data Relevant to the Evaluation of Carcinogenic Risk to Humans

3.1 Carcinogenicity studies in animals

meta-CRESIDINE

Oral administration

Mouse: Groups of 50 male and 50 female B6C3F$_1$ mice, six weeks of age, were administered *meta*-cresidine (of indeterminate purity with at least three impurities detected by gas chromatography) by gavage in corn oil on 5 days per week at two dosage schedules: 0.08 g/kg bw for 32 weeks, followed by 0.02 g/kg bw for 21 weeks and an observation period of 40-41 weeks; or 0.16 g/kg bw for 32 weeks, followed by 0.04 g/kg for 21 weeks and observation periods of 25 weeks for males and 41 weeks for females. The doses were selected on the basis of a range-finding study [see section 3.2(*a*)]. A group consisting of 50 mice of each sex served as untreated controls; groups of 25 mice of each sex received the vehicle only. There were erratic depressions in mean body weight gain throughout the study in all treated males and females that received the high dose. Among male mice, there was a positive association between dosage and mortality ($P < 0.001$): the median survival time for high-dose males was 26 weeks, and that for low-dose males was 52 weeks; these survival times were inadequate for an evaluation of carcinogenicity. There was a significantly higher mortality in high-dose females: by the end of the study, 80-84% of control females, 86% of low-dose females and 60% of the high-dose females were still alive. The incidences of hepatocellular adenomas plus carcinomas in female mice were 0/23 in vehicle controls, 1/50 (1 carcinoma) in low-dose mice and 5/45 (4 carcinomas) in high-dose mice, indicating a significant positive association ($P = 0.027$) between dose and incidence (National Cancer Institute, 1978).

Rat: Groups of 50 male and 50 female Fischer 344 rats were administered the same sample as above of *meta*-cresidine by gavage in corn oil at dosages of 0.08 or 0.16 g/kg bw for 77 weeks and observed for an additional 32-33 weeks. The doses were selected on the basis of a range-finding study [see section 3.2(*a*)]. A group consisting of 50 rats of each sex served as untreated controls, and groups of 25 animals of each sex as vehicle controls. There was a depression in mean body weight gain among high-dose male rats during the first 48 weeks, but thereafter their mean body weight gain was similar to that of controls. Mean body weight gains of low-dose males and of all treated females were similar to those of controls throughout the study. Among males, there was a positive association between dosage and mortality: 64% of the untreated control males, 68% of the vehicle control males, 62% of the low-dose males and 48% of the high-dose males survived to the end of the study. Among females, there was a significantly greater mortality in the high-dose group: 52% of the vehicle controls, 72% of the untreated controls, 57% of the low-dose and 44% of the high-dose females survived to the end of the study. Among male rats, there was a dose-related (P = 0.014) increase in the incidence of transitional-cell carcinomas of the urinary bladder: vehicle controls, 0/25; low-dose rats, 0/45; high-dose rats, 5/44. Among females, transitional-cell carcinomas were found in 0/24 vehicle controls, 1/46 low-dose animals and 2/44 high-dose animals. Hyperplasia of the renal pelvis was found in 26/45 high-dose males and 29/45 high-dose females, but not in untreated or vehicle control or low-dose rats (National Cancer Institute, 1978).

para-CRESIDINE

Oral administration

Mouse: Groups of 50 male and 50 female B6C3F$_1$ mice, six weeks of age, were fed diets containing *para*-cresidine (of indeterminate purity, with at least three impurities detected by thin-layer chromatography) dissolved in acetone. Low-dose males and females were fed 5000 mg/kg of diet for 21 weeks and 1500 mg/kg for 83 additional weeks, followed by an observation period of two weeks. High-dose males and females mice received diets containing 10000 mg/kg for 21 weeks and 3000 mg/kg for 71 weeks (males) or 83 weeks (females); females were observed for an additional two weeks. The doses were selected on the basis of a range finding study [see section 3.2(*a*)]. A group consisting of 50 animals of each sex served as matched controls. Mean body weight gain was depressed among treated males and females as compared with that of controls throughout the study. Mortality was dose-related and associated with the development of bladder tumours. Of the males, 98% of the controls, 50% of the low-dose mice and 10% of the high-dose mice lived longer than 75 weeks; of the females, 90% of the controls, 78% of the low-dose mice and 28% of the high-dose mice survived past 75 weeks. The incidence of squamous-cell or transitional-cell carcinoma of the urinary bladder was dose-related ($P < 0.001$): 0/50 in control males, 40/42 in low-dose males ($P < 0.001$), 31/31 in high-dose males ($P < 0.001$), 0/45 in control females, 41/46 in low-dose females ($P < 0.001$) and 44/46 in high-dose females ($P < 0.001$). Most of the bladder carcinomas were squamous-cell carcinomas; several metastasized to the lung or peritoneal cavity. A few treated mice also had sarcomas of the urinary bladder. Malignant tumours of the nasal cavity were found in 0/50 control males, 2/47 low-dose males, 1/41 high-dose males, 0/47 control females, 0/47 low-dose females and 1/48 high-dose females. Hepatocellular carcinomas or adenomas were found in 0/45 control females, 14/44 low-dose females (13 carcinomas) ($P < 0.001$) and 6/40 high-dose females (6 carcinomas) ($P = 0.009$); one of these carcinomas (in a low-dose female) metastasized to the lung (National Cancer Institute, 1979).

Rat: Groups of 50 male and 50 female Fischer 344 rats, six weeks of age, were fed diets containing 5000 or 10000 mg/kg *para*-cresidine (same sample as used above) dissolved in acetone for 104 weeks and were observed for an additional one or two weeks. The doses were selected on the basis of a range-finding study [see section 3.2(a)]. A group of 50 animals of each sex served as matched controls. There was a dose-related depression in mean body weight gain throughout the study. By 75 weeks, 94, 96 and 62% of males, and 96, 98 and 76% of females were still alive in the control, low-dose and high-dose groups, respectively. Mortality was associated with the development of tumours of the urinary bladder and nasal cavity. Carcinomas of the urinary bladder, primarily of the transitional-cell type, were found in a dose-related trend in 0/48 control males, 16/48 low-dose males, 41/47 high-dose males, 0/47 control females, 31/49 low-dose females and 41/46 high-dose females. Bladder papillomas were found in 0/48 control males, 14/48 low-dose males, 3/47 high-dose males, 0/47 control females, 6/49 low-dose females and 2/46 high-dose females. A few rats had leiomyosarcomas of the bladder; and most of the treated rats that did not have bladder neoplasms had bladder epithelial hyperplasia. Olfactory neuroblastomas were found in 0/48 control males, 1/50 low-dose males and 21/47 high-dose males ($P < 0.001$), and in 0/50 control females, 0/50 low-dose females and 11/49 high-dose females ($P < 0.001$). A few treated rats had other types of malignant nasal tumours or epithelial hyperplasia of the nasal epithelium. Neoplastic liver nodules, hepatocellular carcinomas or hepato-/cholangiocarcinomas were found in 0/48 control males, 13/49 low-dose males ($P < 0.001$) and 2/46 high-dose males; and neoplastic liver nodules were found in 0/50 control females, 4/48 low-dose females and 0/48 high-dose females (National Cancer Institute, 1979).

3.2 Other relevant biological data

(a) Experimental systems

Toxic effects

meta-CRESIDINE

No LD_{50} values were available to the Working Group.

In seven-week subchronic studies, male and female Fischer 344 rats and $B6C3F_1$ mice were given *meta*-cresidine in corn oil by intubation at doses of up to 300 mg/kg bw (rat) and up to 600 mg/kg bw (mouse) on five days a week. The male rats showed a dose-dependent depression in mean body weight gain of up to 27%; female mice showed depressions of up to 18%. The female rats and male mice showed less pronounced weight reductions. Chronic administration of 160 mg/kg bw per day led to hyperplastic and degenerative lesions of the urinary system in rats, and to dose-dependent nephrotoxicity (especially in males) in mice (National Cancer Institute, 1978).

para-CRESIDINE

The oral LD_{50} of *para*-cresidine in rats was 1450 mg/kg bw (Lewis & Tatken, 1979).

In chronic feeding studies, male and female Fischer 344 rats and B6C3F$_1$ mice exhibited depression of body weight gain with doses as low as 0.5% (mice) and 1% (rats). Chronic feeding of this compound led to epithelial hyperplasia of the urinary bladder and nasal cavity in rats, and to hydronephrosis, cystic hyperplasia of the uterus and splenic lesions in mice (National Cancer Institute, 1979).

Effects on reproduction and prenatal toxicity

No data were available to the Working Group.

Absorption, distribution, excretion and metabolism

No data were available to the Working Group.

Mutagenicity and other short-term tests

meta-Cresidine (same sample as used in the carcinogenicity tests) was mutagenic in neither *Escherichia coli* WP2uvrA nor *Salmonella typhimurium* strains TA1537, TA1538, TA98 or TA100 when tested in the presence or absence of Aroclor-induced or uninduced hamster, rat or mouse liver microsomes (Rosenkranz & McCoy, 1981).

para-Cresidine (same sample as used in the carcinogenicity tests) induced reverse mutations in *S. typhimurium* strains TA1538, TA98 and TA100 in the presence of liver microsomes from Aroclor 1254-induced or uninduced rats, hamsters or mice (Rosenkranz & McCoy, 1981).

(b) Humans

No data were available to the Working Group.

3.3 Case reports and epidemiological studies of carcinogenicity in humans

No data were available to the Working Group.

4. Summary of Data Reported and Evaluation

4.1 Experimental data

meta- and *para*-Cresidine (technical grades) were tested in mice and rats by dietary administration. In the only experiment in rats with *meta*-cresidine, it produced an increased incidence of transitional-cell carcinomas of the urinary bladder in males. The results of the study in mice were inconclusive. *para*-Cresidine produced malignant tumours of the urinary bladder in both mice and rats, olfactory neuroblastomas in rats of both sexes and liver tumours in male rats; it also produced nasal cavity tumours in male mice and liver-cell tumours in female mice.

Technical-grade *meta*-cresidine was not mutagenic to *Salmonella typhimurium* or *Escherichia coli*. *para*-Cresidine was mutagenic to *S.typhimurium* after metabolic activation.

4.2 Human data

Although *meta*-cresidine is produced commercially, the extent to which humans are exposed to it is unknown.

para-Cresidine has been produced commercially since 1926. Its use as a dye intermediate could result in occupational exposure, and such exposure may have increased markedly in recent years as a result of the greater production of *para*-cresidine-based food dyes.

No case report or epidemiological study was available to the Working Group.

4.3 Evaluation

The available data were inadequate to evaluate the carcinogenicity of *meta*-cresidine in experimental animals. No evaluation of the carcinogenicity of *meta*-cresidine to humans could be made.

There is *sufficient evidence*[1] for the carcinogenicity of *para*-cresidine in experimental animals. In the absence of data on humans, *para*-cresidine should be regarded, for practical purposes, as if it presented a carcinogenic risk to humans.

---

[1] See preamble, p. 18.

5. References

Egan, H., Fishbein, L., Castegnaro, M., O'Neill, I.K. & Bartsch, H., eds (1981) *Environmental Carcinogens - Selected Methods of Analysis*, Vol. 4, *Some Aromatic Amines and Azodyes in the General and Industrial Environment (IARC Scientific Publications No. 40)*, Lyon, International Agency for Research on Cancer

Hawley, G.G., ed. (1977) *The Condensed Chemical Dictionary*, 9th ed., New York, Van Nostrand-Reinhold, p. 560

IARC (1975) *IARC Monographs on the Evaluation of Carcinogenic Risk of Chemicals to Man*, Vol. 8, *Some aromatic azo Compounds*, Lyon, pp. 41-51

Lewis, R.J., Sr & Tatken, R.L., eds (1979) *1978 Registry of Toxic Effects of Chemical Substances*, Cincinnati, OH, US Department of Health, Education, & Welfare, p. 116

National Cancer Institute (1978) *Bioassay of m-Cresidine for Possible Carcinogenicity (Tech. Rep. Ser. No. 105; DHEW Publ. No. (NIH) 78-1355)*, Washington DC, US Government Printing Office

National Cancer Institute (1979) *Bioassay of p-Cresidine for Possible Carcinogenicity (Tech. Rep. Ser. No. 142; DHEW Publ. No. (NIH) 79-1397)*, Washington DC, US Government Printing Office

Rosenkranz, H.S. & McCoy, E.C. (1981) meta-*Cresidine*, para-*Cresidine* (Contract No. I-CP-65855), Bethesda, MD, National Cancer Institute, Division of Cancer Cause and Prevention, pp. 1-10, 871-879

The Sherwin-Williams Co. (1980) *p-Cresidine (Technical Bulletin 131)*, Cleveland, OH

The Society of Dyers & Colourists (1971) *Colour Index*, 3rd ed., Vol. 4, Bradford, Yorkshire, Lund Humphries, pp. 4086, 4358, 4706

The Society of Dyers & Colourists (1975) *Colour Index*, revised 3rd ed., Vol. 6, Bradford, Yorkshire, Perkin House, pp. 6405-6406

The Society of Dyers & Colourists (1976) *Colour Index*, revised 3rd ed., *Additions and Amendments*, No. 20, Bradford, Yorkshire, Perkin House, p. 585

The Society of Dyers & Colourists (1977) *Colour Index*, revised 3rd ed., *Additions and Amendments*, No. 25, Bradford, Yorkshire, Perkin House, p. 683

US Environmental Protection Agency (1980) *Chemical in Commerce Information System (CICIS)*, Washington DC, Office of Pesticides and Toxic Substances, Chemical Information Division

US International Trade Commission (1976) *Synthetic Organic Chemicals, US Production and Sales, 1974 (USITC Publication 776)*, Washington DC, US Government Printing Office, pp. 54-55, 65, 68-70

US International Trade Commission (1977a) *Imports of Benzenoid Chemicals and Products, 1976 (USITC Publication 828)*, Washington DC, US Government Printing Office, p. 21

US International Trade Commission (1977b) *Synthetic Organic Chemicals, US Production and Sales, 1976 (USITC Publication 833)*, Washington DC, US Government Printing Office, pp. 77, 92

US International Trade Commission (1978) *Synthetic Organic Chemicals, US Production and Sales, 1977 (USITC Publication 920)*, Washington DC, US Government Printing Office, pp. 96, 112

US International Trade Commission (1979) *Synthetic Organic Chemicals, US Production and Sales, 1978* (*USITC Publication 1001*), Washington DC, US Government Printing Office, pp. 86, 111

US International Trade Commission (1980a) *Imports of Benzenoid Chemicals and Products, 1979* (*USITC Publication 1083*), Washington DC, US Government Printing Office, p. 24

US International Trade Commission (1980b) *Synthetic Organic Chemicals, US Production and Sales, 1979* (*USITC Publication 1099*), Washington DC, US Government Printing Office, pp. 29, 49, 68-69, 72, 92

US Tariff Commission (1927) *Census of Dyes and Other Synthetic Organic Chemicals, 1926* (*Tariff Information Series No. 35*), Washington DC, US Government Printing Office, p. 27

Weast, R.C., ed. (1979) *CRC Handbook of Chemistry and Physics*, 60th ed., Cleveland, OH, Chemical Rubber Co., p. C-520

2,4-DIAMINOANISOLE and 2,4-DIAMINOANISOLE SULPHATE

These substances were considered by a previous Working Group, in June 1977 (IARC, 1978). Since that time new data have become available, and these have been incorporated into the monograph and taken into consideration in the present evaluation.

1. Chemical and Physical Data

2,4-DIAMINOANISOLE

1.1 Synonyms and trade names

Chem. Abstr. Services Reg. No.: 615-05-4
Chem. Abstr. Name: 1,3-Benzenediamine, 4-methoxy-
IUPAC Systematic Name: 4-Methoxy-*meta*-phenylenediamine
Synonyms: C.I. 76050; C.I. Oxidation Base 12; 2,4-DAA; 2,4-diaminoanisol; 2,4-diaminoanisole base; *meta*-diaminoanisole 1,3-diamino-4-methoxybenzene; 2,4-diamino-1-methoxybenzene; *para* methoxy-*meta*-phenylenediamine; 4-MMPD
Trade names: Furro L; Pelagol DA; Pelagol Grey L; Pelagol L

1.2 Structural and molecular formulae and molecular weight

$C_7H_{10}N_2O$ Mol. wt: 138.2

1.3 Chemical and physical properties of the pure substance

From Weast (1979), unless otherwise specified
(a) *Description*: Needles (when crystallized from diethyl ether)
(b) *Melting-point*: 67-68°
(c) *Spectroscopic data*: Infra-red, nuclear magnetic resonance and ultra-violet spectral data have been reported (Sadtler Research Laboratories, Inc., undated).
(d) *Solubility*: Soluble in ethanol and hot diethyl ether

1.4 Technical products and impurities

2,4-Diaminoanisole is available in Japan with a minimum purity of 98.5%.

2,4-DIAMINOANISOLE SULPHATE

1.1 Synonyms and trade names

Chem. Abstr. Services Reg. No.: 39156-41-7 (6219-67-6 for the salt with an unknown ratio of sulphuric acid to diamine)

Chem. Abstr. Name: 1,3-benzenediamine, 4-methoxy, sulfate (1:1)

IUPAC Systematic Name: 4-Methoxy-*meta*-phenylenediamine sulphate

Synonyms: C.I. 76051; C.I. Oxidation Base 12A; 2,4-DAA sulphate; 2,4-diaminoanisol sulphate; 1,3-diamino-4-methoxybenzene sulphate: 2,4-diamino-1-methoxybenzene sulphate; *para*-methoxy-*meta*-phenylenediamine sulphate; 4-MMPD sulphate

Trade names: BASF Ursol SLA; Durafur Brown MN; Fouramine BA; Fourrine 76; Fourrine SLA; Furro SLA; Nako TSA; Pelagol BA; Pelagol Grey SLA; Pelagol SLA; Renal SLA; Ursol SLA; Zoba SLE

1.2 Structural and molecular formulae and molecular weight

$C_7H_{10}N_2O \cdot H_2SO_4$ Mol. wt: 263.3

1.3 Chemical and physical properties of the pure substance

(a) *Solubility*: Soluble in water and ethanol; insoluble in sodium hydroxide (The Society of Dyers & Colourists, 1971a)

(b) *Spectroscopic data*: Infra-red, nuclear magnetic resonance and ultra-violet spectral data have been reported (Sadtler Research Laboratories, Inc., undated).

1.4 Technical products and impurities

No data were available to the Working Group.

2. Production, Use, Occurrence and Analysis

2.1 Production and use

(a) Production

2,4-Diaminoanisole was first prepared by the reduction of 2,4-dinitroanisole with iron and acetic acid in 1913 (Richter, 1933). It is produced commercially in Japan by methoxylation of 2,4-dinitro-1-chlorobenzene followed by reduction with iron.

In the Federal Republic of Germany, one company produces 2,4-diaminoanisole and another the sulphate. Commercial production of 2,4-diaminoanisole in Japan was started in 1974; one company is estimated to have produced several thousand kg in 1979.

Commercial production of 2,4-diaminoanisole in the US was first reported in 1933 (US Tariff Commission, 1934) and for the last time in 1940 (US Tariff Commission, 1941); production of 2,4-diaminoanisole sulphate was first reported in 1967 (US Tariff Commission, 1969) and last in 1971 (US Tariff Commission, 1973). In 1979, both compounds were included in the Toxic Substances Control Act Chemical Substance Inventory (US Environmental Protection Agency, 1979), indicating that they had been manufactured, imported or processed for commercial purposes in the US since 1 January 1975. At the present time, one US company is believed to produce 2,4-diaminoanisole sulphate in commercial quantities.

Imports of 2,4-diaminoanisole through principal US customs districts amounted to 8.8 thousand kg in 1979 (US International Trade Commission, 1980), down from 16.2 thousand kg in 1974 (US International Trade Commission, 1976).

(b) Use

In 1977, the estimated annual consumption of 2,4-diaminoanisole in the US was 13.6 thousand kg (National Cancer Institute, 1978). In the past, most of the 2,4-diaminoanisole and 2,4-diaminoisole sulphate produced in the US appears to have been used as components of hair dyes. 2,4-Diaminoanisole is reported to have been employed in this way for over 60 years (US Food & Drug Administration, 1979). In January 1978, the National Institute for Occupational Safety & Health (1978) reported that approximately 75% of the then current oxidation ('permanent') hair-dye formulations contained 2,4-diaminoanisole at levels of 0.05-2.0%, depending on the shade. Formulations for producing brown and blonde shades have been reported to contain significantly less 2,4-diaminoanisole than those for black shades (US Food & Drug Administration, 1979).

A regulation in the US requiring a warning label on all hair dyes containing 2,4-diaminoanisole or its sulphate was to have become effective in April 1980 (US Food & Drug Administration, 1979); however, enforcement of this regulation was subsequently stayed by a US district court and remanded for reconsideration (Anon., 1980). Industry spokesmen stated that these chemicals had been removed from such products before the regulation was to have taken effect (Piellisch, 1980).

2,4-Diaminoanisole is also believed to be used as an intermediate in the manufacture of one commercially significant dye, C.I. Basic Brown 2. One company reported production of an undisclosed amount (see preamble, p. 20) of this dye in 1978 (US International Trade Commission, 1979). In 1971, C.I. Basic Brown 2 was reported to be used mainly in dyeing acrylic fibres, although it was also present in a variety of other products, including shoe polishes (The Society of Dyers & Colourists, 1971b).

2,4-Diaminoanisole and its sulphate may serve as oxidation bases in the dyeing of furs (The Society of Dyers & Colourists, 1971c).

2,4-Diaminoanisole is prohibited in Italy and in the Federal Republic of Germany by application of a 'safeguard clause' in the laws of the member states of the European Communities relating to cosmetic products. The Scientific Advisory Committee for cosmetics, however, recommended that 2,4-diaminoanisole could be permitted for use temporarily, in view of its limited resorption through the skin. Thus, *meta*-phenylendiamines (including 2,4-diaminoanisole) and their salts are permitted for use in cosmetics in the European Communities as oxidizing colouring agents for hair dyes, only if the concentration in the finished product is less than 6% (calculated as the free base) (Commission of the European Communities, 1976).

In Japan, 2,4-diaminoanisole is used in the production of pharmaceuticals.

2.2 Occurrence

(a) Natural occurrence

Neither 2,4-diaminoanisole nor 2,4-diaminoanisole sulphate is known to occur as a natural product.

(b) Occupational exposure

The National Institute for Occupational Safety & Health (1978) in the US estimated that approximately 400000 workers were potentially exposed occupationally to 2,4-diaminoanisole. Hairdressers and cosmetologists comprised the largest portion of this group; a relatively small number of fur dyers were probably exposed to higher levels.

2.3 Analysis

An IARC manual (Egan et al., 1981) gives selected methods for the analysis of aromatic amines. Typical methods for the analysis of 2,4-diaminoanisole and 2,4-diaminoanisole sulphate in various matrices are summarized in Table 1.

Table 1. Methods for the analysis of 2,4-diaminoanisole and 2,4-diaminoanisole sulphate

Sample matrix	Sample preparation	Assay procedure[a]	Limit of detection	Reference
Hair dyes	Separate on chromatography column, using pentane and methanol to elute	HPLC-UV	not given	Turchetto et al. (1978)
	Emulsify; extract with glacial acetic acid; elute with butanol, ethanol, water and glacial acetic acid	TLC	not given	Zelazna & Legatowa (1971)
	Dilute in aqueous sodium sulphite and ethanol; develop, initially with basic solution, then with acidic solution	TLC	not given	Kottemann (1966)
Amine mixtures	Dissolve in isopropanol and hydrochloric acid; elute with aqueous acetate buffer; develop with N,N-dimethyl-para-aminobenzaldehyde in ethanol and glacial acetic acid	TLC	not given	Lepri et al. (1974)
	Elute with glacial acetic acid, butanol and water	TLC	not given	Kurteva & Takeva (1971)

[a] Abbreviations: HPLC-UV, high-performance liquid chromatography with ultra-violet detection; TLC, thin-layer chromatography

3. Biological Data Relevant to the Evaluation of the Carcinogenic Risk to Humans

3.1 Carcinogenicity studies in animals

(a) Oral administration

Mouse: Groups of 50 male and 50 female B6C3F$_1$ mice, six weeks of age, were fed diets containing 1200 or 2400 mg/kg 2,4-diaminoanisole sulphate (of indeterminate purity, with at least one impurity detected by thin-layer chromatography) for 78 weeks and were observed for a further 18-19 weeks. The doses were selected on the basis of a range-finding study [see section 3.2(a)]. Groups of 50 animals of each sex served as matched controls for each dose group. Mean body weight gains of treated and control animals were similar throughout the study, and survival was comparable among treated and control mice: by the end of the study, 84, 78, 92 and 82% of males and 74, 76, 76 and 78% of females were still alive in the low-dose control, high-dose control, low-dose and high-dose groups, respectively. Among the males, follicular-cell adenomas of the thyroid were seen in 1/47 low-dose controls, 0/40 high-dose controls, 0/46 low-dose and 11/45 high-dose ($P < 0.001$) mice; one low-dose male had a follicular-cell carcinoma. Follicular-cell hyperplasia of the thyroid was found in 12/45 high-dose males, hyperplasia in 1/46 low-dose males and papillary or follicular-cell hyperplasia in 2/47 low-dose male controls. Among the females, follicular-cell adenomas were found in 0/43 low-dose controls, 0/41 high-dose controls, 0/42 low-dose and 6/45 high-dose mice ($P = 0.017$); follicular-cell carcinomas were found in 2/45 high-dose mice. Thus, follicular-cell adenomas or carcinomas were found in 8/45 high-dose females ($P = 0.004$). Follicular-cell hyperplasia of the thyroid was found in 1/45 high-dose females, papillary or follicular-cell hyperplasia in 11/42 low-dose females and papillary or adenomatous hyperplasia in 3/44 high-dose female controls. Pigment granules were found in thyroid follicular cells of most high-dose mice. The incidence of malignant lymphomas in treated females was statistically higher than that in the matched controls; however, other control groups of the same hybrid strain in the same laboratory had higher incidences than those of the two treated groups (National Cancer Institute, 1978).

Rat: Groups of 50 male and 50 female Fischer 344 rats, six weeks of age, were fed 2,4-diaminoanisole sulphate (same sample as used above) at concentrations of 5000 mg/kg of diet for 78 weeks, or 1250 mg/kg of diet for 10 weeks and 1200 mg/kg of diet for 68 weeks, followed by a 29-week observation period. Groups of 50 (49 male high-dose controls) animals of each sex served as matched controls for each dose group. Mean body weight gains of high-dose male and female rats were depressed as compared with those of controls throughout most of the study. Mortality of treated and control male rats was similar by the end of the study: 54, 61, 60 and 54% of males and 46, 74, 58 and 44% of females were still alive in the low-dose control, high-dose control, low-dose and high-dose groups, respectively. Malignant tumours of the follicular cells of the thyroid gland (adenocarcinomas, carcinomas, papillary adenocarcinomas and cystadenocarcinomas) were found in 2/35 low-dose male controls, 0/48 high-dose male controls, 2/47 low-dose males and 17/49 high-dose males (P

= 0.001) and in 2/38 low-dose female controls, 1/45 high-dose female controls, 1/46 low-dose females and 10/49 high-dose females (P = 0.006). Eight high-dose males and three high-dose females but none of the controls had multiple follicular-cell tumours. C-cell or follicular-cell hyperplasia of the thyroid was found in a few high-dose rats; and follicular-cell adenomas were found in some high-dose animals with or without thyroid carcinomas. Tumours of thyroid C-cell origin were found in significantly increased incidences in male rats, but not in female rats. Thus, C-cell adenomas or carcinomas were found in 1/35 low-dose male controls, 1/48 high-dose male controls, 4/47 low-dose and 10/49 high-dose males (P = 0.004); the incidence of C-cell carcinomas alone in treated rats was not significant. In males, squamous-cell carcinomas, basal-cell carcinomas or sebaceous adenocarcinomas were found in 0/36 low-dose controls, 0/48 high-dose controls, 2/48 low-dose and 7/49 high-dose animals (P = 0.007). Preputial or clitoral gland adenomas, papillomas or carcinomas wre found in 0/36 low-dose control males, 0/48 high-dose control males, 2/48 low-dose males, 8/49 high-dose males (P < 0.003), 0/39 low-dose female controls, 3/50 high-dose female controls, 5/49 low-dose females (P = 0.049) and 8/49 high-dose females. The incidence of carcinomas alone was not significant in animals of either sex. Squamous-cell carcinomas or sebaceous adenocarcinomas of the Zymbal gland were found in 0/36 low-dose male controls, 0/48 high-dose male controls, 1/48 low-dose males and 8/49 high-dose males (P = 0.003); and sebaceous adenocarcinomas were found in 0/39 low-dose female controls, 0/50 high-dose female controls, 0/49 low-dose and 7/49 high-dose females (P = 0.006). The incidences of mammary fibroadenomas in low-dose females and of uterine adenocarcinomas in high-dose females were significantly increased when compared with those in matched controls but not when compared with those in low-dose controls; in addition, the incidences of these tumours were within the range of those in controls (National Cancer Institute, 1978; Ward et al., 1979).

Groups of 40-60 female Fischer 344 rats, six weeks of age, were fed 2,4-diaminoanisole sulphate (purity not given, probably technical-grade) at concentrations of 0, 1200, 2400 or 5000 mg/kg of diet for up to 82-86 weeks. Another 15 rats were fed diets containing 5000 mg/kg for 10 weeks and observed up to 87 weeks. Mean body weights of high-dose rats were depressed compared with those of controls. Thyroid goitre was found in 6/6 high-dose rats sacrificed at four weeks and in 2/2 rats from each treatment group sacrificed at two and four months. [The effective number of animals in each group was the number necropsied at 87-94 weeks and the number that died near the end of the study]. By 87-94 weeks, follicular-cell adenomas or carcinomas or C-cell carcinomas of the thyroid were found in 1/37 controls, 2/47 low-dose rats, 3/33 mid-dose rats, 31/40 high-dose rats (21 with adenomas) and 3/12 rats that received the high dose for only 10 weeks. Mammary adenocarcinomas were found in 0/37 controls, 0/47 low-dose rats, 5/33 mid-dose rats and 3/40 high-dose rats; mammary adenomas were found in only 1/33 mid-dose and 1/47 low-dose rats. Carcinomas (squamous-cell, sebaceous-cell or sebaceous-squamous-cell) of the clitoral gland were found in 0/37 controls, 8/47 low-dose, 15/33 mid-dose and 9/40 high-dose rats and in 1/12 animals that received the high dose for only 10 weeks. Pituitary adenomas were found in 0/25 controls, 1/29 low-dose rats, 2/20 mid-dose rats and 3/26 high-dose rats and in 1/10 rats that received the high dose for 10 weeks (Evarts & Brown, 1980).

(b) Skin application

Mouse: Two groups of 50 male and 50 female six-to-eight-week-old Swiss-Webster mice received weekly or fortnightly applications of 0.05 ml of a freshly prepared 1:1 mixture of a hair-dye formulation, containing among its constituents 0.38% 2,4-diaminoanisole sulphate, 3% 2,5-diaminotoluene sulphate and 1.5% *para*-phenylenediamine, with 6% hydrogen peroxide. The mixture was painted on shaved skin in the midscapular region for 18 months, at which time the surviving animals were killed. The incidence of lung tumours did not differ significantly from that in untreated control animals. No thyroid lesion was reported in treated mice, although thyroid glands were examined histologically (Burnett *et al.*, 1975). [The Working Group noted a high incidence of lung tumours in mice treated only with the vehicle dye base, which made up the greatest part of the hair-dye formulation.]

Rat: Two experimental preparations in a carboxymethyl cellulose gel were tested: formulation 1 contained 0.75% 2,4-diaminoanisole and 3% 2,5-diaminotoluene (both as sulphates), and formulation 2 contained no added dye intermediate. Each formulation was mixed with an equal volume of 6% hydrogen peroxide immediately before use, and 0.5 ml of the mixture was applied to the dorsal skin. Two groups, each of 50 male and 50 female Sprague-Dawley rats, three months old, were treated twice weekly for two years with formulation 1 or left untreated; a third group of 25 male and 25 female rats, three months old, were treated with formulation 2. No significant differences were detected in tumour types or incidences between the experimental and control groups. Thyroid glands were not examined histologically, and thus no lesions were reported (Kinkel & Holzmann, 1973). [The Working Group noted the absence of information about survival times and a high incidence of spontaneous tumours in untreated and vehicle controls.]

3.2 Other relevant biological data

(a) Experimental systems

Toxic effects

The oral LD_{50} of 2,4-diaminoanisole sulphate in an oil-in-water emulsion in rats was >4000 mg/kg bw; the i.p. LD_{50} in dimethylsulphoxide was 372 mg/kg bw (Burnett *et al.*, 1977).

In four-week subchronic feeding studies, followed by a two-week observation period, male and female $B6C3F_1$ mice and Fischer 344 rats were fed diets containing up to 0.58% 2,4-diaminoanisole sulphate. One male rat fed each dosage and one male mouse fed the highest dose died. No gross abnormalities were noted in either rats or mice during the test, except that thyroid goitre was found in 6/6 rats that received high doses and were sacrificed at four weeks (see section 3.1 and Ward *et al.*, 1979). The only compound-related organ lesions observed in a chronic feeding study with up to 0.5% (rats) or 0.24% (mice) 2,4-diaminoanisole sulphate were diffuse hyperplasia in the liver and C-cell hyperplasia in the thyroid of female rats (National Cancer Institute, 1978).

Effects on reproduction and prenatal toxicity

No studies were available in which 2,4-diaminoanisole was tested alone.

Three commercially available hair-dye formulations, containing 0.02, 2 or 4% 2,4-diaminoanisole sulphate and several aromatic amine derivatives among their constituents, were tested for teratogenicity in groups of 20 mated female Charles River CD rats. Each formulation was applied topically to a shaved site in the dorsoscapular region at a dose of 2 ml/kg bw on days l, 4, 7, 10, 13, 16 and 19 of gestation; just prior to its use, each formulation was mixed with an equal volume of 6% hydrogen peroxide. The mothers were killed on day 20 of gestation. There was no significant increase in soft-tissue anomalies in the living fetuses; but significant skeletal changes were seen in nine of 169 live fetuses in three of 20 litters of dams given the formulation containing the highest concentration of 2,4-diaminoanisole sulphate (4%). On comparison with three control groups, this finding was statistically significant ($P < 0.05$ to $P < 0.01$) (Burnett *et al.*, 1976).

Absorption, distribution, excretion and metabolism

Following i.p. administration of 50 mg/kg bw ^{14}C-labelled 2,4-diaminoanisole to rats, 85% of the radioactivity was excreted in the urine and 9% in the faeces after 48 hours. The major metabolites were 4-acetylamino-2-aminoanisole, 2,4-diacetylaminoanisole, 2,4-diacetylaminophenol, 5-hydroxy-2,4-diacetylaminoanisole and 2-methoxy-5-(glycolamido)-acetanilide or its isomer. These metabolites were excreted in the urine both free and as glucuronides and sulphates (Grantham *et al.*, 1979). Pretreatment with 3-methylcholanthrene, β-naphthoflavone and 2,3,7,8-tetrachlorodibenzo-*para*-dioxin reduced the amounts of unconjugated metabolites and increased the amounts of more water-soluble metabolites. No significant changes in metabolite excretion were observed following pretreatment with phenobarbital (Grantham & Benjamin, 1980).

Dybing *et al.* (1979a,b) showed that the metabolism and covalent binding of 2,4-diaminoanisole *in vitro* and *in vivo* are cytochrome-P450-dependent. Incubation of rat liver and kidney microsomes with radiolabelled 2,4-diaminoanisole in the presence of NADPH and oxygen led to the formation of products that were bound covalently to microsomal protein. Inhibitors of cytochrome P450 *in vivo* and *in vitro* decreased the binding, pretreatment with phenobarbital increased binding; and pretreatment with β-napthoflavone had no effect. More ring-labelled than methyl-labelled 2,4-diaminoanisole was bound; when the hydrogens in the methyl group were replaced by deuterium, both the binding and the mutagenicity of 2,4-diaminoanisole increased. Liver microsomes catalysed irreversible binding to endogenous microsomal RNA; no binding to exogenously added DNA was detected (Dybing *et al.*, 1979a).

When 10-200 mg/kg bw ^{3}H-2,4-diaminoanisole were injected into rats, the label was bound covalently to liver and kidney proteins. The binding was increased by pretreating the animals with phenobarbital or β-naphthoflavone, but was decreased by pretreatment with cobaltous chloride. Irreversibly bound material was localized preferentially in the microsomal fraction. No covalent binding to hepatic RNA or DNA was detected (Dybing et al., 1979b).

Mutagenicity and other short-term tests

2,4-Diaminoanisole was mutagenic to *Salmonella typhimurium* in the presence of microsomes derived from induced or uninduced mice (Dybing & Thorgeirsson, 1977; Aune & Dybing, 1979), rats (Ames et al., 1975; Dybing & Aune, 1977; Yoshikawa et al., 1977; Aune et al., 1980a; Prival et al., 1980; Reddy et al., 1980; de Giovanni-Donnelly, 1981), rabbits (Aune et al., 1980b) or humans (Dybing et al., 1979c).

2,4-Diaminoanisole sulphate produced mitotic recombination in *Saccharomyces cerevisiae* only when growing yeast were exposed to the chemical for a prolonged period (72 h) in the absence of a microsomal preparation (Mayer & Goin, 1980).

2,4-Diaminoanisole sulphate induced mutations at the thymidine kinase locus of L5178Y mouse lymphoma cells over a narrow concentration range in the absence of metabolic activation (Palmer et al., 1977); it also induced mutations in *Drosophila melanogaster* (Blijleven, 1977)

2,4-Diaminoanisole did not increase the incidence of micronuclei in rats at a dose of 1000 mg/kg bw per 24 h (Hossack & Richardson, 1977), or in mice (no dosage given) (Heddle & Bruce, 1977). No increases were reported in the frequency of morphologically abnormal sperm in treated mice (Heddle & Bruce, 1977).

2,4-Diaminoanisole and its sulphate were negative in a dominant lethal assay in which rats were given 40 mg/kg bw per day three times a week for 10 weeks (Sheu & Green, 1979) or 20 mg/kg bw three times a week for eight weeks (Burnett et al., 1977)].

(b) Humans

No data were available to the Working Group.

3.3 Case reports and epidemiological studies of carcinogenicity in humans

No data were available to the Working Group. See, however, Appendix 1.

2,4-DIAMINOANISOLE and its SULPHATE

4. Summary of Data Reported and Evaluation

4.1 Experimental data

2,4-Diaminoanisole sulphate (technical-grade) was tested by dietary administration in one experiment in mice and in two experiments in one strain of rats. Benign or malignant tumours of the thyroid gland were induced in rats and mice with the highest dose tested; tumours of the skin and of the preputial, clitoral and Zymbal glands were also induced in rats. It was also tested in a hair-dye formulation by skin application in mice and rats, but the studies were considered to be inadequate for evaluation.

2,4-Diaminoanisole or its sulphate was mutagenic for *Salmonella typhimurium*, cultured mouse lymphoma cells and *Drosophila melanogaster*. The sulphate produced mitotic recombination in yeast. Neither 2,4-diaminoanisole nor its sulphate induced micronuclei, sperm abnormalities, or dominant lethal mutations in rodents.

4.2 Human data

2,4-Diaminoanisole and its sulphate have been used in hair dyes for over 60 years. In addition to the consumer and occupational exposure resulting from this use, past use of 2,4-diaminoanisole as a chemical intermediate and as a fur-dye component could have resulted in other occupational exposure.

No case report or epidemiological study was available to the Working Group, but see Appendix 1.

4.3 Evaluation

There is *sufficient evidence*[1] for the carcinogenicity of 2,4-diaminoanisole sulphate in experimental animals.

In the absence of epidemiological studies relating specifically to 2,4-diaminoanisole, it should be regarded, for practical purposes, as if it presented a carcinogenic risk to humans.

For a consideration of the possible carcinogenic risk of hair dyes containing 2,4-diaminoanisole, see Appendix 1.

---

[1] See preamble, p. 18.

5. References

Ames, B.N., Kammen, H.O. & Yamasaki, E. (1975) Hair dyes are mutagenic: identification of a variety of mutagenic ingredients. *Proc. natl Acad. Sci. (USA), 72*, 2423-2427

Anon. (1980) Court stays FDA hair dye regulation. *Chemical Engineering News*, 29 September, p. 20

Aune, T. & Dybing, E. (1979) Mutagenic activation of 2,4-diaminoanisole and 2-aminofluorene *in vitro* by liver and kidney fractions from aromatic hydrocarbon responsive and nonresponsive mice. *Biochem. Pharmacol., 28*, 2791-2797

Aune, T., Dybing, E. & Nelson, S.D. (1980a) Mutagenic activation of 2,4-diaminoanisole and 2-aminofluorene by isolated rat liver nuclei and microsomes. *Chem.-biol. Interactions, 31*, 35-49

Aune, T., Dybing, E. & Thorgeirsson, S.S. (1980b) Developmental pattern of 3-methylcholanthrene-inducible mutagenic activation of N-2-fluorenylacetamide, 2-fluorenamine, and 2,4-diaminoanisole in the rabbit. *J. natl Cancer Inst., 64*, 765-769

Blijleven, W.G.H. (1977) Mutagenicity of four hair dyes in *Drosophila melanogaster. Mutat. Res., 48*, 181-186

Burnett, C., Lanman, B., Giovacchini, R., Wolcott, G., Scala, R. & Keplinger, M. (1975) Long-term toxicity studies on oxidation hair dyes. *Food Cosmet. Toxicol., 13*, 353-357

Burnett, C., Goldenthal, E.I., Harris, S.B., Wazeter, F.X., Strausburg, J., Kapp, R. & Voelker, R. (1976) Teratology and percutaneous toxicity studies on hair dyes. *J. Toxicol. environ. Health, 1*, 1027-1040

Burnett, C., Loehr, R. & Corbett, J. (1977) Dominant lethal mutagenicity study on hair dyes. *J. Toxicol. environ. Health, 2*, 657-662

Commission of the European Communities (1976) Council Directive of 27 July 1976 on the approximation of the laws of the Member States relating to cosmetic products. *Off. J. Eur. Communities, L262*, 184-185

Dybing, E. & Aune, T. (1977) Hexachlorobenzene induction of 2,4-diaminoanisole mutagenicity *in vitro. Acta pharmacol. toxicol., 40*, 575-583

Dybing, E. & Thorgeirsson, S.S. (1977) Metabolic activation of 2,4-diaminoanisole, a hair-dye component. I. Role of cytochrome P-450 metabolism in mutagenicity *in vitro. Biochem. Pharmacol., 26*, 729-734

Dybing, E., Aune, T. & Nelson, S.D. (1979a) Metabolic activation of 2,4-diaminoanisole, a hair-dye component. II. Role of cytochrome P-450 metabolism in irreversible binding *in vitro. Biochem. Pharmacol, 28*, 43-50

Dybing, E., Aune, T. & Nelson, S.D. (1979b) Metabolic activation of 2,4-diaminoanisole, a hair-dye component. III. Role of cytochrome P-450 metabolism in irreversible binding *in vivo. Biochem. Pharmacol., 28*, 51-55

Dybing, E., von Bahr, C., Aune, T., Glaumann, H., Levitt, D.S. & Thorgeirsson, S.S. (1979c) *In vitro* metabolism and activation of carcinogenic aromatic amines by subcellular fractions of human liver. *Cancer Res., 39*, 4206-4211

Egan, H., Fishbein, L., Castegnaro, M., O'Neill, I.K. & Bartsch, H., eds (1981) *Environmental Carcinogens - Selected Methods of Analysis*, Vol. 4, *Some Aromatic Amines and Azodyes in the General and Industrial Environment (IARC Scientific Publications No. 40)*, Lyon, International Agency for Research on Cancer

Evarts, R.P. & Brown, C.A. (1980) 2,4-Diaminoanisole sulfate: early effect on thyroid gland morphology and late effect on glandular tissue of Fischer 344 rats. *J. natl Cancer Inst., 65*, 197-204

de Giovanni-Donnelly, R. (1981) The comparative response of *Salmonella typhimurium* strains TA1538, TA98 and TA100 to various hair-dye components. *Mutat. Res., 91*, 21-25

Grantham, P.H. & Benjamin, T. (1980) *The effect of cytochrome P-450 inducers on the in vivo metabolism of 2,4-diaminoanisole, a hair dye intermediate* (Abstract No. 19). In: *Abstracts of Papers. Society of Toxicology Incorporated, Nineteenth Annual Meeting, Washington DC*, Washington DC, Society of Toxicology Inc., p. A7

Grantham, P.H., Benjamin, T., Tahan, L.C., Roller, P.P., Miller, J.R. & Weisburger, E.K. (1979) Metabolism of the dyestuff intermediate 2,4-diaminoanisole in the rat. *Xenobiotica, 9*, 333-341

Heddle, J.A. & Bruce, W.R. (1977) *Comparison of tests for mutagenicity or carcinogenicity using assays for sperm abnormalities, formation of micronuclei, and mutation in* Salmonella. In: Hiatt, H.H., Watson, J.D. & Winsten, J.A., eds, *Origins of Human Cancer*, Book C, Cold Spring Harbor, NY, Cold Spring Harbor Laboratory, pp. 1549-1557

Hossack, D.J.N. & Richardson, J.C. (1977) Examination of the potential mutagenicity of hair dye constituents using the micronucleus test. *Experientia, 33*, 377-378

IARC (1978) *IARC Monographs on the Evaluation of the Carcinogenic Risk of Chemicals to Man, Vol. 16, Some Aromatic Amines and Related Nitro Compounds - Hair Dyes, Colouring Agents and Miscellaneous Industrial Chemicals*, Lyon, pp. 51-62

Kinkel, H.J. & Holzmann, S. (1973) Study on long-term percutaneous toxicity and carcinogenicity of hair dyes (oxidizing dyes) in rats. *Food Cosmet. Toxicol., 11*, 641-648

Kottemann, C.M. (1966) Two-dimensional thin layer chromatographic procedure for the identification of dye intermediates in arylamine oxidation hair dyes. *J. Assoc. off. anal. Chem., 49*, 954-959

Kurteva, R. & Takeva, S. (1971) Thin-layer chromatography of aromatic diamines and their *N*-derivatives (Bulg.). *God. Nauchnoizsled. Inst. Khim. Prom., 9*, 53-57 [*Chem. Abstr., 80*, 33618n]

Lepri, L., Desideri, P.G. & Coas, V. (1974) Chromatographic and electrophoretic behaviour of primary aromatic amines on anion-exchange thin layers. *J. Chromatogr., 90*, 331-339

Mayer, V.W. & Goin, C.J. (1980) Induction of mitotic recombination by certain hair-dye chemicals in *Saccharomyces cerevisiae*. *Mutat. Res., 78*, 243-252

National Cancer Institute (1978) *Bioassay of 2,4-Diaminoanisole Sulfate for Possible Carcinogenicity (Tech. Rep. Ser. No. 84; DHEW Publ. No. (NIH) 78-1334)*, Washington DC, US Government Printing Office

National Institute for Occupational Safety & Health (1978) *2,4-Diaminoanisole (4-Methoxy-m-phenylenediamine) in Hair and Fur Dyes (Current Intelligence Bulletin 19; DHEW (NIOSH) Publ. No. 78-111)*, Rockville, MD

Palmer, K.A., Denunzio, A. & Green, S. (1977) The mutagenic assay of some hair dye components, using the thymidine kinase locus of L5178Y mouse lymphoma cells. *J. environ. Pathol. Toxicol., 1*, 87-91

Piellisch, R. (1980) FDA and industry in battle over hair dye labeling rule. *Chemical Marketing Reporter*, 23 June, pp. 43, 44, 47

Prival, M.J., Mitchell, V.D. & Gomez, Y.P. (1980) Mutagenicity of a new hair dye ingredient: 4-ethoxy-*m*-phenylenediamine. *Science, 207*, 907-908

Reddy, T.V., Benjamin, T., Grantham, P.H., Weisburger, E.K. & Thorgeirsson, S.S. (1980) Mutagenicity of urine from rats after administration of 2,4-diaminoanisole, a hair dye component: the effect of microsomal enzyme inducers (Abstract No. 266). *Proc. Am. Assoc. Cancer Res., 21*, 66

Richter, F., ed. (1933) *Beilsteins Handbuch der Organischen Chemie*, 4th ed., 1st Suppl., Vol. 13, Syst. No. 1854, Berlin, Springer, p. 204

Sadtler Research Laboratories, Inc. (undated) *Sadtler Standard Spectra*, Philadelphia, PA

Sheu, C.-J.W. & Green, S. (1979) Dominant lethal assay of some hair-dye components in random-bred male rats. *Mutat. Res., 68*, 85-98

The Society of Dyers & Colourists (1971a) *Colour Index*, 3rd ed., Vol. 4, Bradford, Yorkshire, Lund Humphries, p. 4644

The Society of Dyers & Colourists (1971b) *Colour Index*, 3rd ed., Vol. 1, Bradford, Yorkshire, Lund Humphries, p. 1683

The Society of Dyers & Colourists (1971c) *Colour Index*, 3rd ed., Vol. 3, Bradford, Yorkshire, Lund Humphries, p. 3262

Turchetto, L., Cuozzo, V., Terracciano, M., Papetti, P., Percaccio, G. & Quercia, V. (1978) Analytical study of some hair dyes (Ital.). *Boll. Chim. Farm., 117*, 475-478 [*Chem. Abstr., 90*, 174500m]

US Environmental Protection Agency (1979) *Toxic Substances Control Act (TSCA) Chemical Substance Inventory, Initial Inventory*, Vol. 1, Washington DC, Office of Toxic Substances, pp. 65, 280

US Food & Drug Administration (1979) Cosmetic product warning statements; coal tar hair dyes containing 4-methoxy-*m*-phenylenediamine (2,4-diaminoanisole) or 4-methoxy-*m*-phenylenediamine sulfate (2,4-diaminoanisole sulfate). *Fed. Regist., 44*, 59509-59522

US International Trade Commission (1976) *Imports of Benzenoid Chemicals and Products, 1974 (USITC Publication 762)*, Washington DC, US Government Printing Office, p. 22

US International Trade Commission (1979) *Synthetic Organic Chemicals, US Production and Sales, 1978 (USITC Publication 1001)*, Washington DC, US Government Printing Office, p. 98

US International Trade Commission (1980) *Imports of Benzenoid Chemicals and Products, 1979 (USITC Publication 1083)*, Washington DC, US Government Printing Office, p. 24

US Tariff Commission (1934) *Production and Sales of Dyes and Other Synthetic Organic Chemicals, 1933 (Report No. 89, Second Series)*, Washington DC, US Government Printing Office, p. 9

US Tariff Commission (1941) *Synthetic Organic Chemicals, US Production and Sales, 1940 (Report No. 148, Second Series)*, Washington DC, US Government Printing Office, p. 11

US Tariff Commission (1969) *Synthetic Organic Chemicals, US Production and Sales, 1967 (TC Publication 295)*, Washington DC, US Government Printing Office, p. 81

US Tariff Commission (1973) *Synthetic Organic Chemicals, US Production and Sales, 1971 (TC Publication 614)*, Washington DC, US Government Printing Office, p. 43

Ward, J.M., Stinson, S.F., Hardisty, J.F., Cockrell, B.Y. & Hayden, D.W. (1979) Neoplasms and pigmentation of thyroid glands in F344 rats exposed to 2,4-diaminoanisole sulfate, a hair dye component. *J. natl Cancer Inst., 62*, 1067-1073

Weast, R.C., ed. (1979) *CRC Handbook of Chemistry and Physics*, 60th ed., Cleveland, OH, Chemical Rubber Co., p. C-156

Yoshikawa, K., Uchino, H., Tateno, N. & Kurata, H. (1977) Mutagenic activities of the samples prepared with raw material of hair dye (Jpn.). *Eisei Shikensho Hokoku* (*Bull. natl Inst. hyg. Sci. Tokyo*), *95*, 15-24

Zelazna, K. & Legatowa, B. (1971) Identification of basic dyes in emulsified hair dyes by thin layer chromatography (Pol.). *Rocz. Panstw. Zakl. Hig.*, *22*, 427-430 [*Chem. Abstr.*, *76*, 6623w]

4,4'-METHYLENEBIS(N,N-DIMETHYL)BENZENAMINE

1. Chemical and Physical Data

1.1 Synonyms and trade names

Chem. Abstr. Services Reg. No.: 101-61-1
Chem. Abstr. Name: Benzenamine, 4,4'-methylenebis(N,N-dimethyl)-
IUPAC Systematic Name: 4,4'-Methylenebis(N,N-dimethylaniline)
Synonyms: 4,4'-Bis(dimethylamino)diphenylmethane; *para,para'*-bis(dimethylamino)-diphenylmethane; bis[4-(dimethylamino)phenyl]methane; bis[4-(N,N–dimethylamino)phenyl]methane; bis[*para*-(dimethylamino)phenyl]methane; bis [*para*-N,N-(dimethylamino)phenyl]methane; methane base; Michler's Base, Michler's hydride; Michler's methane; tetrabase; tetramethyldiaminodiphenylmethane; 4,4'-tetramethyldiaminodiphenylmethane; *para,para'*-tetramethyldiaminodiphenylmethane; N,N,N',N'-tetramethyl-4,4'-diaminodiphenylmethane; N,N,N',N'-tetramethyl-*para,para'*-diaminodiphenylmethane

1.2 Structural and molecular formulae and molecular weight

$(CH_3)_2N$—⟨phenyl⟩—CH_2—⟨phenyl⟩—$N(CH_3)_2$

$C_{17}H_{22}N_2$ Mol. wt: 254.4

1.3 Chemical and physical properties of the pure substance

From Windholz (1976), unless otherwise specified
(a) Description: Lustrous leaflets
(b) Boiling-point: 390° (sublimes without decomposition)
(c) Melting-point: 90-91°
(d) Solubility: Insoluble in water; soluble in acids and hot ethanol; very soluble in diethyl ether, benzene and carbon disulphide (Weast, 1979)
(e) Spectroscopic data: Infra-red, nuclear magnetic resonance, ultra-violet (Sadtler Research Laboratories, Inc., undated) and mass spectral data (Mass Spectrometry Data Centre, 1974) have been published.

1.4 Technical products and impurities

No data were available to the Working Group.

2. Production, Use, Occurrence and Analysis

2.1 Production and use

(a) Production

4,4'-Methylenebis(N,N-dimethyl)benzenamine can be prepared from N,N-dimethylaniline by: (a) heating it with a 40% aqueous formaldehyde solution and concentrated aqueous hydrochloric acid; (b) reacting it with diacetyl peroxide; or (c) reacting it with *tert*-butyl perbenzoate (Windholz, 1976). The first method is probably that used for commercial production.

4,4'-Methylenebis(N,N-dimethyl)benzenamine has been produced commercially in the US since at least 1921 (US Tariff Commission, 1922). Two US companies reported commercial production of an undisclosed amount (see preamble, p. 20) in 1979 (US International Trade Commission, 1980). In 1978, three US companies reported production of this chemical, but separate production data were not published (US International Trade Commission, 1979). Total US production by three companies in 1977 amounted to 453 thousand kg (US International Trade Commission, 1978), down sharply from the 845 thousand kg produced by three companies in 1974 (US International Trade Commission, 1976).

4,4'-Methylenebis(N,N-dimethyl)benzenamine is believed to be produced by one company in the Federal Republic of Germany. It is not produced commercially in Japan.

(b) Use

4,4'-Methylenebis(N,N-dimethyl)benzenamine is believed to be used primarily as a dye intermediate. The Society of Dyers & Colourists (1971) report that six dyes and one pigment can be prepared from it. In 1979, only one of the dyes, Basic Yellow 2, was being produced commercially by two companies in the US (US International Trade Commission, 1980). This dye, which is the hydrochloride salt of auramine (see IARC, 1972), can also be produced from Michler's ketone [4,4'-bis(dimethylamino)benzophenone]. Separate US production data for Basic Yellow 2 were last reported in 1972, when three companies produced a total of 183 thousand kg (US Tariff Commission, 1974). Commercial production of another dye, Basic Orange 14, in the US was last reported by one company in 1977 (US International Trade Commission, 1978). Commercial production in the US of another dye, Solvent Yellow 34, which is the free base of auramine, was last reported by two companies in 1975 (US International Trade Commission, 1977).

The hydrochloride of 4,4'-methylenebis(*N*,*N*-dimethyl)benzenamine is reportedly used as a reagent for the determination of lead (Windholz, 1976).

2.2 Occurrence

4,4'-Methylenebis(*N*,*N*-dimethyl)benzenamine has not been reported to occur as a natural product. No data on its occurrence in the environment were available to the Working Group.

2.3 Analysis

An IARC manual (Egan *et al.*, 1981) gives selected methods for the analysis of aromatic amines. No information on quantitative methods of analysis for 4,4'-methylenebis(*N*,*N*-dimethyl)benzenamine were available to the Working Group.

3. Biological Data Relevant to the Evaluation of Carcinogenic Risk to Humans

3.1 Carcinogenicity studies in animals

Oral administration

Mouse: Groups of 50 male and 50 female B6C3F$_1$ mice, six weeks of age, were fed diets containing 1250 or 2500 mg/kg 4,4'-methylenebis(*N*,*N*-dimethyl)benzenamine (technical grade, purity not established, but no impurities detected by thin-layer chromatography, nuclear magnetic resonance or infra-red spectral examination) for 78 weeks; animals were observed for an additional 13 weeks. The doses were selected on the basis of a range-finding study [see section 3.2(a)]. A group of 20 mice of each sex served as matched controls and were kept under observation for 91 weeks. By the end of the study, 100, 94 and 100% of males and 90, 92 and 84% of females were still alive in the control, low-dose and high-dose groups, respectively. Statistically significant ($P = 0.025$), dose-related increases in the incidences of hepatocellular adenomas and carcinomas were observed in treated mice as follows: hepatocellular adenomas and carcinomas in 5/20 control males, 12/50 low-dose males, 22/48 high-dose males, 1/19 control females, 19/49 low-dose females ($P = 0.005$) and 23/48 high-dose females ($P = 0.001$). The incidences of hepatocellular carcinomas were as follows: males - 3/20, 9/50 and 6/48; females - 0/19, 1/49 and 1/48, in the three groups, respectively (National Cancer Institute, 1979).

Rat: Two groups of 50 male and 50 female Fischer 344 rats, six weeks of age, were fed diets containing 375 or 750 mg/kg 4,4'-methylenebis(*N*,*N*-dimethyl)benzenamine (same sample as used above) for 59 weeks and were observed for an additional 45 weeks. The doses were selected on the basis of a range-finding study [see section 3.2(a)]. A group of 20 rats of each sex served as matched controls and were kept under observation for 104 weeks. By the end of the study, 80, 78 and 88% of males and 85, 82 and 74% of females were still alive in the control, low-dose and high-dose groups, respectively. Statistically significant ($P < 0.001$), dose-related increases in the incidences of thyroid follicular-cell adenomas and carcinomas were observed in treated groups as follows: in 1/18 control males, 4/50 low-dose males, 34/46 high-dose males ($P < 0.001$), 0/20 control females, 4/46 low-dose females and 36/45 high-dose females ($P < 0.001$). The incidences of follicular-cell carcinomas were as follows: males - 1/18, 4/50 and 21/46 ($P = 0.002$); females - 0/20, 3/46 and 23/45 ($P < 0.001$), in the three groups, respectively (National Cancer Institute, 1979; Murthy, 1980).

3.2 Other relevant biological data

(a) Experimental systems

Toxic effects

The oral LD_{50} of 4,4'-methylenebis-(N,N-dimethyl)benzenamine in mice was 3160 mg/kg (Lewis & Tatken, 1979).

Dietary administration for four weeks of up to 11 380 mg/kg of this compound produced no deaths and no inhibition of growth in $B6C3F_1$ mice. During chronic studies with levels of 1250 and 2500 mg/kg of diet, no decrease in body weight gain occurred in animals of either sex during the first 30 weeks of the experiment; however, compound-dependent non-neoplastic proliferative lesions of the thyroid were observed. In similar studies in Fischer 344 rats, administration for four weeks of 3155 mg/kg of diet did not affect survival of either males or females, but there was a severe compound-dependent depression of weight gain in animals of both sexes. Chronic studies with levels of 750 and 375 mg/kg of diet did not result in growth inhibition; however, compound-dependent non-neoplastic proliferative lesions of the thyroid were detected in animals of both sexes (National Cancer Institute, 1979).

Effects on reproduction and prenatal toxicity

No data were available to the Working Group.

Absorption, distribution, excretion and metabolism

Following injection of randomly labelled ^{3}H-4,4'-methylenebis(N,N-dimethyl)-benzenamine to rats, radioactivity was found bound to liver nucleic acids (Schribner *et al.*, 1980).

Mutagenicity and other related short-term tests

4,4'-Methylenebis(*N,N*-dimethyl)benzenamine preferentially inhibited the growth of DNA repair-deficient *Escherichia coli* even in the absence of mammalian microsomes (Rosenkranz & Poirier, 1979).

Doses of up to 400 µg/plate were not found to be mutagenic to *Salmonella typhimurium* in the presence of liver microsomes from Aroclor-induced or uninduced rats (Rosenkranz & Poirier, 1979; Simmon, 1979a; Schribner *et al.*, 1980). A subsequent study with the sample used in the carcinogenicity tests indicated borderline mutagenicity in *Salmonella* strains TA98 and TA100, in the presence of hepatic microsomes from rats or mice, even with a low concentration (10 µg/plate) (Dunkel & Simmon, 1980; Rosenkranz & McCoy, 1981), but not in *E. coli* WP2*uvr* (Rosenkranz & McCoy, 1981).

4,4'-Methylenebis(*N,N*-dimethyl)benzenamine did not induce mitotic recombination in *Saccharomyces cerevisiae* when tested in the presence and absence of liver microsomes from Aroclor-induced rats (Simmon, 1979b).

It gave a positive result in a host-mediated assay in mice with *S. typhimurium* as the genetic indicator and with i.m. administration of 125 mg/kg bw of the chemical. When *Saccharomyces cerevisiae* was used as the genetic indicator, oral administration of 1600 mg/kg bw gave negative results (Simmon *et al.*, 1979).

This compound induced morphological transformation of hamster embryo cells (Pienta *et al.*, 1977).

(b) Humans

No data were available to the Working Group.

3.3 Case reports and epidemiological studies of carcinogenicity in humans

No data were available to the Working Group.

4. Summary of Data Reported and Evaluation

4.1 Experimental data

4,4'-Methylenebis(*N,N*-dimethyl)benzenamine was tested in one experiment in mice and in one experiment in rats by dietary administration. It produced liver-cell adenomas in female mice and thyroid follicular-cell carcinomas in male and female rats given the highest dose.

4,4'-Methylenebis(*N,N*-dimethyl)benzenamine was positive in *Escherichia coli* repair tests. It was mutagenic to *Salmonella typhimurium*, both *in vitro* and in the murine host-mediated assay. It did not induce mitotic recombination in yeast, but caused morphological transformation in hamster embryo cells.

4.2 Human data

4,4'-Methylenebis(*N,N*-dimethyl)benzenamine has been produced commercially since at least 1921. Its use as a dye intermediate could result in occupational exposure.

No case report or epidemiological study was available to the Working Group.

4.3 Evaluation

There is *limited evidence*[1] for the carcinogenicity of 4,4'-methylenebis(*N,N*-dimethyl)benzenamine in experimental animals.

No evaluation of the carcinogenicity of 4,4'-methylenebis(*N,N*-dimethyl)benzenamine to humans could be made.

---

[1] See preamble pp. 18-19.

5. References

Dunkel, V.C. & Simmon, V.F. (1980) *Mutagenic activity of chemicals previously tested for carcinogenicity in the National Cancer Institute Bioassay Program.* In: Montesano, R., Bartsch, H. & Tomatis, L., eds, *Molecular and Cellular Aspects of Carcinogen Screening Tests (IARC Scientific Publications No. 27)*, Lyon, International Agency for Research on Cancer, pp. 283-302

Egan, H., Fishbein, L., Castegnaro, M., O'Neill, I.K. & Bartsch, H., eds (1981) *Environmental Carcinogens - Selected Methods of Analysis*, Vol. 4, *Some Aromatic Amines and Azodyes in the General and Industrial Environment (IARC Scientific Publications No. 40)*, Lyon, International Agency for Research on Cancer

IARC (1972) *IARC Monographs on the Evaluation of Carcinogenic Risk of Chemicals to Man*, Vol. 1, Lyon, pp. 69-73

Lewis, R.J., Sr & Tatken, R.L., eds (1979) *1978 Registry of Toxic Effects of Chemical Substances*, Cincinnati, OH, US Department of Health, Education, & Welfare, p. 113

Mass Spectrometry Data Centre (1974) *Eight Peak Index of Mass Spectra*, Reading, UK, Atomic Weapons Research Establishment

Murthy, A.S.K. (1980) Morphology of the neoplasms of the thyroid gland in Fischer 344 rats treated with 4,4'-methylene-bis-(N,N-dimethyl)-benzenamine. *Toxicol. Lett., 6*, 391-397

National Cancer Institute (1979) *Bioassay of 4,4'-Methylenebis(N,N- dimethyl)benzenamine for Possible Carcinogenicity (Tech. Rep. Ser. No. 186; DHEW Publ. No. (NIH) 79-1742)*, Washington DC, US Government Printing Office

Pienta, R.J., Poiley, J.A. & Lebhertz, W.B., III (1977) Morphological transformation of early passage golden Syrian hamster embryo cells derived from cryopreserved primary cultures as a reliable *in vitro* bioassay for identifying diverse carcinogens, *Int. J. Cancer, 19*, 642-655

Rosenkranz, H.S. & McCoy, E.C. (1981) *4,4'-Methylenebis(N,N-dimethyl)aniline* (Contract No. 1-CP-65855), Bethesda, MD, National Cancer Institute, Division of Cancer Cause and Prevention, pp. 385-604

Rosenkranz, H.S. & Poirier, L.A. (1979) Evaluation of the mutagenicity and DNA-modifying activity of carcinogens and noncarcinogens in microbial systems. *J. natl Cancer Inst., 62*, 873-892

Sadtler Research Laboratories, Inc. (undated) *Sadtler Standard Spectra*, Philadelphia, PA

Schribner, J.D., Koponen, G., Fisk, S.R. & Woodworth, B. (1980) Binding of the dye intermediates Mischler's ketone and methane base to rat liver nucleic acids and lack of mutagenicity in *Salmonella typhimurium. Cancer Lett., 9*, 117-121

Simmon, V.F. (1979a) *In vitro* mutagenicity assays of chemical carcinogens and related compounds with *Salmonella typhimurium. J. natl Cancer Inst., 62*, 893-899

Simmon, V.F. (1979b) *In vitro* assays for recombinogenic activity of chemical carcinogens and related compounds with *Saccharomyces cerevisiae* D3. *J. natl Cancer Inst., 62*, 901-909

Simmon, V.F., Rosenkranz, H.S., Zeiger, E. & Poirier, L.A. (1979) Mutagenic activity of chemical carcinogens and related compounds in the intraperitoneal host-mediated assay. *J. natl Cancer Inst., 62*, 911-918

The Society of Dyers & Colourists (1971) *Colour Index*, 3rd ed., Vol. 4, Bradford, Yorkshire, Perkin House, p. 4703

US International Trade Commission (1976) *Synthetic Organic Chemicals, US Production and Sales, 1974* (*USITC Publication 776*), Washington DC, US Government Printing Office, pp. 22, 39

US International Trade Commission (1977) *Synthetic Organic Chemicals, US Production and Sales, 1975* (*USITC Publication 804*), Washington DC, US Government Printing Office, p. 72

US International Trade Commission (1978) *Synthetic Organic Chemicals, US Production and Sales, 1977* (*USITC Publication 920*), Washington DC, US Government Printing Office, pp. 48, 74, 108

US International Trade Commission (1979) *Synthetic Organic Chemicals, US Production and Sales, 1978* (*USITC Publication 1001*), Washington DC, US Government Printing Office, p. 68

US International Trade Commission (1980) *Synthetic Organic Chemicals, US Production and Sales, 1979* (*USITC Publication 1099*), Washington DC, US Government Printing Office, pp. 50, 77

US Tariff Commission (1922) *Census of Dyes and other Synthetic Organic Chemicals, 1921* (*Tariff Information Series No. 26*), Washington DC, US Government Printing Office, p. 25

US Tariff Commission (1974) *Synthetic Organic Chemicals, US Production and Sales, 1972* (*TC Publication 681*), Washington DC, US Government Printing Office, pp. 59, 73

Weast, R.C., ed. (1979) *CRC Handbook of Chemistry and Physics*, 60th ed., Cleveland, OH, Chemical Rubber Co., p. C-372

Windholz, M., ed. (1976) *The Merck Index*, 9th ed., Rahway, NJ, Merck & Co., p. 1189

1,5-NAPHTHALENEDIAMINE

1. Chemical and Physical Data

1.1 Synonyms and trade names

Chem. Abstr. Services Reg. No.: 2243-62-1
Chem. Abstr. Name and IUPAC Systematic Name: 1,5-Naphthalenediamine
Synonyms: 1,5-Diaminonaphthalene; 1,5-naphthylenediamine

1.2 Structural and molecular formulae and molecular weight

$C_{10}H_{10}N_2$ Mol. wt: 158.2

1.3 Chemical and physical properties of the pure substance

From Weast (1979), unless otherwise specified
(a) Description: Colourless crystals (Hawley, 1977)
(b) Boiling-point: Sublimes
(c) Melting-point: 190°
(d) Density: 1.4
(e) Solubility: Soluble in hot water, ethanol and diethyl ether; very soluble in chloroform, hot ethanol and hot diethyl ether
(f) Spectroscopic data: Infra-red and ultra-violet spectral data have been reported (Sadtler Research Laboratories, Inc., undated).

1.4 Technical products and impurities

No data were available to the Working Group.

2. Production, Use, Occurrence and Analysis

2.1 Production and use

(a) Production

1,5-Naphthalenediamine can be prepared by the reduction of 1,5-dinitronaphthalene (Sandridge & Staley, 1978) or by ammonolysis of 1,5-dihydroxynaphthalene. Both methods are believed to be used for its commercial production in Japan.

1,5-Naphthalenediamine is believed to be produced by two companies in the Federal Republic of Germany. It has been produced commercially in Japan since 1957; in 1979, two companies produced an estimated 50 thousand kg.

No evidence was found that 1,5-naphthalenediamine has ever been produced in commercial quantities in the US. Two thousand kg were imported through principal US customs districts in 1979 (US International Trade Commission, 1980).

(b) Use

1,5-Naphthalenediamine is believed to be used almost exclusively as an intermediate for the manufacture of 1,5-naphthalene diisocyanate and organic dyes. In Japan, an estimated 75% is consumed in the production of the isocyanate and 25% in dye synthesis.

1,5-Naphthalene diisocyanate, the subject of an earlier monograph (IARC, 1979), is used in Japan and western Europe in the production of polyurethane elastomers. The Society of Dyers & Colourists (1971) reports that 1,5-naphthalenediamine can serve as an oxidation base and that one dye can be prepared from it. No evidence was found that it is presently used commercially in these two applications. The nature of the dyes presently being produced in commercial quantities from 1,5-naphthalenediamine is not known.

2.2 Occurrence

1,5-Naphthalenediamine has not been reported to occur as a natural product. No data on its occurrence in the environment were available to the Working Group.

2.3 Analysis

An IARC manual (Egan et al., 1981) gives selected methods for the analysis of aromatic amines. No information on quantitative methods of analysis for 1,5-naphthalenediamine were available to the Working Group.

3. Biological Data Relevant to the Evaluation of Carcinogenic Risk to Humans

3.1 Carcinogenicity studies in animals

Oral administration

Mouse: Groups of 50 male and 50 female B6C3F$_1$ mice, approximately seven weeks of age, were fed diets containing 1000 or 2000 mg/kg 1,5-naphthalenediamine (probably no more than 89% pure, with at least one unspecified impurity detected by thin-layer chromatography) for 103 weeks. The doses were selected on the basis of a range-finding study [see section 3.2(a)]. Groups of 50 mice of each sex served as matched controls. All animals in the study received food and water *ad libitum*, and all were treated for parasites with 3 g/l piperazine adipate added for three days per week to the drinking-water for two weeks prior to treatment with the test chemical. The observation periods were 105-106 weeks for treated mice and 109 weeks for controls. There was no significant association between dose of 1,5-naphthalenediamine and mortality in animals of either sex; 58-82% of treated mice and 60-66% of controls survived the observation period. Statistically significant increases in tumour incidence were observed for the following neoplasms: (a) a dose-related increase ($P = 0.005$) in C-cell carcinomas of the thyroid gland in females: controls, 0/44; low-dose, 1/49; high-dose, 6/45 ($P = 0.014$); (b) dose-related increases ($P < 0.001$ and $P = 0.003$) in neoplasms of the thyroid gland (follicular-cell adenomas, papillary adenomas and papillary adenomas plus papillary cystadenomas) in 0/38 male controls, 8/46 low-dose males ($P = 0.006$), 16/43 high-dose males ($P < 0.001$), 2/44 female controls, 17/49 low-dose females ($P < 0.001$) and 14/45 high-dose females ($P = 0.001$); (c) an increase in hepatocellular carcinomas in females: controls, 1/46; low-dose, 25/49 ($P < 0.001$); high-dose, 16/46 ($P < 0.001$); and (d) an increase in alveolar/bronchiolar adenomas and carcinomas in females: controls, 0/49; low-dose, 10/48 ($P = 0.001$); high-dose, 5/46 ($P = 0.024$) (National Cancer Institute, 1978).

Rat: Groups of 50 male and 50 female Fischer 344 rats, approximately seven weeks of age, were fed diets containing 500 or 1000 mg/kg 1,5-naphthalenediamine (same sample as used above) for 103 weeks. The doses were selected on the basis of a range-finding study [see section 3.2(a)]. Groups of 25 rats of each sex served as matched controls. All animals under study received food and water *ad libitum*, and all were treated for parasites with piperazine adipate added for three days to the drinking-water (followed by three days of plain tap-water and three subsequent days of piperazine adipate) two weeks prior to treatment with the test chemical. The observation periods were 106-107 weeks for treated rats and 109-110 weeks for controls. There was no significant association between dose of 1,5-naphthalenediamine and mortality of animals of either sex: 74-80% of treated rats and 64-68% of controls survived the observation period. A statistically significant, dose-related increase ($P = 0.003$) in the incidence of adenomas plus carcinomas of the clitoral gland was observed: controls, 1/24; low-dose, 3/50; high-dose, 13/50 ($P = 0.021$) (National Cancer Institute, 1978). [The Working Group noted that the increase in the incidence of clitoral gland tumours was only marginally significant, and that histological section of this organ was performed only when it showed gross abnormality.]

3.2 Other relevant biological data

(a) Experimental systems

Toxic effects

No LD_{50} values were available to the Working Group.

In eight-week subchronic feeding studies, male and female Fischer 344 rats and $B6C3F_1$ mice received up to 3.0% 1,5-naphthalenediamine in the diet. Some deaths were observed in treated groups fed 0.3% or more. Mean body weight gain was depressed by 3-22%. No compound-related lesions were observed in chronic studies with 1,5-naphthalenediamine in rats and mice (highest dose, 0.1% in rats and 0.2% in mice) (National Cancer Institute, 1978).

Effects on reproduction and prenatal toxicity

No data were available to the Working Group.

Absorption, distribution, excretion and metabolism

No data were available to the Working Group.

Mutagenicity and other short-term tests

1,5-Naphthalenediamine (same sample as used in the carcinogenicity tests) was mutagenic to *Salmonella typhimurium* strain TA100 without metabolic activation (Dunkel & Simmon, 1980).

(b) Humans

No data were available to the Working Group.

3.3 Case reports and epidemiological studies of carcinogenicity in humans

No data were available to the Working Group.

4. Summary of Data Reported and Evaluation

4.1 Experimental data

1,5-Naphthalenediamine (technical grade) was tested in one experiment in mice and in one experiment in rats by dietary administration. It produced adenomas of the thyroid in male mice and carcinomas and adenomas of the thyroid and lungs and carcinomas of the liver in female mice. The experiment in rats was inadequate for evaluation.

1,5-Naphthalenediamine (technical-grade) was mutagenic to *Salmonella typhimurium*.

4.2 Human data

1,5-Naphthalenediamine has been produced commercially since at least 1957. Its use as an intermediate in the manufacture of 1,5-naphthalene diisocyanate and of dyes could result in occupational exposure.

No case report or epidemiological study was available to the Working Group.

4.3 Evaluation

There is *limited evidence* [1] for the carcinogenicity of 1,5-naphthalenediamine in experimental animals.

No evaluation of the carcinogenicity of 1,5-naphthalenediamine to humans could be made.

---

[1] See preamble, pp. 18-19.

5. References

Dunkel, V.C. & Simmon, V.F. (1980) *Mutagenic activity of chemicals previously tested for carcinogenicity in the National Cancer Institute Bioassay Program.* In: Montesano, R., Bartsch, H. & Tomatis, L., eds, Molecular and Cellular Aspects of Carcinogen Screening Tests *(IARC Scientific Publication No. 27), Lyon, International Agency for Research on Cancer, pp. 283-302*

Egan, H., Fishbein, L., Castegnaro, M., O'Neill, I.K. & Bartsch, H., eds (1981) *Environmental Carcinogens - Selected Methods of Analysis*, Vol. 4, *Some Aromatic Amines and Azodyes in the General and Industrial Environment (IARC Scientific Publications No. 40)*, Lyon, International Agency for Research on Cancer

Hawley, G.G., ed. (1977) *The Condensed Chemical Dictionary*, 9th ed., New York, Van Nostrand-Reinhold, p. 601

IARC (1979) *IARC Monographs on the Evaluation of the Carcinogenic Risk of Chemicals to Humans*, Vol. 19, *Some Monomers, Plastics and Synthetic Elastomers, and Acrolein*, Lyon, pp. 311-340

National Cancer Institute (1978) *Bioassay of 1,5-Naphthalenediamine for Possible Carcinogenicity (Tech. Rep. Ser. No. 143; DHEW Publ. No. (NIH) 78-1398)*, Washington DC, US Government Printing Office

Sadtler Research Laboratories, Inc. (undated) *Sadtler Standard Spectra*, Philadelphia, PA

Sandridge, R.L. & Staley, H.B. (1978) *Amines by reduction.* In: Kirk, R.E. & Othmer, D.F., eds, *Encyclopedia of Chemical Technology*, 3rd ed., Vol. 2, New York, John Wiley & Sons, p. 365

The Society of Dyers & Colourists (1971) *Colour Index*, 3rd ed., Vol. 4, Bradford, Yorkshire, Perkin House, p. 4782

US International Trade Commission (1980) *Imports of Benzenoid Chemicals and Products, 1979 (USITC Publication 1083)*, Washington DC, US Government Printing Office, p. 25

Weast, R.C., ed. (1979) *CRC Handbook of Chemistry and Physics*, 60th ed., Cleveland, OH, Chemical Rubber Co., p. C-385

5-NITRO-*ortho*-ANISIDINE

1. Chemical and Physical Data

1.1 Synonyms and trade names

Chem. Abstr. Services Reg. No.: 99-59-2
Chem. Abstr. Name: Benzenamine, 2-methoxy-5-nitro-
IUPAC Systematic Name: 5-Nitro-*ortho*-anisidine
Synonyms: 2-Amino-1-methoxy-4-nitrobenzene; 2-amino-4-nitroanisole; *ortho*-anisidine nitrate; Azoic Diazo Component 13, Base; 2-methoxy-5-nitroaniline; 3-nitro-6-methoxyaniline; 5-nitro-2- methoxyaniline
Trade names: Azoamine Scarlet K; Azogene Ecarlate R

1.2 Structural and molecular formulae and molecular weight

$C_7H_8N_2O_3$ Mol. wt: 168.2

1.3 Chemical and physical properties of the pure substance

From Weast (1979), unless otherwise specified
(a) *Description*: Orange-red needles
(b) *Melting-point*: 118°
(c) *Density*: d^{156} 1.2068
(d) *Spectroscopic data*: Infra-red, nuclear magnetic resonance, ultra-violet (Sadtler Research Laboratories, Inc., undated) and mass spectral data (Mass Spectrometry Centre, 1974) have been published.
(e) *Solubility*: Soluble in hot water, hot diethyl ether and ethyl acetate; very soluble in ethanol, acetone, hot benzene and acetic acid; slightly soluble in petroleum ether

1.4 Technical products and impurities

5-Nitro-*ortho*-anisidine is available in western Europe as a yellow powder containing 97% active ingredient and with a melting-point of 116-118°.

2. Production, Use, Occurrence and Analysis

2.1 Production and use

(a) Production

5-Nitro-*ortho*-anisidine was first synthesized in 1911. It can be (a) obtained as a by-product during the manufacture of 4-nitro-*ortho*-anisidine, or (b) produced by partial reduction of 2,4-dinitroanisole or (c) produced by nitration of *ortho*-anisidine (The Society of Dyers & Colourists, 1971). At least one European producer uses the second method, and the last method is believed to be used for its commercial production in Japan.

5-Nitro-*ortho*-anisidine has been produced commercially in western Europe since at least 1920. It is believed to be produced at present by one company in the UK and one company in France.

It has been produced in the US since at least 1937 (US Tariff Commission, 1938). In 1979, two US companies reported commercial production of this chemical under its synonym, Azoic Diazo Component 13, base (US International Trade Commission, 1980a); the production in 1977 of one of these companies was reported as 4.5-45.4 thousand kg (US Environmental Protection Agency, 1980). In 1979, 13.3 thousand kg 5-nitro-*ortho*-anisidine and 36.3 thousand kg Azoic Diazo Component 13, base were imported through principal US customs districts (US International Trade Commission, 1980b).

Commercial production of 5-nitro-*ortho*-anisidine in Japan was started prior to 1945; total production in 1979 by two companies is estimated to have been 400 thousand kg.

(b) Use

5-Nitro-*ortho*-anisidine is believed to be used almost exclusively as an intermediate in the manufacture of organic dyes and pigments. The Society of Dyers & Colourists (1971) reported that it can be used as an azoic dye component (under the name Azoic Diazo Component 13, base), and that two organic pigments and one organic dye can be prepared from it.

For the production of azoic dyes, it is converted to its stabilized diazonium salt, known as Azoic Diazo Component 13, salt, and this salt is condensed with one of several Azoic Coupling Components to produce the desired colour. Two US companies reported commercial production of an undisclosed amount (see preamble, p. 20) of this diazonium salt in 1979 (US International Trade Commission, 1980a); separate data on total US production were last reported in 1976, when 119.4 thousand kg were produced by four companies (US International Trade Commission, 1977).

Of the two pigments and one dye that can be prepared from 5-nitro-*ortho*-anisidine, only Pigment Red 23 has been produced commercially in the US. In 1979, total US production of this pigment (which is produced by condensation of the diazonium salt of 5-nitro-*ortho*-anisidine with 3-hydroxy-3'-nitro-2-naphthanilide) by 11 companies amounted to 151 thousand kg (US International Trade Commission, 1980a).

2.2 Occurrence

5-Nitro-*ortho*-anisidine has not been reported to occur as a natural product. No data on its occurrence in the environment were available to the Working Group.

2.3 Analysis

An IARC manual (Egan *et al.*, 1981) gives selected methods for the analysis of aromatic amines. No information on quantitative methods of analysis for 5-nitro-*ortho*-anisidine were available to the Working Group.

3. Biological Data Relevant to the Evaluation of Carcinogenic Risk to Humans

3.1 Carcinogenicity studies in animals

Oral administration

Mouse: Two groups of 50 male and 50 female B6C3F$_1$ mice, six weeks of age, were fed diets containing 5-nitro-*ortho*-anisidine (no impurities detected by thin-layer chromatography) by two treatment schedules: the first group was given 8000 mg/kg for 78 weeks (dose A), and the second group was given 16000 mg/kg for 15 weeks and 4000 mg/kg for 63 weeks (dose B). Both groups were observed for an additional 18 weeks. The doses were selected on the basis of a range-finding study [see section 3.2(*a*)]. Two groups (A and B) of 50 mice of each sex served as matched controls and were kept under observation for 96 weeks. Five males and five females from each of the four groups were sacrificed at week 49 or 78. By the end of the study, 86, 80, 84 and 88% of males and 72, 76, 32 and 66% of females were still alive in the A and B control groups and in the A and B dose groups, respectively. A statistically significant increase in the incidence of hepatocellular carcinomas was observed: males - 12/50 in A controls, 6/48 in B controls, 25/48 in A dose (P = 0.004), 3/47 in B dose; females - 2/47 in A control, 1/50 in B control, 0/41 in A dose, 8/43 in B dose (P = 0.008) (National Cancer Institute, 1978). [The Working Group noted that control and treated mice were received in separate shipments and that those that received dose A were obtained from a different commercial source than their controls. In addition, it was noted that the incidences of liver tumours in treated mice were in the range observed in historical controls.]

Rat: Two groups of 50 male and 50 female Fischer 344 rats, six weeks of age, were fed diets containing 4000 or 8000 mg/kg 5-nitro-*ortho*-anisidine (same sample as used above) for 78 weeks and were observed for an additional 24-28 weeks. The doses were selected on the basis of a range-finding study [see section 3.2(a)]. Two groups (one for each dose group), of 49-50 males and 50 females each, served as matched controls and were kept under observation for 108-109 weeks; five males and five females from each control group were sacrificed at week 78-80. Treated animals had significantly greater mortality than their respective controls: up to 78% of the males in each group survived at least 70 weeks; of the females, 88% of low-dose controls, 86% of high-dose controls, 74% of low-dose rats and 58% of high-dose rats were still alive at 85 weeks. Statistically significant increases in the incidences of the following tumours were observed in the control groups and in the low- and high-dose groups, respectively. Skin carcinomas (several types): males, 1/48, 0/48, 30/50 ($P < 0.001$), 40/48 ($P < 0.001$); Zymbal gland or skin of ear: males, 1/48, 0/48, 2/50, 10/48 ($P = 0.001$); females, 0/49, 0/50, 3/49, 7/46 ($P = 0.004$); mammary adenocarcinoma: females, 0/49, 0/50, 10/49 ($P = 0.001$), 4/46 ($P = 0.049$); preputial gland adenoma or carcinoma: 2/48, 0/48, 2/50, 5/48 ($P = 0.028$); clitoral gland adenoma or carcinoma: 1/49, 2/50, 12/49 ($P = 0.001$), 14/46 ($P < 0.001$) (National Cancer Institute, 1978). [The Working Group noted that control and treated rats were received in separate shipments and that low-dose rats were obtained from a different commercial source than their controls.]

3.2 Other relevant biological data

(a) Experimental systems

Toxic effects

The oral LD_{50} of 5-nitro-*ortho*-anisidine in rats was 704 mg/kg bw (Lewis & Tatken, 1979).

Administration of up to 4000 mg/kg of diet for 7 weeks to male and female $B6C3F_1$ mice led to no more than an 8% reduction in body weight gain. Concentrations of 4000 and 8000 mg/kg of the compound in the diet led to depressions in weight gain. In a group that received 16000 mg/kg, decreased to 4000 mg/kg at 15 weeks, non-neoplastic lesions were observed, including degenerative and hyperplastic changes in the liver and lymphocytic inflammatory infiltration of the thyroid in male mice, and myelofibrosis in the bones of female mice (National Cancer Institute, 1978).

Administration of diets containing up to 4000 mg/kg 5-nitro-*ortho*-anisidine to male and female Fischer 344 rats for seven weeks produced depressions in weight gain of no more than 10%. Feeding of 8000 mg/kg of diet resulted in liver degeneration, testicular degeneration and sclerosis of the gastric mucosa in males, and pigmentation of the thyroid follicles and haematopoiesis in the spleen in animals of both sexes (National Cancer Institute, 1978).

Effects on reproduction and prenatal toxicity

No data were available to the Working Group

Absorption, distribution, excretion and metabolism

No data were available to the Working Group.

Mutagenicity and other short-term tests

5-Nitro-*ortho*-anisidine induced reverse mutations in *Salmonella typhimurium* strain TA98 in the absence of metabolic activation (Chiu *et al.*, 1978).

(b) Humans

No data were available to the Working Group.

3.3 Case reports and epidemiological studies of carcinogenicity in humans

No data were available to the Working Group.

4. Summary of Data Reported and Evaluation

4.1 Experimental data

5-Nitro-*ortho*-anisidine was tested for carcinogenicity in one experiment in mice and in one experiment in rats by dietary administration. It produced skin carcinomas in male rats, carcinomas of the Zymbal gland or skin of the ear in male and female rats, mammary adenocarcinomas in female rats and adenomas or carcinomas of the preputial and clitoral glands in male and female rats, respectively. The experiment in mice was inadequate for evaluation.

5-Nitro-*ortho*-anisidine was mutagenic to *Salmonella typhimurium*.

4.2 Human data

5-Nitro-*ortho*-anisidine has been produced commercially since at least 1937. Its principal use as a dye intermediate could result in occupational exposure.

No case report or epidemiological study was available to the Working Group.

4.3 Evaluation

There is *limited evidence* [1] for the carcinogenicity of 5-nitro-*ortho*-anisidine in experimental animals.

No evaluation of the carcinogenicity of 5-nitro-*ortho*-anisidine to humans could be made.

---

[1] See preamble, pp. 18-19.

5. References

Chiu, C.W., Lee, L.H., Wang, C.Y. & Bryan, G.T. (1978) Mutagenicity of some commercially available nitro compounds for *Salmonella typhimurium*. *Mutat. Res.*, *58*, 11-22

Egan, H., Fishbein, L., Castegnaro, M., O'Neill, I.K. & Bartsch, H., eds (1981) *Environmental Carcinogens - Selected Methods of Analysis*, Vol. 4, *Some Aromatic Amines and Azodyes in the General and Industrial Environment (IARC Scientific Publications No. 40)*, Lyon, International Agency for Research on Cancer

Lewis, R.J., Sr & Tatken, R.L., eds (1979) *1978 Registry of Toxic Effects of Chemical Substances*, Rockville, MD, US Department of Health, Education, & Welfare, p. 116

Mass Spectrometry Data Centre (1974) *Eight Peak Index of Mass Spectra*, Reading, UK, Atomic Weapons Research Establishment

National Cancer Institute (1978) *Bioassay of 5-nitro-o-anisidine for Possible Carcinogenicity (Tech. Rep. Ser. No. 127; DHEW Publ. No. (NIH) 78-1382)*, Washington DC, US Government Printing Office

Sadtler Research Laboratories, Inc. (undated) *Sadtler Standard Spectra*, Philadelphia, PA

The Society of Dyers & Colourists (1971) *Colour Index*, 3rd ed., Vol. 4, Bradford, Yorkshire, Perkin House, pp. 4350, 4706

US Environmental Protection Agency (1980) *Chemicals in Commerce Information System (CICIS)*, Washington DC, Office of Pesticides & Toxic Substances, Chemical Information Division

US International Trade Commission (1977) *Synthetic Organic Chemicals, US Production and Sales, 1976 (USITC Publication 833)*, Washington DC, US Government Printing Office, pp. 76, 87

US International Trade Commission (1980a) *Synthetic Organic Chemicals, US Production and Sales, 1979 (USITC Publication 1099)*, Washington DC, US Government Printing Office, pp. 75, 76, 103, 109

US International Trade Commission (1980b) *Imports of Benzenoid Chemicals and Products, 1979 (USITC Publication 1083)*, Washington DC, US Government Printing Office, pp. 26, 72

US Tariff Commission (1938) *Dyes and Other Synthetic Organic Chemicals in the United States, 1937 (Report No. 132, Second Series)*, Washington DC, US Government Printing Office, p. 17

Weast, R.C., ed. (1979) *CRC Handbook of Chemistry and Physics*, 60th ed., Cleveland, OH, Chemical Rubber Co., p. C-117

2,2',5,5'-TETRACHLOROBENZIDINE

1. Chemical and Physical Data

1.1 Synonyms and trade names
Chem. Abstr. Services Reg. No.: 15721-02-5
Chem. Abstr. Name: (1,1'-Biphenyl)-4,4'-diamine, 2,2',5,5'-tetrachloro-
IUPAC Systematic Name: 2,2',5,5'-Tetrachlorobenzidine
Synonyms: 3,3',6,6'-Tetrachlorobenzidine; 2,2',5,5'-tetrachloro-4,4'-diaminodiphenyl

1.2 Structural and molecular formulae and molecular weight

$C_{12}H_8Cl_4N_2$ Mol. wt: 322.0

1.3 Chemical and physical properties of the pure substance

No data were available to the Working Group.

1.4 Technical products and impurities

2,2',5,5'-Tetrachlorobenzidine is available in Japan as a light-brown powder containing 98% active ingredient and a maximum of 1% ash, with a minimal melting-point of 133°.

2. Production, Use, Occurrence and Analysis

2.1 Production and use

(a) Production

2,2',5,5'-Tetrachlorobenzidine is produced in Japan by reduction of 2,5-dichloronitrobenzene to the corresponding tetrachloro-substituted hydrazobenzene, and benzidine rearrangement of the product.

It has been produced in Japan since 1965; in 1979, one company produced an estimated 20 thousand kg. It is also believed to be produced by one company in the Federal Republic of Germany. This chemical is not produced in commercial quantities in the US; imports through principal US customs districts were last reported in 1978, when they amounted to only 65 kg (US International Trade Commission, 1979).

(b) Use

2,2',5,5'-Tetrachlorobenzidine is believed to be used almost exclusively as an intermediate in the manufacture of organic pigments. According to The Society of Dyers & Colourists (1977), two pigments can be prepared from it. No evidence was found that these compounds, Pigment Yellow 81 and Pigment Yellow 113, have ever been produced commercially in the US.

2.2 Occurrence

2,2',5,5'-Tetrachlorobenzidine has not been reported to occur as a natural product. No data on its occurrence in the environment were available to the Working Group.

2.3 Analysis

An IARC manual (Egan *et al.*, 1981) gives selected methods for the analysis of aromatic amines. No information on quantitative methods of analysis of 2,2',5,5'-tetrachlorobenzidine were available to the Working Group.

3. Biological Data Relevant to the Evaluation of Carcinogenic Risk to Humans

3.1 Carcinogenicity studies in animals

Oral administration

Mouse: A group of 50 weanling female dd mice were fed a pelleted diet with 25% protein (procedure unspecified) containing 1000 mg/kg 2,2',5,5'-tetrachlorobenzidine (purity unspecified) for at least 300 days. A group of 25 female dd mice served as controls. Twenty-five treated and 20 control animals survived more than 300 days after the onset of treatment; these were the only animals necropsied. Histological analyses were made only on those organs that showed gross pathological changes. Of the 25 treated animals that survived, 16 developed tumours: 14 malignant lymphomas, 2 mammary carcinomas and 1 mammary fibrosarcoma. No tumours were found in the controls (Yoshimoto, 1978). [The Working Group noted the small number of control animals and the incomplete necropsies and histological analyses.]

Rat: A group of 30 male and 30 female weanling Wistar rats were fed a pelleted diet with 25% protein (procedure unspecified) containing 3000 mg/kg 2,2',5,5'-tetrachlorobenzidine (purity unspecified) for at least 300 days. A group of 15 male and 10 female rats served as controls. Twenty-two male and 23 female treated rats and 11 male and 10 female control rats survived more than 300 days; these were the only animals necropsied. Histological analyses were made only on those organs that showed gross pathological changes. Bladder carcinomas were found in 3/22 male and 1/23 female treated rats but in none of the controls (Yoshimoto, 1978). [The Working Group noted the small number of control animals and the incomplete necropsies and histological analyses.]

3.2 Other relevant biological data

(a) Experimental systems

Toxic effects

No data were available to the Working Group.

Effects on reproduction and prenatal toxicity

No data were available to the Working Group.

Absorption, distribution, excretion and metabolism

No data were available to the Working Group.

Mutagenicity and other short-term tests

2,2',5,5'-Tetrachlorobenzidine induced reverse mutations in *Salmonella typhimurium* strains TA98 and TA100 when tested in the presence of a rat liver activation system (Yoshimoto, 1978).

(b) Humans

No data were available to the Working Group.

3.3 Case reports and epidemiological studies of carcinogenicity in humans

No data were available to the Working Group.

4. Summary of Data Reported and Evaluation

4.1 Experimental data

2,2',5,5'-Tetrachlorobenzidine (of unknown purity) was tested for carcinogenicity by dietary administration in mice and rats. The available data were insufficient for evaluation.

2,2',5,5'-Tetrachlorobenzidine was mutagenic to *Salmonella typhimurium* with metabolic activation.

4.2 Human data

2,2',5,5'-Tetrachlorobenzidine has been produced commercially since at least 1965. Its use as an intermediate in the manufacture of organic pigments could result in occupational exposure.

No case report or epidemiological study was available to the Working Group.

4.3 Evaluation

The available data were inadequate for an evaluation of the carcinogenicity of 2,2',5,5'-tetrachlorobenzidine in experimental animals.

No evaluation of the carcinogenicity of this compound to humans could be made.

5. References

Egan, H., Fishbein, L., Castegnaro, M., O'Neill, I.K. & Bartsch, H., eds (1981) *Environmental Carcinogens - Selected Methods of Analysis*, Vol. 4, *Some Aromatic Amines and Azodyes in the General and Industrial Environment* (*IARC Scientific Publications No. 40*), Lyon, International Agency for Research on Cancer

The Society of Dyers & Colourists (1977) *Colour Index*, revised 3rd ed., *Additions and Amendments*, No. 22, Bradford, Yorkshire, Perkin House, p. 621

US International Trade Commission (1979) *Imports of Benzenoid Chemicals and Products, 1978* (*USITC Publication 990*), Washington DC, US Government Printing Office, p. 29

Yoshimoto, S. (1978) Carcinogenicity and mutagenicity of tetrachloro benzidine. *Jikeikai med. J., 25*, 123-128

4,4'-THIODIANILINE

This substance was considered by a previous Working Group, in June 1977 (IARC, 1978). Since that time new data have become available, and these have been incorporated into the monograph and taken into consideration in the present evaluation.

1. Chemical and Physical Data

1.1 Synonyms and trade names

Chem. Abstr. Services Reg. No.: 139-65-1
Chem. Abstr. Name: Benzenamine, 4,4'-thiobis-
IUPAC Systematic Name: 4,4'-Thiodianiline
Synonyms: Bis(4-aminophenyl)sulphide; 4,4'-diaminodiphenyl sulphide; *para,para'*-diaminodiphenyl sulphide; 4,4'-diaminophenyl sulphide; di(*para*-aminophenyl)-sulphide; thioaniline; 4,4'-thiobis(aniline); *para,para'*-thiodianiline; thiodi-*para*-phenylenediamine

1.2 Structural and molecular formulae and molecular weight

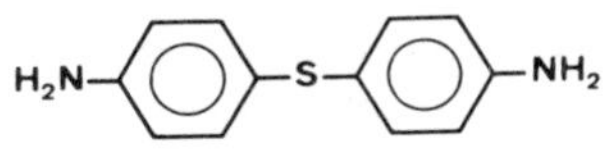

$H_2N-\langle\bigcirc\rangle-S-\langle\bigcirc\rangle-NH_2$

$C_{12}H_{12}N_2S$ Mol. wt: 216.3

1.3 Chemical and physical properties of the pure substance

From Weast (1979)
(a) *Description*: Needles
(b) *Melting-point*: 108-109°
(c) *Spectroscopic data*: Infra-red, nuclear magnetic resonance, ultra-violet (Sadtler Research Laboratories, Inc., undated) and mass spectral data (Mass Spectrometry Data Centre, 1974) have been published.
(d) *Solubility*: Slightly soluble in hot water; very soluble in ethanol, diethyl ether and hot benzene

1.4 Technical products and impurities

No data were available to the Working Group.

2. Production, Use, Occurrence and Analysis

2.1 Production and use

(a) Production

4,4'-Thiodianiline was prepared by Merz & Weith in 1871 by boiling sulphur with aniline for several days (Prager *et al.*, 1930). No information was available on the methods used for its commercial production.

Production of 4,4'-thiodianiline in the US was first reported in 1941-1943 (US Tariff Commission, 1945); currently, one company is believed to produce it. 4,4'-Thiodianiline is also produced in the USSR. No evidence was found that it is produced in commercial quantities in western Europe or Japan.

(b) Use

4,4'-Thiodianiline appears to be used almost exclusively as a dye intermediate. The Society of Dyers & Colourists (1971) reported that three dyes can be prepared from it. Only one of these, Mordant Yellow 16, has been produced commercially in the US; one company reported production of an undisclosed amount (see preamble, p. 20) of this dye in 1979 (US International Trade Commission, 1980).

The USSR has established a ceiling value of 1 mg/m^3 of air for occupational exposure to 4,4'-thiodianiline (International Labour Office, 1977).

2.2 Occurrence

4,4'-Thiodianiline is not known to occur as a natural product. No data on its occurrence in the environment were available to the Working Group.

2.3 Analysis

An IARC manual (Egan *et al.*, 1981) gives selected methods for the analysis of aromatic amines.

Thin-layer chromatography, with four different solvent systems, has been used to separate and identify a group of aromatic diamines, including 4,4'-thiodianiline, and N-benzamides (Krasnov et al., 1970); and paper chromatography has been used to separate and identify a group of aromatic para,para'-diamines, including 4,4'-thiodianiline (Gasparic & Snobl, 1971). A group of aromatic amines, including this compound, has been separated and identified by gas chromatography (Kazinik et al., 1971a,b).

3. Biological Data Relevant to the Evaluation of Carcinogenic Risk to Humans

3.1 Carcinogenicity studies in animals

Oral administration

Mouse: Groups of 35 male and 35 female $B6C3F_1$ mice, 46-47 days of age, were fed diets containing 2500 or 5000 mg/kg 4,4'-thiodianiline (purity probably greater than 97%, with at least one unspecified impurity detected by gas chromatography) on five days per week for 77-79 weeks. The doses were selected on the basis of a range-finding study in male Swiss mice [see section 3.2(a)]. A group of 14 mice of each sex served as matched controls. All animals under study received food and water *ad libitum*. Mean body weight gain of treated male and female mice was markedly reduced, and a positive dose-related trend in mortality was observed: by 52 weeks, 97% of the low-dose animals and 83% of the high-dose animals of both sexes were still alive, as compared with 86% of female and 100% of male controls. All low- and high-dose animals had died by the end of 77-79 weeks of treatment, while the controls were kept for observation for 91 weeks. Statistically significant increases in tumour incidences were observed for several neoplasms: (a) There was a dose-related increase ($P < 0.001$) in hepatocellular carcinomas in females: 0/12 controls, 32/34 low-dose ($P < 0.001$) and 30/31 high-dose animals ($P < 0.001$); in males, hepatocellular carcinomas were seen in 1/13 controls, 32/34 low-dose ($P < 0.001$) and 22/24 high-dose animals ($P < 0.001$). Such tumours metastasized to the lungs in 9-14% of females and to the kidneys in 3% of low-dose males and high-dose females. (b) There was a dose-related increase ($P < 0.001$) in follicular-cell carcinomas of the thyroid gland in females: 0/11 controls, 3/33 low-dose and 15/30 high-dose animals ($P = 0.002$); and in males, these tumours were seen in 0/14 controls, 15/33 low-dose ($P < 0.001$) and 20/23 high-dose animals ($P < 0.001$). Pulmonary metastases were found in some males. When the incidences of follicular-cell adenoma and carcinoma were grouped for analysis, a significant increase was observed for animals of each sex and at each dose level. In addition, the high incidence of thyroid follicular-cell hyperplasia was considered to be related to the treatment (National Cancer Institute, 1978; Cueto & Chu, 1979).

Rat: In an experiment designed to screen carcinogens that act on the mammary gland of young female Sprague-Dawley rats, 20 animals, 40 days of age, were given 10 doses of 40 mg/animal 4,4'-thiodianiline in 1 ml sesame oil every three days by gastric intubation (total dose, 400 mg); a further group of 10 females received a total dose of 300 mg/rat; 140 control rats received the vehicle alone. All surviving animals were killed after nine months. Of 12 surviving rats given 400 mg and which were autopsied, three developed mammary carcinomas; of those that received 300 mg/animal, 1/8 developed a mammary carcinoma. Among 132 controls autopsied, of which 127 survived nine months, three had mammary carcinomas and one a mammary fibroadenoma (Griswold et al., 1968), [The Working Group noted the inadequate duration of the experiment.]

Groups of 35 male and 35 female Fischer 344 rats, 47-48 days of age, were fed diets containing 1500 or 3000 mg/kg 4,4'-thiodianiline (purity probably greater than 97%, with at least one unspecified impurity detected by gas chromatography) on five days per week for 68-72 weeks. The doses were selected on the basis of a range-finding study in male Sprague-Dawley rats [see section 3.2(a)]. A group of 15 rats of each sex served as matched controls. All animals under study received food and water *ad libitum*. Mean body weight gain of treated male and female rats was markedly reduced, and a positive, dose-related trend in mortality was observed: by 52 weeks, 51-91% of the treated rats and 100% of controls were still alive. The low-dose animals had all died by 68-72 weeks of treatment and the high-dose animals by 68-69 weeks; the controls were observed up to 104 weeks. Statistically significant increases in tumour incidences were observed for several neoplasms, all found in treated animals only. (a) An increase occurred in follicular-cell carcinomas of the thyroid gland in females: 24/33 low-dose ($P < 0.001$), 32/32 high-dose ($P < 0.001$); and in males: 28/33 low-dose ($P < 0.001$), 32/33 high-dose ($P < 0.001$). In most cases, these tumours were bilateral, and infiltration of neoplastic cells was frequent. Pulmonary metastases were found in 34-52% of animals in the different groups; metastases to other sites were rare. (b) There was an increased incidence of adenocarcinomas of the uterus: 31/33 low-dose ($P < 0.001$), 23/32 high-dose ($P < 0.001$). Pulmonary metastases were observed in 44% of the low-dose group; metastases to other sites were rare. (c) An increase was observed in the incidence of squamous-cell papillomas or carcinomas of the ear canal in males: 15/33 low-dose ($P = 0.001$), 8/33 high-dose ($P = 0.037$). These tumours metastasized to the lungs in 2/66 animals. (d) An increase occurred in the incidence of hepatocellular carcinomas in males: 21/33 low-dose ($P < 0.001$), 10/33 high-dose ($P = 0.014$). The carcinomas metastasized to the lungs in 21% of animals in the low-dose group and in 3% of those in the high-dose group (National Cancer Institute, 1978; Cueto & Chu, 1979).

3.2 Other relevant biological data

(a) *Experimental data*

Toxic effects

The oral LD_{50} of 4,4'-thiodianiline in rats has been reported as 1100 mg/kg bw (Lewis & Tatken, 1979).

In 45-day subchronic feeding studies with male Sprague-Dawley rats and male Swiss mice, all animals that received diets containing 3% (rats) or 2.5% (mice) or more 4,4'-thiodianiline died. In chronic studies with Fischer 344 rats and B6C3F$_1$ mice, squamous metaplasia of alveolar and bronchiolar epithelium, nodular hyperplasia of the liver and hyperplasia of the bile ducts were observed (National Cancer Institute, 1978).

Effects on reproduction and prenatal toxicity

Oral administration of 50 mg/kg bw 4,4'-thiodianiline to mice on days 1-5 of pregnancy slightly reduced fetal implantation; doses ⩾100 mg/kg bw prevented implantation (Kamboj & Kar, 1966).

Absorption, distribution, excretion and metabolism

No data were available to the Working Group.

Mutagenicity and other short-term tests

4,4'-Thiodianiline induced reverse mutations in *Salmonella typhimurium* strains TA98 and TA100 when tested in the presence of a liver activation system prepared from Aroclor 1254-induced rats (Lavoie *et al.*, 1979).

(b) Humans

No data were available to the Working Group.

3.3 Case reports and epidemiological studies of carcinogenicity in humans

No data were available to the Working Group.

4. Summary of Data Reported and Evaluation

4.1 Experimental data

4,4'-Thiodianiline was tested adequately for carcinogenicity in one experiment in mice and in one experiment in rats by dietary administration. It was carcinogenic for animals of both sexes of both species. In mice, it produced hepatocellular carcinomas and carcinomas or adenomas of the thyroid gland in animals of both sexes. In rats, it produced metastatic thyroid gland carcinomas in animals of both sexes, hepatocellular carcinomas and ear-canal papillomas or carcinomas in males and adenocarcinomas of the uterus in females.

4,4'-Thiodianiline was mutagenic to *Salmonella typhimurium* with metabolic activation.

4.2 Human data

4,4'-Thiodianiline was first produced commercially in the early 1940s. Its use as a dye intermediate could lead to occupational exposure.

No case report or epidemiological study was available to the Working Group.

4.3 Evaluation

There is *sufficient evidence*[1] for the carcinogenicity of 4,4'-thiodianiline in experimental animals.

In the absence of data on humans, 4,4'-thiodianiline should be regarded, for practical purposes, as if it presented a carcinogenic risk to humans.

---

[1] See preamble, p. 18.

5. References

Cueto, C., Jr & Chu, K.C. (1979) *Carcinogenicity of dapsone and 4,4'-thiodianiline*. In: Deichmann, W.B., ed., *Toxicology and Occupational Medicine*, Vol. 4, Amsterdam, Elsevier, pp. 99-108

Egan, H., Fishbein, L., Castegnaro, M., O'Neill, I.K. & Bartsch, H., eds (1981) *Environmental Carcinogens - Selected Methods of Analysis, Vol. 4, Some Aromatic Amines and Azodyes in the General and Industrial Environment (IARC Scientific Publications No. 40)*, Lyon, International Agency for Research on Cancer

Gasparic, J. & Snobl, D. (1971) Identification of organic compounds. LXXIII. Paper chromatography of aromatic p,p'-diamines (Czech.). *Sb. Ved. Pr., Vys. Sk. Chemickotechnol., Pardubice, 25,* 33-40 [*Chem. Abstr., 75,* 147558v]

Griswold, D.P., Jr, Casey, A.E., Weisburger, E.K. & Weisburger, J.H. (1968) The carcinogenicity of multiple intragastric doses of aromatic and heterocyclic nitro or amino derivatives in young female Sprague-Dawley rats. *Cancer Res., 28,* 924-933

IARC (1978) *IARC Monographs on the Evaluation of the Carcinogenic Risk of Chemicals to Humans, Vol. 16, Some Aromatic Amines and Related Nitro Compounds - Hair Dyes, Colouring Agents and Miscellaneous Industrial Chemicals*, Lyon, pp. 343-348

International Labour Office (1977) *Occupational Exposure Limits for Airborne Toxic Substances (Occupational Safety & Health Series No. 37)*, Geneva, pp. 86-87

Kamboj, V.P. & Kar, A.B. (1966) Antiimplantation effect of some aromatic sulfur derivatives. *Indian J. exp. Biol., 4,* 120-121 [*Chem. Abstr., 65,* 7847a]

Kazinik, E.M., Gudkova, G.A. & Shcheglova, T.A. (1971a) Gas-liquid chromatographic analysis of some high-boiling aromatic amines. *Zh. anal. Khim., 26,* 154-157

Kazinik, E.M., Gudkova, G.A., Mesh, L.Y., Shcheglova, T.A. & Ivanov, A.V. (1971b) Gas-chromatographic analysis of high-boiling diamines - diphenyl, diphenyl sulfide, and diphenyl sulfone derivatives. *Zh. anal. Khim., 26,* 1920-1923

Krasnov, E.P., Logunova, V.I. & Shumakova, V.D. (1970) *Thin-layer chromatography of aromatic diamines and N-benzamides* (Russ.). In: Pakshver, A.B., ed., *Synthetic Polymer Fibres*, Moscow, Khimiya, pp. 292-294 [*Chem. Abstr., 74,* 112598k]

Lavoie, E., Tulley, L., Fow, E. & Hoffmann, D. (1979) Mutagenicity of aminophenyl and nitrophenyl ethers, sulfides and disulfides. *Mutat. Res., 67,* 123-131

Lewis, R.J., Sr & Tatken, R.L., eds (1979) *1978 Registry of Toxic Effects of Chemical Substances*, Cincinnati, OH, US Department of Health, Education, & Welfare, p. 114

Mass Spectrometry Data Centre (1974) *Eight Peak Index of Mass Spectra*, 2nd ed., Reading, UK, Atomic Weapons Research Establishment

National Cancer Institute (1978) *Bioassay of 4,4'-Thiodianiline for Possible Carcinogenicity (Tech. Rep. Ser. No. 47; DHEW Publ. No. (NIH) 78-847)*, Washington DC, US Government Printing Office

Prager, B., Jacobson, P., Schmidt, P. & Stern, D., eds (1930) *Beilsteins Handbuch der Organischen Chemie*, 4th ed., Vol. 13, Syst. No. 1853, Berlin, Springer, pp. 535-536

Sadtler Research Laboratories, Inc. (undated) *Sadtler Standard Spectra*, Philadelphia, PA

The Society of Dyers & Colourists (1971) *Colour Index*, 3rd ed., Vol. 4, Bradford, Yorkshire, Lund Humphries, p. 4704

US International Trade Commission (1980) *Synthetic Organic Chemicals, US Production and Sales, 1979 (USITC Publication 1099)*, Washington DC, US Government Printing Office, p. 93

US Tariff Commission (1945) *Synthetic Organic Chemicals, US Production and Sales, 1941-43 (Report No. 153, Second Series)*, Washington DC, US Government Printing Office, p. 20

Weast, R.C., ed. (1979) *CRC Handbook of Chemistry and Physics*, 60th ed., Cleveland, OH, Chemical Rubber Co., p. C-505

ortho-**TOLUIDINE** and ortho-**TOLUIDINE HYDROCHLORIDE**

These substances were considered by a previous Working Group, in June 1977 (IARC, 1978a). Since that time new data have become available, and these have been incorporated into the monograph and taken into consideration in the present evaluation.

1. Chemical and Physical Data

ortho-TOLUIDINE

1.1 Synonyms and trade names

Chem. Abstr. Services Reg. No.: 95-53-4
Chem. Abstr. Name: Benzenamine, 2-methyl-
IUPAC Systematic Name: o-Toluidine
Synonyms: 1-Amino-2-methylbenzene; 2-amino-1-methylbenzene; 2-aminotoluene; ortho-aminotoluene; C.I. Azoic Brown 29 (Component); C.I. 37077; 1-methyl-2-aminobenzene; 2-methyl-1-aminobenzene; 2-methylaniline; ortho-methylaniline; 2-methylbenzenamine; ortho-methylbenzenamine; ortho-toluidin; 2-toluidine; ortho-tolylamine

1.2 Structural and molecular formulae and molecular weight

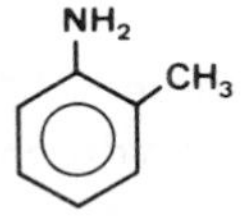

C_7H_9N Mol. wt: 107.2

1.3 Chemical and physical properties of the pure substance

From Weast (1979), unless otherwise specified
(a) *Description*: Light-yellow liquid (Windholz, 1976)
(b) *Boiling-point*: 200.2°
(c) *Melting-point*: -14.7° (β-form)
(d) *Density*: d_4^{20} 0.9984
(e) *Refractive index*: n_D^{20} 1.5725

(f) Spectroscopic data: Infra-red, nuclear magnetic resonance, mass and ultra-violet spectral data have been tabulated (Grasselli, 1973).

(g) Solubility: Slightly miscible with water; miscible with carbon tetrachloride, diethyl ether and ethanol

(h) Volatility: Vapour pressure is 1 mm at 44° and 10 mm at 81.4°.

(i) Stability: Flash-point, 87°; turns reddish brown in air and light (Windholz, 1976). Technical grade decomposes exothermically above 350° (E.I. du Pont de Nemours Co., 1979).

1.4 Technical products and impurities

ortho-Toluidine is available in the US as a technical grade (with or without added stabilizer), with the following specifications: a minimum of 99.5% *ortho*-toluidine, a maximum of 0.4% *meta*-toluidine, a maximum of 0.1% *para*-toluidine, a maximum of 0.1% moisture and a maximal optical absorbance at 450 nm of 0.100 (unstabilized) or 0.025 (stabilized) (E.I. du Pont de Nemours & Co., 1979). It is available in western Europe in several grades containing minima of 97%, 99% or more active ingredient (with *para*-toluidine as the principal impurity). In Japan, it can be obtained as a colourless to pale-yellow liquid containing a minimum of 99% active ingredient and with a maximum of 0.2% moisture content.

ortho-TOLUIDINE HYDROCHLORIDE

1.1 Synonyms and trade names

Chem. Abstr. Services Reg. No.: 636-21-5
Chem. Abstr. Name: Benzenamine, 2-methyl-, hydrochloride
IUPAC Systematic Name: *o*-Toluidine hydrochloride
Synonyms: 1-Amino-2-methylbenzene hydrochloride; 2-amino-1-methylbenzene hydrochloride; 2-aminotoluene hydrochloride; *ortho*-aminotoluene hydrochloride; 1-methyl-2-aminobenzene hydrochloride; 2-methyl-1-aminobenzene hydrochloride; 2-methylaniline hydrochloride; *ortho*-methylaniline hydrochloride; 2-methylbenzenamine hydrochloride; *ortho*-methylbenzenamine hydrochloride; *ortho*-toluidin hydrochloride; 2-toluidine hydrochloride; *ortho*-tolylamine hydrochloride

1.2 Structural and molecular formulae and molecular weight

$C_7H_9N \cdot HCl$ Mol. wt: 143.6

1.3 Chemical and physical properties of the pure substance

From Weast (1979), unless otherwise specified
- *(a) Description*: Monoclinic prisms (from cold water)
- *(b) Boiling-point*: 242.2°
- *(c) Melting-point*: 215°
- *(d) Spectroscopic data*: Infra-red, nuclear magnetic resonance, ultra-violet (Sadtler Research Laboratories, Inc., undated) and mass spectral data (Mass Spectrometry Data Centre, 1974) have been published.
- *(e) Solubility*: Very soluble in water; soluble in ethanol; insoluble in benzene and diethyl ether

1.4 Technical products and impurities

No data were available to the Working Group.

2. Production, Use, Occurrence and Analysis

2.1 Production and use

(a) Production

ortho-Toluidine was prepared by Muspratt & Hofmann in 1844 by the reduction of 1-methyl-2-nitrobenzene with alcoholic ammonium sulphide (Prager *et al.*, 1929). It is produced commercially in the US, western Europe and Japan by iron reduction or catalytic hydrogenation of this nitro compound (Sandridge & Staley, 1978). It can be prepared as part of a mixture with *para*-toluidine by the reduction of crude nitrotoluene (Hawley, 1977).

ortho-Toluidine was first produced commercially in the UK in 1880. It is believed that at least seven companies in western Europe currently produce it: three in the Federal Republic of Germany, three in the UK and one in France. In addition, two companies in the UK are believed to produce mixed toluidines.

ortho-Toluidine has been produced commercially in the US for over 50 years (US Tariff Commission, 1922), and commercial production of the hydrochloride was first reported in 1956 (US Tariff Commission, 1957). In 1979, two US companies reported production of an undisclosed amount (see preamble, p. 20) of *ortho*-toluidine, and one of the companies also reported production of an undisclosed amount of mixed toluidines (US International Trade Commission, 1980a). Production of *ortho*-toluidine by one of these companies in 1977 was 0.454-4.54 million kg (US Environmental Protection Agency, 1980a). Commercial production of *ortho*-toluidine hydrochloride was last reported in the US in 1975, when one company reported production of an undisclosed amount (US International Trade Commission, 1977a); one US company is believed to produce this chemical currently.

Imports of *ortho*-toluidine through principal US customs districts in 1979 were 1.37 million kg (US International Trade Commission, 1980a), a much larger quantity than had been reported in previous years. Imports of the hydrochloride were last reported in 1975, when one thousand kg were imported (US International Trade Commission, 1977b).

Commercial production of *ortho*-toluidine in Japan was started prior to 1945. Total production in 1979 by four companies is estimated to have been 1.8 million kg; and 88 thousand kg were imported. The hydrochloride is not produced in commercial quantities in Japan.

(b) Use

The principal commercial use of *ortho*-toluidine and its hydrochloride is believed to be as intermediates in the manufacture of a variety of dyes. *ortho*-Toluidine has also been reported to be used as an intermediate for rubber chemicals, pharmaceuticals and pesticides (E.I. du Pont de Nemours & Co., 1979).

The Society of Dyers & Colourists (1971) reported that 93 dyes and pigments can be prepared from *ortho*-toluidine and its derivatives. Sixteen of these were reported to be produced commercially in the US in 1979. Separate data were published for only one dye made directly from *ortho*-toluidine, i.e., Direct Red 72: total production by five companies amounted to 266 thousand kg. Separate production data on two pigments based on *ortho*-toluidine derivatives were also reported: six companies reported a total of 41 thousand kg of Pigment Red 17 (made from the *ortho*-toluidine derivative, 3-hydroxy-2-naphtho-*ortho*-toluidide), and 15 companies reported a total of 1.85 million kg of Pigment Yellow 14 (made from the *ortho*-toluidine derivative, *ortho*-acetoacetotoluidide). Other commercially important dyes based on *ortho*-toluidine and its derivatives are Scarlet Red, or Solvent Red 24 (see IARC, 1975a) (four producers); and Solvent Red 26 (two producers) (US International Trade

Commission, 1980b). Of lesser commercial importance in the US are the following dyes, all of which have been the subject of previous IARC monographs: magenta (also known as Fuchsine and Basic Violet 14) (IARC, 1974); *ortho*-aminoazotoluene [4-(*ortho*-tolylazo)-*ortho*-toluidine; Solvent Yellow 3] (IARC, 1975b); and Oil Orange SS (Solvent Orange 2) (IARC, 1975c). Another dye derived from *ortho*-toluidine, which has not been produced commercially in the US for many years, is Yellow OB (Solvent Yellow 6) (IARC, 1975d).

The only commercially significant rubber-processing chemicals in the US which are believed to be derived from *ortho*-toluidine are two accelerators: 1,3-di-*ortho*-tolylguanidine and its salt with dicatechol borate.

ortho-Toluidine has been reported to be involved in the manufacture of the hypnotic agent methaqualone and the anaesthetic aptocaine. It can also be used in the synthesis of diacetylaminoazotoluene (see IARC, 1975e), which is not of commercial significance in the US but has reportedly been employed in human and veterinary medicine.

The only pesticide derived from *ortho*-toluidine that is produced in commercial quantities in the US (*via* 2-methylcyclohexylamine) is 1-(2-methylcyclohexyl)-3-phenylurea (common name: siduron), which has limited application in controlling weeds in lawns and turf.

ortho-Toluidine is believed to be used as a chemical intermediate in the production of the following chemicals that are produced in commercial quantities in the US: *meta*-[(4-amino-3-tolyl)azo]benzenesulphonic acid; 7-[(4-amino-*ortho*-tolyl)azo]-1,3-naphthalenedisulphonic acid;*para*-chloro-*ortho*-toluidine (see IARC, 1978b); di-*ortho*-tolylthiourea (used as a mineral flotation agent); *ortho*-tolylbiguanide (an antibacterial agent used in bath soaps); 2-*ortho*-toluidinoethanol; *ortho*-toluidinomethanesulphonic acid; and tolyltriazole (a corrosion inhibitor) (US International Trade Commission, 1980b).

ortho-Toluidine may also be used to produce special dyes for colour photography, and as an ingredient in clinical laboratory reagents for determining glucose and haemoglobin. Other reported uses are in printing of textiles blue-black and in making colours fast to acids (Windholz, 1976).

ortho-Toluidine is believed to be employed in many of the same applications in western Europe, i.e., in the production of dyes, pesticides, pharmaceuticals, pigments, rubber chemicals, and textile auxiliaries, and for other miscellaneous products.

Approximately 50% of the *ortho*-toluidine consumed in Japan in 1979 was in the production of dyes; 20% was used to produce pigments, 20% for the production of antioxidants, and the remaining 10% for miscellaneous applications.

Permissible levels of *ortho*-toluidine in the working environment have been established by regulation or recommended guidelines in at least 15 countries. These standards are listed in Table 1. The American Conference of Governmental Industrial Hygienists (1980) published a notice of intended change in the recommended threshold limit value-time weighted average for dermal exposure to *ortho*-toluidine to 9 mg/m^3 of air (2 ppm) for any eight-hour work day or 40-hour work week. This trial limit will remain as an intended change for at least two years before reconsideration for addition to the adopted list.

Table 1. National occupational exposure limits for ortho-toluidine[a]

Country	Year	Concentration (mg/m^3)	(ppm)	Interpretation[b]	Status
Australia	1973	22	5	TWA	Guideline
Belgium	1974	22	5	TWA	Guideline
Bulgaria	1971	3	–	Ceiling	Regulation
Czechoslovakia	1976	5	–	TWA	Regulation
		20	–	Ceiling	
Finland	1975	22	5	TWA	Regulation
German Democratic Republic	1973	20	–	Max. average ceiling (30 min)	Regulation
		10	–	TWA	
Federal Republic of Germany	1979	22	5	TWA	Guideline
Italy	1975	10	–	TWA	Guideline
The Netherlands	1973	22	5	TWA[c]	Guideline
Poland	1976	3	–	Ceiling	Regulation
Romania	1975	3	–	TWA	Regulation
		5	–	Ceiling	
Switzerland	1976	22	5	TWA	Regulation
US	1980[d]	22	5	TWA	Regulation
USSR	1977	3	–	Ceiling	Regulation
Yugoslavia	1971	1.2[e]	5	Ceiling	Regulation

[a] From International Labour Office (1977)]
[b] TWA - time-weighted average
[c] Skin irritant notation added
[d] From US Occupational Safety & Health Administration (1980)
[e] Value appears to be in error: should probably be 22

Since the US Environmental Protection Agency has identified *ortho*-toluidine hydrochloride as a toxic waste, as of 19 November 1980, persons who generate, transport, treat, store or dispose of this compound must comply with the regulations of the Federal hazardous waste management programme (US Environmental Protection Agency, 1980b).

Toluidines and their salts (including *ortho*-toluidine and its hydrochloride) are not permitted to be used in cosmetics in the European Economic Communities (Commission of the European Communities, 1976).

2.2 Occurrence

(a) Occupational exposure

The concentration of *ortho*-toluidine measured in the air of the working environment in an *ortho*-toluidine manufacturing plant in the USSR was reported to be in the range of 0.5-28.6 mg/m^3 (Khlebnikova *et al.*, 1970).

(b) Water

ortho-Toluidine has been reported in surface water samples taken from certain rivers in the Federal Republic of Germany at levels of 0.3-1 μg/kg (Neurath *et al.*, 1977). It has also been identified in sea water and in chemical plant effluents in the US (Shackelford & Keith, 1976).

(c) Food and beverages

Unspecified isomers of toluidine were found in samples of kale (1.1 mg/kg) and carrots (7.2 mg/kg) (Neurath *et al.*, 1977). *ortho*-Toluidine has been identified in the volatile aroma components of black tea (Vitzthum *et al.*, 1975).

(d) Tobacco and tobacco smoke

ortho-Toluidine has been found in tobacco smoke in concentrations of 23.3-213 ng/cigarette (Patrianakos & Hoffmann, 1979) and in steam volatiles from the distillation of one type of tobacco leaves (Latakia) (Irvine & Saxby, 1969).

(e) Other

ortho-Toluidine has been reported to be a component of the tar produced by low-temperature carbonization of coal (Pichler & Hennenberger, 1969). It has also been detected in the gasoline fraction of hydrocracked Arlan petroleum (Ben'kovskii *et al.*, 1972).

This compound has been found in humans as a metabolite of aptocaine, a local anaesthetic (Beckett & Vutthikongsirigool, 1976).

It is also an impurity (less than 0.5%) in formulated corrosion inhibitors sold under the trade name Rodine (Anon., 1980).

2.3 Analysis

An IARC manual (Egan et al., 1981) gives selected methods for the analysis of aromatic amines, including *ortho*-toluidine. Typical methods for its analysis in various matrices are summarized in Table 2.

Table 2. Methods for the analysis of ortho-toluidine

Sample matrix	Sample preparation	Assay procedure[a]	Limit of detection	Reference
Bulk chemical	Dissolve in ethanol; add hydrochloric acid and *para*-dimethyl-aminocinnamaldehyde; add water	S[b]	0.1 mg/l	Qureshi & Khan (1976)
Amine mixtures	Dissolve in hydrochloric acid and propanol; elute with aqueous methanol; develop with N,N-dimethyl-*para*-aminobenzaldehyde in ethanol and glacial acetic acid	TLC	not given	Lepri et al. (1979)
	Dissolve in aqueous hydrochloric acid	^{1}H-NMR	not given	Yasuda & Kakiyama (1974)
	Inject into column packed with aminopropyltrimethoxysilane and copper (II) ion on Partisil-10; elute with methanol and cyclohexane	HPLC-FL	not given	Chow & Grushka (1977)

Sample matrix	Sample preparation	Assay procedure[a]	Limit of detection	Reference
	Dissolve in methanol; elute with benzene and acetic acid or with acetic acid, ethyl acetate and diisopropyl ether	TLC	not given	Yasuda (1971)
Air	Draw air through silica gel adsorbent; elute with 95% ethanol containing 0.1% heptanol	GC-FID	not given	Wood & Anderson (1975)
Water	Extract with benzene or hexane; derivatize with perfluoromethoxy-propionic anhydride	GC-ECD	2-3 mg/l	Kulikova et al. (1979)
	Add phosphate buffer sodium hydroxide and 4-amino-antipyrine; add potassium ferricyanide	C[b]	0.1 mg/l	El-Dib et al. (1975)

[a] Abbreviations: S, spectrophotometry; TLC, thin-layer chromatography; ^{1}H-NMR, proton nuclear magnetic resonance spectroscopy; HPLC-FL, high-performance liquid chromatography with fluorometric detection; GC-FID, gas chromatography with flame ionization detection; GC-ECD, gas chromatography with electron-capture detection; C, colorimetry

[b] **Not specific for *ortho*-toluidine**

3. Biological Data Relevant to the Evaluation of Carcinogenic Risk to Humans

3.1 Carcinogenicity studies in animals[1]

(a) Oral administration

Mouse: Groups of 50 male and 50 female B6C3F$_1$ hybrid mice, six weeks of age, were fed diets containing 1000 or 3000 mg/kg *ortho*-toluidine hydrochloride (purity probably at least 99%, as determined by high-performance liquid chromatography, with at least one unspecified contaminant) for 102-103 weeks. The doses were selected on the basis of a range-finding study [see section 3.2(a)]. A group of 20 mice of each sex served as matched controls. All animals under study received food and water *ad libitum*. No significant dose-related trend in mortality was seen for animals of either sex: by the end of the study, 75, 86 and 68% of males and 95, 78 and 86% of females were still alive in the control, low-dose and high-dose groups, respectively. Statistically significant ($P = 0.001$ and $P = 0.004$), dose-related increases in tumour incidences were observed for several neoplasms. (a) The incidences of hepatocellular carcinomas or adenomas were increased in female mice: in 0/20 controls, 4/49 low-dose and 13/50 high-dose animals ($P = 0.007$). (b) Haemangiosarcomas at all sites were found in increased incidence in male mice: in 17/19 controls, 1/50 low-dose and 10/50 high-dose animals (National Cancer Institute, 1979).

Groups of 25 male and 25 female random-bred albino CD-1 mice, six to eight weeks old, were fed diets containing *ortho*-toluidine hydrochloride (97-99% pure) at two dose levels: 16000 mg/kg of diet for three months and then 8000 mg/kg of diet for a further 15 months; or 32000 mg/kg of diet for three months and then 16000 mg/kg of diet for a further 15 months. The initial high doses were close to the maximum tolerated concentrations established in a subchronic toxicity study [see section 3.2(a)]. Groups of 25 mice of each sex served as matched controls and groups of 102 females and 99 males as pooled controls. All animals were observed for 21 months. [No data on dynamics in growth or survival were given.] Only animals that lived six or more months were examined histologically. Significant increases were observed in the incidences of vascular tumours, i.e., haemangiosarcomas and haemangiomas of the abdominal viscera. Among the females: 0/15 in matched controls, 9/102 in pooled controls, 5/18 in the low-dose group ($P < 0.05$ when compared with either matched or pooled controls) and 9/21 in the high-dose group ($P < 0.025$ when compared with all controls). Among the males: 0/14 in matched controls, 5/99 in pooled controls, 5/14 in the low-dose group ($P < 0.025$ when compared with all controls) and 9/11 in the high-dose group ($P < 0.025$ when compared with all controls) (Weisburger *et al.*, 1978).

Rat: Groups of 50 male and 50 female Fischer 344 rats, six weeks of age, were fed diets containing 3000 or 6000 mg/kg *ortho*-toluidine hydrochloride (purity probably at least 99%,

---

[1] The Working Group was aware of a completed but as yet unpublished study by s.c. administration to rats (IARC, 1981).

as determined by high-pressure liquid chromatography, with at least one unspecified contaminant) for 101 or 104 weeks. The doses were selected on the basis of a range-finding study [see section 3.2(a)]. A group of 20 rats of each sex served as matched controls. All animals under study received food and water *ad libitum*. A significant ($P < 0.001$), positive, dose-related trend in mortality was observed for animals of each sex; all high-dose males had died by 100 weeks. Statistically significant ($P < 0.001$), dose-related increases in tumour incidences were observed for several neoplasms. (a) The combined incidence of sarcomas, fibrosarcomas, angiosarcomas and osteosarcomas of multiple organs was increased: among females, 0/20 in controls, 3/50 in the low-dose group, 21/49 in the high-dose group ($P < 0.001$); and among males, 0/20 in controls, 15/50 in the low-dose group ($P = 0.003$) and 37/49 in the high-dose group ($P < 0.001$). (b) The combined incidence of sarcomas, angiosarcomas and osteosarcomas of the spleen was increased in females: 0/20 in controls, 9/49 in the low-dose group, ($P = 0.036$) and 12/49 in the high-dose group ($P = 0.01$). (c) Transitional-cell carcinomas or papillomas of the urinary bladder were increased in females: 0/20 in controls, 10/45 in the low-dose group ($P = 0.018$) and 22/47 in the high-dose group ($P < 0.001$). Metastases were rare; hyperplasia of the transitional-cell epithelium of the bladder was also seen [see section 3.2(a)]. (d) Fibromas of the subcutaneous tissue in the integumentary system were increased in males: 0/20 in controls, 28/50 in the low-dose group ($P < 0.001$) and 27/49 in the high-dose group ($P < 0.001$). (e) Fibroadenomas of the mammary gland were increased in females: 6/20 in controls, 20/50 in the low-dose group and 35/49 in the high-dose group ($P = 0.002$). The combined incidences of fibroadenomas and adenomas of the mammary gland were: 7/20 in controls, 20/50 in the low-dose group, and 35/49 in the high-dose group ($P = 0.006$). (g) Mesotheliomas of multiple organs or of the tunica vaginalis were increased in males: 0/20 in controls, 17/50 in the low-dose group ($P = 0.001$) and 9/49 in the high-dose group ($P = 0.036$) (National Cancer Institute, 1979).

Groups of 25 male Charles River CD rats, six to eight weeks old, were fed diets containing *ortho*-toluidine hydrochloride (97-99% pure) at two dose levels: 8000 mg/kg of diet for three months and then 4000 mg/kg of diet for a further 15 months; or 16000 mg/kg of diet for three months and then 8000 mg/kg of diet for a further 15 months. The initial high doses were close to the maximum tolerated concentrations established in a subchronic toxicity study [see section 3.2(a)]. A group of 25 rats served as matched controls and a group of 111 rats as pooled controls. All animals were observed for 24 months. [No data on dynamics in growth and survival were given]. Only animals that lived six or more months were examined histologically. Statistically significant increases in tumour incidences were reported for several neoplasms. (a) Subcutaneous fibromas and fibrosarcomas were increased: 0/16 in matched controls, 18/111 in pooled controls, 18/23 in the low-dose group, and 21/24 in the high-dose group ($P < 0.025$). (b) The incidence of multiple tumours, i.e., fibromas, fibrosarcomas, bladder carcinomas, pituitary or adrenal adenomas, was also increased: 3/16 in matched controls, 14/111 in pooled controls, 6/23 in the low-dose group (difference not significant) and 8/24 in the high-dose group ($P < 0.025$ when compared with pooled controls only). In addition, there was a certain increase (although not significant) in the incidence of transitional-cell carcinomas of the urinary bladder: 0/16, 5/111, 3/23 and 4/24 in the four groups, respectively (Weisburger *et al.*, 1978).

(b) Subcutaneous and/or intramuscular administration

Various species: Fifteen rabbits and 10 guinea-pigs were injected subcutaneously with 1.0 and 0.5 ml, respectively, of a 2% solution of *ortho*-toluidine in olive oil six times a week. Rabbits that survived more than 100 days developed papillomas in the bladder (number of animals not specified). Guinea-pigs did not live long enough to develop papillomas (Morigami & Nisimura, 1940). Papillomas in the bladder were found in 1/2 rats, 4/5 rabbits and 5/8 guinea-pigs injected subcutaneously with *ortho*-toluidine (dose and duration unspecified) in olive oil or alcohol (Satani *et al.*, 1941). [The Working Group noted the inadequate reporting of these experiments.]

3.2 Other relevant biological data

(a) Experimental systems

Toxic effects

The single oral LD_{50} of *ortho*-toluidine hydrochloride in rats when given by stomach tube was 2950 mg/kg bw (Lindstrom *et al.*, 1969); the oral LD_{50} of the free base in rats was 900 mg/kg bw when undiluted compound was administered by intubation (Jacobson, 1972). Oral LD_{50}s of *ortho*-toluidine in oil were 520, 670 and 840 mg/kg bw in mice, rats and rabbits, respectively (Lunkin, 1967).

Toxic effects of *ortho*-toluidine include methaemoglobinaemia, reticulocytosis and anaemia in rats (Lunkin, 1967) and methaemoglobinaemia in mice (Nomura, 1977). In rats, it produced keratosis and metaplasia in the epithelium of the bladder (Ekman & Strombeck, 1947).

In seven-week subchronic feeding studies, groups of $B6C3F_1$ mice of each sex were fed diets containing up to 50000 mg/kg *ortho*-toluidine hydrochloride. All animals showed a dose-dependent depression in mean body weight gain of up to 37%. The spleens of animals that received 50000 mg/kg had pigment deposition. In similar studies with Fischer 344 rats, males and females given 50000 mg/kg of diet had dose-dependent depressions in mean body weight gain of up to 49 and 58%, respectively, as compared with controls. All animals given doses of less than 50000 mg/kg survived to the end of the study. Renal and splenic pigmentation were observed in male and female rats that received 12500 mg/kg (National Cancer Institute, 1979).

Effects on reproduction and prenatal toxicity

No data were available to the Working Group.

Absorption, distribution, excretion and metabolism

Following the s.c. administration of a single dose of 400 mg/kg bw *ortho*-[methyl-^{14}C]-toluidine hydrochloride to male Fischer 344 rats, 56% of the ^{14}C was recovered in the 24-hour urine, 2.3% in the faeces and 1% as exhaled $^{14}CO_2$. After 48 hours, 83.9% of the ^{14}C appeared in the urine, 3.3% in the faeces, and 1.4 was exhaled. Ether-extractable urinary compounds were identified as: *ortho*-toluidine (5.1% of the dose); azoxytoluene (0.2%); *ortho*-nitrosotoluene (≤0.1%); *N*-acetyl-*ortho*-toluidine (0.2%); *N*-acetyl-*ortho*-aminobenzyl alcohol (0.3%); 4-amino-*meta*-cresol (0.6%); N-acetyl-4-amino-*meta*-cresol (0.3%); anthranilic acid (0.3%); and *N*-acetylanthranilic acid (0.3%). Acid-conjugated urinary metabolites (51% of the dose) were identified as sulphates of 4-amino-*meta*-cresol (27.8%), *N*-acetyl-4-amino-*meta*-cresol (8.5%) and 2-amino-*meta*-cresol (2.1%), and glucuronides of 4-amino-*meta*-cresol (2.6%), *N*-acetyl-4-amino-m*eta*-cresol (2.8%) and *N*-acetyl-*ortho*-aminobenzyl alcohol. Evidence for the presence of a double-acid conjugate of 4-amino-*meta*-cresol was also obtained (Son *et al.*, 1980).

Mutagenicity and other short-term tests

Using a modified *Escherichia coli* DNA repair test (*pol A-/A +*), Rosenkranz and Poirier (1979) showed that 0.5 μg/ml *ortho*-toluidine was mutagenic when tested in the absence of an exogenous metabolic activation system. The compound induced prophage λ in *E. coli* K 12 without metabolic activation (Galkiewicz *et al.*, 1979).

Levels of up to 1000 μg/plate did not induce reverse mutations in *Salmonella typhimurium* strains TA1535, TA1537, TA1538, TA98 or TA100 when tested in the presence or absence of a rat liver activation system (McCann *etal.*, 1975; Garner & Nutman, 1977; Rosenkranz & Poirier, 1979; Simmon, 1979). It induced reverse mutations in *S. typhimurium* strain TA98 only when tested in the presence of both norharman and a liver activation system prepared from rats induced with polychlorinated biphenyls (Nagao *et al.*, 1977, 1978).

A urine concentrate from male Sprague-Dawley rats given 300 mg/kg bw *ortho*-toluidine orally was mutagenic to *S. typhimurium* strain TA98 when tested in the presence of a liver activation system from rats treated with polychlorinated biphenyls (Tanaka *et al.*, 1980).

No single-strand breaks were observed in the DNA of V79 Chinese hamster cells incubated with 0.3-10 mM *ortho*-toluidine for two hours in the presence of Aroclor-induced rat liver microsomes (Zimmer *et al.*, 1980).

(b) Humans

No data were available to the Working Group.

3.3 Case reports and epidemiological studies of carcinogenicity in humans

Case & Pearson (1954) and Case et al. (1954) reported a historical cohort study of bladder tumours among workers in the British chemical industry engaged in the manufacture of dyestuff intermediates. Groups were defined in terms of exposure to aniline, benzidine, 1- and 2-naphthylamine, auramine and magenta. Many aniline workers were also exposed to toluidines produced in the same plant. The 21 firms in the survey included the major British producers of aniline, and the groups at risk were observed from 1921-1952. There were 812 men classified as having had exposure to aniline but not to the other suspected carcinogens. The death certificate of one of these mentioned a bladder tumour; 0.52 was expected at national rates. [Exposure of workers to *ortho*-toluidine is not specified in this paper but can be inferred from knowledge of aniline dye production processes, see 2.1*(b)*, p. 158.]

Uebelin & Pletscher (1954) studied the occurrence of bladder tumours in workers in a factory in Switzerland producing dyestuff intermediates. Among 300 workers exposed to 2-naphthylamine and/or benzidine alone or with other aromatic amines for unspecified periods during 1924-1953, 97 cases were found. A further 650 workers had been exposed only to aniline or other aromatic amines but not to 2-naphthylamine or benzidine; three of these men developed bladder tumours, but their exact exposures were not known with certainty. The authors distinguished a subgroup of 35 men who prepared *para*-chloro-*ortho*-toluidine from *ortho*-toluidine; no bladder tumours were found among these men. [Insufficient details were provided concerning person-years at risk or follow-up to permit evaluation of the significance of these observations.]

Gropp (1958) described 98 cases of bladder tumours that occurred from 1903-1955 among workers at a German factory, and Oettel (1959, 1967) and Oettel *et al.* (1968) reported both these and some further cases. Most of Gropp's cases were reported as having had exposure to 2-naphthylamine alone or with other amines, including benzidine. However, 11 cases were reported as having had exposure only to aniline/toluidine. These 11 cases were all diagnosed between 1905 and 1935 and had their first exposure between 1874 and 1905; they had longer latent periods and later ages at onset than the cases associated with exposure to 2-naphthylamine and benzidine. [The population at risk is not known.]

Bladder cancer has been reported in workers exposed to 1-naphthylamine (Vigliani & Barsotti, 1961) or benzidine (Vigliani & Barsotti, 1961; Zavon *et al.*, 1973) and other aromatic amines, who were also exposed to *ortho*-toluidine.

Khlebnikova *et al.* (1970) reported bladder tumours in workers engaged in the production of *ortho*-toluidine and/or *para*-toluidine in the USSR during the 1960s. The concentration range of *ortho*-toluidine measured in the air of the working environment was 0.5-28.6 mg/m^3. Two cases of bladder tumours (papillomas) were found when 75 out of 81 current toluidine workers were examined cystoscopically. The 81 workers comprised 19 women and 62 men, aged 22-55; 35 were operators, 18 were fitters, and 10 were cleaners. Nine had worked in toluidine production for less than one year, and 31 for 1-5 years; the remaining 41 workers had had longer exposure. Most had been exposed to both *ortho-* and *para*-toluidine. One worker with a bladder tumour, aged 26, had been exposed only to *para*-toluidine and only for 20 months; the other, aged 49, had worked in contact with both isomers for 23 years. Six other cases of bladder tumours, of which four were carcinomas (one was found to be a papilloma and one a multiple papilloma), had been found earlier, upon cystoscopic examination of 16 former workers who had worked with the toluidines for periods ranging from 12-17 years. [The information contained in the paper with regard to person-years of exposure and to other substances to which these workers may have had prior or concomitant exposure is insufficient to evaluate the significance of these observations.]

Lipkin (1972), in a report also from the USSR, referred to 27 cases of bladder cancer occurring in individuals exposed occupationally to *ortho-* and *para*-toluidine, and to 21 cases in workers exposed to both these chemicals and to 1-naphthylamine. Genin *et al.* (1978) referred briefly to 36 people with bladder cancer who had been exposed occupationally to *ortho-* and *para*-toluidine and to 'some other aromatic amino and nitro compounds' in a chemical plant. The cancers were diagnosed in men and women aged between 25 and 61. [No information was given on the sizes of the populations at risk. It is possible, but not stated explicitly, that these cases included some or all of those reported by Khlebnikova *et al.* (1970).]

Rubino *et al.* (1979) reported the results of follow-up of 868 of 906 men first employed in a dyestuffs factory between 1922 and 1970. Mortality was determined over the period 1946 to 1976 and compared with that of the Italian male population as a whole. Thirty-six deaths from urinary bladder cancer were observed, whereas 1.23 were expected. Thirty-one of these deaths occurred in 610 men with exposure to benzidine, 1-naphthylamine, 2-naphthylamine or several of these; but five (with 0.08 expected) occurred in 53 men engaged solely in the manufacture of Fuchsine (magenta; see IARC, 1974) and Safranine-T, involving exposure either to a combination of toluene, *ortho*-nitrotoluene, *ortho*-toluidine and 4,4'-methylene bis(2-methylaniline) (three deaths from bladder cancer) or to a combination of *ortho*-toluidine, 4,4'-methylene bis(2-methylaniline), *ortho*-nitrotoluene, 2,5-diaminotoluene, *ortho*-aminoazotoluene, aniline, Fuchsine and Safranine-T (two deaths). [There is a definite excess of bladder cancer in those involved in manufacture of Fuchsine and Safranine-T, but it cannot be attributed with certainty to exposure to *ortho*-toluidine or to any one of the other compounds involved.]

4. Summary of Data Reported and Evaluation

4.1 Experimental data

ortho-Toluidine hydrochloride was tested in several experiments, but only two experiments in mice and two experiments in rats by oral administration were adequate. It was carcinogenic in mice, producing hepatocellular carcinomas or adenomas in females and haemangiosarcomas at all sites in males of one strain and haemangiosarcomas and haemangiomas of abdominal viscera in animals of both sexes of another strain. It was carcinogenic in rats: in animals of one strain it produced sarcomas of multiple organs in animals of both sexes, fibromas of the subcutaneous tissue and mesotheliomas in males; and sarcomas of the spleen, transitional-cell papillomas and carcinomas of the urinary bladder and mammary gland fibroadenomas and adenomas in females. In male rats of another strain it produced subcutaneous fibromas and fibrosarcomas as well as a slight increase in the incidence of multiple tumours, including transitional-cell carcinomas of the urinary bladder.

ortho-Toluidine gave positive results in *Escherichia coli* DNA-repair and phage-induction assays, but it was not mutagenic in *Salmonella typhimurium* unless tested in the presence of both norharman and a metabolic activation system. Urine concentrates from rats treated with *ortho*-toluidine were mutagenic for *S. typhimurium* with metabolic activation.

4.2 Human data

ortho-Toluidine has been produced commercially since 1880. Its use as an intermediate in the manufacture of dyes and other chemicals as well as its occurrence as an impurity in some formulated corrosion inhibitors could result in occupational exposure.

The epidemiological studies available to the Working Group dealt only with workers who had been exposed to *ortho*-toluidine in combination with other chemicals. The majority were case reports and did not provide estimates of the population at risk. In one follow-up study, five cases of bladder cancer (0.08 expected) occurred in 53 men exposed to *ortho*-toluidine and to other possibly carcinogenic agents in the manufacture of Fuchsine and Safranine T.

4.3 Evaluation

There is *sufficient evidence*[1] for the carcinogenicity of *ortho*-toluidine hydrochloride in experimental animals.

An increased incidence of bladder cancer has been observed in workers exposed to *ortho*-toluidine, but as all were exposed to other possibly carcinogenic chemicals, *ortho*-toluidine cannot be identified specifically as the responsible agent.

ortho-Toluidine should be regarded, for practical purposes, as if it presented a carcinogenic risk to humans.

---

[1] See preamble, p. 18.

5. References

American Conference of Governmental Industrial Hygienists (1980) *TLVs Threshold Limit Values for Chemical Substances and Physical Agents in the Workroom Environment with Intended Changes for 1980*, Cincinnati, OH, pp. 35, 39

Anon. (1980) TSCA substantial risk reports made by four corporations. *Pesticide & Toxic Chemicals News*, 16 July, p. 4

Beckett, A.H. & Vutthikongsirigool, W. (1976) Identification of metabolites of aptocaine and some related compounds. *J. Pharm. Pharmacol.*, *28, Suppl.*, 54P

Ben'kovskii, V.G., Baikova, A.Y., Bulatova, B.T., Lyubopytova, N.S. & Popov, Y.N. (1972) Nitrogenous bases in gasoline from the hydrocracking of Arlan petroleum (Russ.). *Neftekhimiya*, *12*, 454-459 [*Chem. Abstr.*, *77*, 116548d]

Case, R.A.M. & Pearson, J.T. (1954) Tumours of the urinary bladder in workmen engaged in the manufacture and use of certain dyestuff intermediates in the British chemical industry. II. Further consideration of the role of aniline and of the manufacture of auramine and magenta (fuchsine) as possible causative agents. *Br. J. ind. Med.*, *11*, 213-216

Case, R.A.M., Hosker, M.E., McDonald, D.B. & Pearson, J.T. (1954) Tumours of the urinary bladder in workmen engaged in the manufacture and use of certain dyestuff intermediates in the British chemical industry. I. The role of aniline, benzidine, *alpha*-naphthylamine, and *beta*-naphthylamine. *Br. J. ind. Med.*, *11*, 75-104

Chow, F.K. & Grushka, E. (1977) Separation of aromatic amine isomers by high pressure liquid chromatography with a copper(II)-bonded phase. *Anal. Chem.*, *49*, 1756-1761

Commission of the European Communities (1976) Council Directive of 27 July 1976 on the approximation of the laws of the Member States relating to cosmetic products. *Off. J. Eur. Communities*, *L262*, 175

Egan, H., Fishbein, L., Castegnaro, M., O'Neill, I.K. & Bartsch, H., eds (1981) *Environmental Carcinogens - Selected Methods of Analysis, Vol. 4, Some Aromatic Amines and Azodyes in the General and Industrial Environment (IARC Scientific Publications No. 40)*, Lyon, International Agency for Research on Cancer

E.I. du Pont de Nemours & Co. (1979) *o-Toluidine Technical Data Sheet*, Wilmington, DE

Ekman, B. & Stroembeck, J.P. (1947) Demonstration of tumorigenic decomposition products of 2,3-azotoluene. *Acta physiol. scand.*, *14*, 43-50

El-Dib, M.A., Abdel-Rahman, M.O. & Aly, O.A. (1975) 4-Aminoantipyrine as a chromogenic agent for aromatic amine determination in natural water. *Water Res.*, *9*, 513-516

Galkiewicz, E., Pyziak, M., Chomiczewski, J. & Gorski, T. (1979) Mutagenic properties of some chemical compounds as evaluated by induction of the prophage lambda of *Escherichia coli* K12 (Pol.). *Med. Dosw. Mikrobiol.*, *31*, 11-16

Garner, R.C. & Nutman, C.A. (1977) Testing of some azo dyes and their reduction products for mutagenicity using *Salmonella typhimurium* TA 1538. *Mutat. Res.*, *44*, 9-19

Genin, V.A., Pliss, G.B., Pylev, L.N. & Shabad, L.M. (1978) Prevention of occupational urinary bladder tumours in the manufacture of toluidines (Russ.). *Gig. Tr. Prof. Zabol.*, *7*, 10-15

Grasselli, J.G., ed. (1973) *CRC Atlas of Spectral Data and Physical Constants for Organic Compounds*, Cleveland, OH, Chemical Rubber Co., pp. B-942-B-943

Gropp, D. (1958) In: *Zur Atiologie des sogenannten Anilin-Blasenkrebses* (*Etiology of So-called Aniline Bladder Tumours*). Inaugural Dissertation Thesis, Johannes-Gutenberg Universitat, Mainz, FRG

Hawley, G.G., ed. (1977) *The Condensed Chemical Dictionary*, 9th ed., New York, Van Nostrand-Reinhold, p. 870

IARC (1974) *IARC Monographs on the Evaluation of Carcinogenic Risk of Chemicals to Man*, Vol. 4, *Some aromatic amines, hydrazine and related substances, N-nitroso compounds and miscellaneous alkylating agents*, Lyon, pp. 57-64

IARC (1975a) *IARC Monographs on the Evaluation of Carcinogenic Risk of Chemicals to Man*, Vol. 8, *Some aromatic azo compounds*, Lyon, pp. 217-224

IARC (1975b) *IARC Monographs on the Evaluation of Carcinogenic Risk of Chemicals to Man*, Vol. 8, *Some aromatic azo compounds*, Lyon, pp. 61-74

IARC (1975c) *IARC Monographs on the Evaluation of Carcinogenic Risk of Chemicals to Man*, Vol. 8, *Some aromatic azo compounds*, Lyon, pp. 165-171

IARC (1975d) *IARC Monographs on the Evaluation of Carcinogenic Risk of Chemicals to Man*, Vol. 8, *Some aromatic azo compounds*, Lyon, pp. 287-294

IARC (1975e) *IARC Monographs on the Evaluation of Carcinogenic Risk of Chemicals to Man*, Vol. 8, *Some aromatic azo compounds*, Lyon, pp. 113-116

IARC (1978a) *IARC Monographs on the Evaluation of the Carcinogenic Risk of Chemicals to Man*, Vol. 16, *Some Aromatic Amines and Related Nitro Compounds - Hair Dyes, Colouring Agents and Miscellaneous Industrial Chemicals*, Lyon, pp. 349-366

IARC (1978b) *IARC Monographs on the Evaluation of the Carcinogenic Risk of Chemicals to Man*, Vol. 16, *Some Aromatic Amines and Related Nitro Compounds - Hair Dyes, Colouring Agents and Miscellaneous Industrial Chemicals*, Lyon, pp. 277-285

IARC (1981) *Information Bulletin on the Survey of Chemicals Being Tested for Carcinogenicity*, No. 9, Lyon, p. 29

International Labour Office (1977) *Occupational Exposure Limits for Airborne Toxic Substances* (*Occupational Safety & Health Series No. 37*), Geneva, pp. 204-205

Irvine, W.J. & Saxby, M.J. (1969) Steam volatile amines of Latakia tobacco leaf. *Phytochemistry*, *8*, 473-476

Jacobson, K.H. (1972) Acute oral toxicity of mono- and di-alkyl ring- substituted derivatives of aniline. *Toxicol. appl. Pharmacol.*, *22*, 153-154

Khlebnikova, M.I., Gladkova, E.V., Kurenko, L.T., Pshenitsyn, A.V. & Shalin, B.M. (1970) Problems of industrial hygiene and health status of workers engaged in the production of o-toluidine (Russ.). *Gig. Tr. Prof. Zabol.*, *14*, 7-10

Kulikova, G.S., Kirichenko, V.E. & Pashkevich, K.I. (1979) Gas-liquid chromatographic determination of aniline derivatives in water. *Zh. anal. Khim.*, *34*, 790-793

Lepri, L., Desideri, P.G. & Heimler, D. (1979) Soap thin-layer chromatography of sulphonamides and aromatic amines. *J. Chromatogr.*, *169*, 271-278

Lindstrom, H.V., Bowie, W.C., Wallace, W.C., Nelson, A.A. & Fitzhugh, O.G. (1969) The toxicity and metabolism of mesidine and pseudocumidine in rats. *J. Pharmacol. exp. Ther.*, *167*, 223-234

Lipkin, I.L. (1972) *Prevention of occupational urinary bladder tumour diseases among workers in the aniline dye industry in the USSR* (Russ.). In: Shabad, L.M., ed., *Some Results of a*

Study of Pollution of the Environment with Carcinogenic Substances, Moscow, Ministry of Public Health, pp. 107-111

Lunkin, V.N. (1967) Information for the hygienic establishment of the maximum allowable concentration of *para-* and *ortho*-toluidines in inland waters. *Ref. Zh. Otd. Vyp. Farmakol. Khimioter. Sredstva. Toksikol.*, No. 12.54.1096

Mass Spectrometry Data Centre (1974) *Eight Peak Index of Mass Spectra*, 2nd ed., Reading, UK, Atomic Weapons Research Establishment

McCann, J., Choi, E., Yamasaki, E. & Ames, B.N. (1975) Detection of carcinogens as mutagens in the *Salmonella*/microsome test: assay of 300 chemicals. *Proc. natl Acad. Sci. (USA)*, 72, 5135-5139

Morigami, S. & Nisimura, I. (1940) Experimental studies on aniline bladder tumors. *Gann*, 34, 146-147

Nagao, M., Yahagi, T., Honda, M., Seino, Y., Matsushima, T. & Sugimura, T. (1977) Demonstration of mutagenicity of aniline and *o*-toluidine by norharman. *Proc. Jpn. Acad., Ser. B.*, 53, 34-37

Nagao, M., Yahagi, T. & Sugimura, T. (1978) Differences in effects of norharman with various classes of chemical mutagens and amounts of S-9. *Biochem. biophys. Res. Commun.*, 83, 373-378

National Cancer Institute (1979) *Bioassay of o-Toluidine Hydrochloride for Possible Carcinogenicity* (Tech. Rep. Ser. No. 153; DHEW Publ. No. (NIH) 79-1709), Washington DC, US Government Printing Office

Neurath, G.B., Duenger, M., Pein, F.G., Ambrosius, D. & Schreiber, O. (1977) Primary and secondary amines in the human environment. *Food Cosmet. Toxicol.*, 15, 275-282

Nomura, A. (1977) Studies on sulfhemoglobin formation by various drugs. *Nippon Yakurigaku Zasshi*, 73, 423-435

Oettel, H. (1959) *On the question of occupational tumours due to chemicals.* In: *Verhandlungen der Deutschen Gesellschaft fur Pathologie, 43 Tagung* (Proceedings of the German Society of Pathology, 43rd Meeting), Stuttgart, Gustav Fischer, pp. 313-320

Oettel, H. (1967) *Bladder cancer in Germany.* In: Lampe, K.F., ed., *Bladder Cancer, A Symposium*, Birmingham, AL, Aesculapius Publishing Co., pp. 196-199

Oettel, H., Thiess, A.M. & Uhl, C. (1968) Contribution to the problem of occupational lung tumours. Long-term observations at BASF (Ger.). *Zbl. Arbeitsmed. Arbeitsschutz*, 18, 291-303

Patrianakos, C. & Hoffmann, D. (1979) Chemical studies on tobacco smoke. LXIV. On the analysis of aromatic amines in cigarette smoke. *J. anal. Toxicol.*, 3, 150-154

Pichler, H. & Hennenberger, P. (1969) Examination of liquid and gaseous products of tar carbonization. III. Composition of oil, effluents and crude benzene in coke production at high temperature (Ger.). *Brennst.-Chem.*, 50, 341-346

Prager, B., Jacobson, P., Schmidt, P. & Stern, D., eds (1929) *Beilsteins Handbuch der Organischen Chemie*, 4th ed., Vol. 12, Syst. No. 1671-1672, Berlin, Springer, pp. 772-784

Qureshi, M. & Khan, I.A. (1976) Detection and spectrophotometric determination of some aromatic nitrogen compounds with *p*-dimethylaminocinnamaldehyde. *Anal. chim. Acta*, 86, 309-311

Rosenkranz, H.S. & Poirier, L.A. (1979) Evaluation of the mutagenicity and DNA-modifying activity of carcinogens and noncarcinogens in microbial systems. *J. natl Cancer Inst.*, *62*, 873-892

Rubino, G.F., Scansetti, G., Piolatto, G. & Pira, E. (1979) A further contribution to the knowledge of carcinogenic effect of aromatic amines. *Arh. Hig. Rada. Toksikol.*, *30*, Suppl., 627-632

Sadtler Research Laboratories, Inc. (undated) *Sadtler Standard Spectra*, Philadelphia, PA

Sandridge, R.L. & Staley, H.B. (1978) *Amines by reduction*. In: Kirk, R.E. & Othmer, D.F., eds, *Encyclopedia of Chemical Technology*, 3rd ed., Vol. 2, New York, John Wiley & Sons, p. 365

Satani, Y., Tanimura, T., Nishimura, I. & Isikawa, Y. (1941) Clinical and experimental examination of bladder papillomas (Jpn.). *Gann*, *35*, 275-276

Shackelford, W.M. & Keith, L.H. (1976) *Frequency of Organic Compounds Identified in Water (EPA-600/4-76-062)*, Athens, GA, Environmental Research Laboratory, pp. 35, 226

Simmon, V.F. (1979) *In vitro* mutagenicity assays of chemical carcinogens and related compounds with *Salmonella typhimurium*. *J. natl Cancer Inst.*, *62*, 893-899

The Society of Dyers & Colourists (1971) *Colour Index*, 3rd ed., Vol. 4, Bradford, Yorkshire, Lund Humphries, p. 4857

Son, O.S., Everett, D.W. & Fiala, E.S. (1980) Metabolism of o-[methyl-^{14}C]toluidine in the F344 rat. *Xenobiotica*, *10*, 457-468

Tanaka, K.-I., Marui, S. & Mii, T. (1980) Mutagenicity of extracts of urine from rats treated with aromatic amines. *Mutat. Res.*, *79*, 173-176

Uebelin, F. & Pletscher, A. (1954) Etiology and prophylaxis of occupational tumours in the dyestuffs industry (Ger.). *Schweiz. med. Wochenschr.*, *84*, 917-928

US Environmental Protection Agency (1980a) *Chemicals in Commerce Information System (CICIS)*, Washington DC, Office of Pesticides & Toxic Substances, Chemical Information Division

US Environmental Protection Agency (1980b) Hazardous waste management system: identification and listing of hazardous wastes. *US Code Fed. Regul.*, Title 40, Part 261; *Fed. Regist.*, *45*, No. 98, 33084, 33124-33127

US International Trade Commission (1977a) *Synthetic Organic Chemicals, US Production and Sales, 1975 (USITC Publication 804)*, Washington DC, US Government Printing Office, p. 42

US International Trade Commission (1977b) *Imports of Benzenoid Chemicals and Products, 1975 (USITC Publication 806)*, Washington DC, US Government Printing Office, p. 27

US International Trade Commission (1980a) *Imports of Benzenoid Chemical and Products, 1979 (USITC Publication 1083)*, Washington DC, US Government Printing Office, p. 33

US International Trade Commission (1980b) *Synthetic Organic Chemicals, US Production and Sales, 1979 (USITC Publication 1099)*, Washington DC, US Government Printing Office, pp. 3l, 37, 57, 67, 81, 95, 103, 107, 109, 255

US Occupational Safety & Health Administration (1980) Air contaminants. *US Code Fed. Regul.*, Title 29, Part 1910.1000, p. 73

US Tariff Commission (1922) *Census of Dyes and Other Synthetic Organic Chemicals, 1921 (Tariff Information Series No. 26)*, Washington DC, US Government Printing Office, p. 26

US Tariff Commission (1957) *Synthetic Organic Chemicals, US Production and Sales, 1956* (Report No. 200, Second Series), Washington DC, US Government Printing Office, p. 79

Vigliani, E.C. & Barsotti, M. (1961) Environmental tumors of the bladder in some Italian dye-stuff factories. *Med. Lav.*, 52, 241-250

Vitzthum, O.G., Werkhoff, P. & Hubert, P. (1975) New volatile constituents of black tea aroma. *J. agric. Food Chem.*, 23, 999-1003

Weast, R.C., ed. (1979) *CRC Handbook of Chemistry and Physics*, 60th ed., Cleveland, OH, Chemical Rubber Co., pp. C-519, D-209

Weisburger, E.K., Russfield, A.B., Homburger, F., Weisburger, J.H., Boger, E., Van Dongen, C.G. & Chu, K.C. (1978) Testing of twenty-one environmental aromatic amines or derivatives for long-term toxicity or carcinogenicity. *J. environ. Pathol. Toxicol.*, 2, 325-356

Windholz, M., ed. (1976) *The Merck Index*, 9th ed., Rahway, NJ, Merck & Co., p. 1226

Wood, G.O. & Anderson, R.G. (1975) Personal air sampling for vapors of aniline compounds. *Am. ind. Hyg. Assoc. J.*, 36, 538-548

Yasuda, K. (1971) Thin-layer chromatography of aromatic amines on cadmium sulphate-impregnated silica gel thin layers. *J. Chromatogr.*, 60, 144-149

Yasuda, S. & Kakiyama, H. (1974) Determination of the three isomers of cresol, toluic acid, and toluidine by nuclear magnetic resonance (Jpn.). *Bunseki Kagaku*, 23, 615-620 [*Chem. Abstr.*, 81, 130660y]

Zavon, M.R., Hoegg, V. & Bingham, E. (1973) Benzidine exposure as a cause of bladder tumors. *Arch. environ. Health*, 27, 1-7

Zimmer, D., Mazurek, J., Petzold, G. & Bhuyan, B.K. (1980) Bacterial mutagenicity and mammalian cell DNA damage by several substituted anilines. *Mutat. Res.*, 77, 317-326

2,4,5- and 2,4,6-TRIMETHYLANILINE and their HYDROCHLORIDES

1. Chemical and Physical Data

2,4,5-TRIMETHYLANILINE

1.1 Synonyms and trade names

Chem. Abstr. Services Reg. No.: 137-17-7
Chem. Abstr. Name: Benzenamine, 2,4,5-trimethyl-
IUPAC Systematic Name: 2,4,5-Trimethylaniline
Synonyms: 1-Amino-2,4,5-trimethylbenzene; *psi*-cumidine; pseudo-cumidine; 1,2,4-trimethyl-5-aminobenzene

1.2 Structural and molecular formulae and molecular weight

$C_9H_{13}N$ Mol. wt: 135.2

1.3 Chemical and physical properties of the pure substance

From Weast (1979), unless otherwise specified
(a) *Description*: Needles (from water)
(b) *Boiling-point*: 234-235°
(c) *Melting-point*: 68°
(d) *Density*: 0.957
(e) *Solubility*: Highly soluble in water; soluble in ethanol
(f) *Volatility*: Vapour pressure is 1 mm at 68.4° and 10 mm at 109.0°.
(g) *Spectroscopic data*: Infra-red, nuclear magnetic resonance and ultra-violet spectral data have been published (Sadtler Research Laboratories, Inc., undated).

1.4 Technical products and impurities

No data were available to the Working Group.

2,4,5-TRIMETHYLANILINE HYDROCHLORIDE

1.1 Synonyms and trade names
Chem. Abstr. Services Reg. No.: 21436-97-5
Chem. Abstr. Name: Benzenamine, 2,4,5-trimethyl-, hydrochloride
IUPAC Systematic Name: 2,4,5-Trimethylaniline hydrochloride
Synonyms: 1-Amino-2,4,5-trimethylbenzene hydrochloride; *psi*-cumidine hydrochloride; pseudocumidine hydrochloride; 1,2,4-trimethyl-5-aminobenzene hydrochloride

1.2 Structural and molecular formulae and molecular weight

$C_9H_{13}N \cdot HCl$ Mol. wt: 171.7

1.3 Chemical and physical properties of the pure substance

Melting-point: 235° (Weisburger *et al.*, 1978)

1.4 Technical products and impurities

No data were available to the Working Group.

2,4,6-TRIMETHYLANILINE

1.1 Synonyms and trade names
Chem. Abstr. Services Reg. No.: 88-05-1

Chem. Abstr. Name: Benezenamine, 2,4,6-trimethyl-
IUPAC Systematic Name: 2,4,6-Trimethylaniline
Synonyms: Aminomesitylene; 2-aminomesitylene; 2-amino-1,3,5-trimethylbenzene; mesidin; mesidine; mesitylamine

1.2 Structural and molecular formulae and molecular weight

$C_9H_{13}N$ Mol. wt: 135.2

1.3 Chemical and physical properties of the pure substance

From Weast (1979), unless otherwise specified
(a) *Boiling-point*: 232-233°
(b) *Melting-point*: -5°
(c) *Density*: 0.9633
(d) *Refractive index*: n_D^{20} 1.5495
(e) *Spectroscopic data*: Infra-red, nuclear magnetic resonance and ultra-violet spectral data have been published (Sadtler Research Laboratories, Inc., undated).

1.4 Technical products and impurities

2,4,6-Trimethylaniline is available in western Europe; it contains a minimum of 99% active ingredient and has a melting point of -5°.

2,4,6-TRIMETHYLANILINE HYDROCHLORIDE

1.1 Synonyms and trade names

Chem. Abstr. Services Reg. No.: 6334-11-8
Chem. Abstr. Name: Benzenamine, 2,4,6-trimethyl-, hydrochloride

IUPAC Systematic Name: 2,4,6-Trimethylaniline hydrochloride
Synonyms: Aminomesitylene hydrochloride; 2-aminomesitylene hydrochloride; 2-amino-1,3,5-trimethylbenzene hydrochloride; mesidin hydrochloride; mesidine hydrochloride; mesitylamine hydrochloride

1.2 Structural and molecular formulae and molecular weight

$C_9H_{13}N \cdot HCl$ Mol. wt: 171.7

1.3 Chemical and physical properties of the pure substance

Melting-point: 251-252° (Weisburger *et al.*, 1978)

1.4 Technical products and impurities

No data were available to the Working Group.

2. Production, Use, Occurrence and Analysis

2.1 Production and use

2,4,5-TRIMETHYLANILINE

(a) Production

Commercial production of 2,4,5-trimethylaniline was first reported in the US in 1941-1943 (US Tariff Commission, 1945) and was last reported by one US company in 1968 (US Tariff Commission, 1970). It is not produced commercially in western Europe or Japan.

(b) Use

The Society of Dyers & Colourists (1971) reported that one acid dye can be prepared from 2,4,5-trimethylaniline. This dye, Ponceau 3R (Food Red 6), which was the subject of an earlier monograph (IARC, 1975), has not been produced commercially in the US since 1960 and is not believed to be produced in western Europe or Japan.

2,4,5-TRIMETHYLANILINE HYDROCHLORIDE

(a) Production

No evidence was found that 2,4,5-trimethylaniline hydrochloride has ever been produced in commercial quantities in the US, western Europe or Japan.

(b) Use

No data were available to the Working Group.

2,4,6-TRIMETHYLANILINE

(a) Production

2,4,6-Trimethylaniline is produced commercially by the reduction of nitromesitylene.

It is believed to be produced by one company each in France and Switzerland. It was produced by two custom manufacturers in Japan; however, present production is believed to be negligible. No evidence was found that 2,4,6-trimethylaniline has ever been produced in commercial quantities in the US.

(b) Use

2,4,6-Trimethylaniline is not presently used commercially in the US. The Society of Dyers & Colourists (1971, 1975) reported that three dyes can be prepared from it. Only one of these, Acid Blue 129, has been produced in commercial quantities in the US; its production was last reported by one US company in 1971 (US Tariff Commission, 1973).

2,4,6-Trimethylaniline is believed to be used as an intermediate in the manufacture of 2-diethylamino-2',4',6'-acetanilide (trimecaine), which is transformed to its hydrochloride salt for use as a local anaesthetic (Windholz, 1976). Although a Swedish company holds a US patent on synthesis of this chemical, no evidence was found that it is marketed in the US or in western Europe. Wade (1977) reported that trimecaine hydrochloride is used as a local anaesthetic in the USSR and eastern Europe.

2,4,6-TRIMETHYLANILINE HYDROCHLORIDE

(a) Production

No evidence was found that 2,4,6-trimethylaniline hydrochloride has ever been produced commercially in the US, western Europe or Japan.

(b) Use

No data were available to the working Group.

2.2 Occurrence

(a) Natural occurrence

2,4,5-Trimethylaniline, 2,4,6-trimethylaniline and their hydrochloride salts have not been reported to occur as natural products.

(b) Tobacco and tobacco smoke

2,4,6-Trimethylaniline has been detected in the steam volatiles from the distillation of the leaves of one type of tobacco (Latakia) (Irvine & Saxby, 1969).

2.3 Analysis

An IARC manual (Egan *et al.*, 1981) gives selected methods for the analysis of aromatic amines. No data on quantitative methods of analysis for 2,4,5-trimethylaniline or its hydrochloride salt were available to the Working Group.

2,4,6-Trimethylaniline has been separated and identified by paper chromatography, using a variety of solvent systems for development and various spray reagents for detection (Reio, 1974). No data were available on methods of analysis for 2,4,6-trimethylaniline hydrochloride.

3. Biological Data Relevant to the Evaluation of Carcinogenic Risk to Humans

3.1 Carcinogenicity studies in animals

2,4,5-TRIMETHYLANILINE and 2,4,5-TRIMETHYLANILINE HYDROCHLORIDE

Oral administration

Mouse: Two groups, each of 25 male and 25 female CD-1 mice, six to eight weeks old, were fed 1000 or 2000 mg/kg of diet 2,4,5-trimethylaniline hydrochloride (97-99% pure) for 18 months and were observed for an additional three months. An untreated, matched control group of 25 males and 25 females were kept under observation for 21 months; pooled control groups were also used in evaluating the results. Only animals that survived six months or more were autopsied. Statistically significant increases in incidence were observed for the following tumours, in matched and pooled control groups and in low- and high-dose groups, in sequence: hepatomas: males, 3/18, 7/99, 9/14 ($P < 0.025$), 19/21 ($P < 0.025$); females, 0/20, 1/102, 6/15 ($P < 0.025$), 14/22 ($P < 0.025$); lung tumours: males, 5/18, 24/99, 11/14 ($P < 0.025$), 10/21 ($P < 0.05$, with pooled controls only); females, 6/20, 32/102, 11/15 ($P < 0.025$), 12/22 ($P < 0.05$, with pooled controls only) (Weisburger *et al.*, 1978). [The Working Group noted that the poor survival and the scant detail in the reporting make evaluation of the study difficult.]

Groups of 50 male and 50 female B6C3F$_1$ mice, six weeks of age, were fed diets containing 50 or 100 mg/kg 2,4,5-trimethylaniline (purity not established; at least one impurity found by gas chromatography) for 101 weeks. The doses were selected on the basis of a range-finding study [see section 3.2(*a*)]. A group of 20 animals of each sex served as matched controls and were kept under observation for 101 weeks. By the end of the study, 80, 86 and 76% of males and 85, 78 and 90% of females were still alive in the control, low-dose and high-dose groups, respectively. A statistically significant increase in the incidence of hepatocellular carcinomas was observed: among males, 5/20 in controls, 26/50 in the low-dose group ($P = 0.035$) and 27/50 in the high-dose group ($P = 0.025$); among females, 0/20 in controls, 18/49 in the low-dose group ($P = 0.001$) and 40/50 in the high-dose group ($P < 0.001$) (National Cancer Institute, 1979).

Rat: Two groups of 25 male CD rats, six to eight weeks old, were fed diets containing 1000 or 2000 mg/kg 2,4,5-trimethylaniline hydrochloride (97-99% pure) for 18 months and were observed for an additional six months. An untreated, matched control group of 25 males were kept under observation for 24 months; pooled control groups were also used in evaluating the results. Only animals that survived six months or more were autopsied. The incidences of subcutaneous and liver tumours in the matched and pooled controls and in the low- and high-dose groups, in sequence, were as follows: subcutaneous fibromas or fibrosarcomas in 4/22, 18/111, 6/17 ($P < 0.05$, with pooled controls only) and 1/25 (a lipoma); liver tumours in 2/22, 2/111, 3/17 ($P < 0.025$, with pooled controls only) and 2/25 (Weisburger *et al.*, 1978).

Groups of 50 male and 50 female Fischer 344 rats, six weeks of age, were fed diets containing 200 or 800 mg/kg 2,4,5-trimethylaniline (purity not established; at least one impurity found by gas chromatography) for 101 weeks. The doses were selected on the basis of a range-finding study [see section 3.2(a)]. A group of 20 animals of each sex served as matched controls and were kept under observation for 101 weeks. By the end of the study 80, 74 and 86% of males and 70, 84 and 84% of females were still alive in the control, low-dose and high-dose groups, respectively. Statistically significant ($P < 0.003$ to $P < 0.001$), dose-related increases in incidences of the following types of tumour were observed in control, low-dose and high-dose animals in sequence: lung carcinomas plus adenomas - males, 1/20 (1 carcinoma), 0/49, 7/50 (2 carcinomas); females, 0/20, 3/43 (2 carcinomas), 11/50 (2 carcinomas) ($P = 0.017$); hepatocellular carcinomas - males, 0/19, 3/50, 11/50 ($P = 0.020$); females, 0/20, 0/49, 9/50 ($P = 0.039$) (National Cancer Institute, 1979).

2,4,6-TRIMETHYLANILINE and 2,4,6-TRIMETHYLANILINE HYDROCHLORIDE

Oral administration

Mouse: Two groups, each of 25 male and 25 female CD-1 mice, six to eight weeks old, were fed diets containing 2,4,6-trimethylaniline hydrochloride (97-99% pure) at levels of 500 or 1000 mg/kg of diet for three months and 300 or 600 mg/kg of diet for 15 months; they were observed for an additional three months. An untreated, matched control group of 25 males and 25 females were kept under observation for 21 months; pooled control groups were also used in evaluating the results. Only animals that survived six months or more were autopsied. Statistically significant increases in the incidences of the following types of tumours were observed in matched controls, pooled controls, low-dose and high-dose groups, in sequence: hepatomas - males, 0/14, 7/99, 5/15 ($P < 0.05$), 9/13 ($P < 0.025$); females, 1/15, 1/102, 1/12, 12/16 ($P < 0.025$); vascular tumours - males, 2/14, 5/99, 1/15, 4/13 ($P < 0.025$, with pooled controls only) (Weisburger *et al.*, 1978). [The Working Group noted that the poor survival and the scant detail in the reporting make evaluation of the study difficult.]

Rat: Two groups of 25 male CD rats, six to eight weeks old, were fed diets containing 2,4,6-trimethylaniline hydrochloride (97-99% pure) at levels of 250 or 500 mg/kg of diet for three months and 125 or 250 mg/kg of diet for 15 months; they were observed for an additional six months. An untreated, matched control group of 25 males were kept under observation for 24 months; pooled control groups were also used in evaluating the results. Only animals that survived six months or more were autopsied. Statistically significant increases in the incidences of the following types of tumours were observed in matched controls, pooled controls, low-dose and high-dose groups, in sequence: liver tumours - 0/16, 2/111, 4/20 ($P < 0.025$, with pooled controls only), 8/21 ($P < 0.025$); lung tumours - 0/16, 2/111, 5/20 ($P < 0.05$), 8/21 ($P < 0.025$); stomach tumours - 0/16, 2/111, 0/20, 3/21 ($P < 0.05$, with pooled controls only) (Weisburger *et al.*, 1978). [The Working Group noted the poor survival of the control group and the inadequate reporting of the data.]

3.2 Other relevant biological data

(a) Experimental systems

Toxic effects

The single oral LD_{50} of 2,4,5-trimethylaniline hydrochloride in rats when administered by stomach tube was 1585 mg/kg bw; that of 2,4,6-trimethylaniline hydrochloride was 660 mg/kg bw (Lindstrom *et al.*, 1969).

In six-month subacute toxicity studies, oral doses of up to 5000 mg/kg bw 2,4,5-trimethylaniline hydrochloride produced a dose-dependent depression of mean body weight gain in male and female rats. Doses of up to 1000 mg/kg bw 2,4,6-trimethylaniline hydrochloride depressed body weight gain after six months' feeding. A dose-related increase in mean liver weight was observed with both compounds (Lindstrom *et al.*, 1969).

Toxic effects of 2,4,6-trimethylaniline in rats include anaemia and methaemoglobinaemia (Bowie *et al.*, 1965).

Effects on reproduction and prenatal toxicity

No data were available to the Working Group.

Absorption, distribution, excretion and metabolism

After an oral dose of 100 mg/kg bw 2,4,6-trimethylaniline hydrochloride or 200 mg/kg bw 2,4,5-trimethylaniline hydrochloride to male rats, 30% of the 2,4,6-trimethylaniline and 15% of the 2,4,5-trimethylaniline were recovered from the urine after acid hydrolysis. No other metabolites were identified, except trace amounts of carboxylic acid and quinone derivatives (Lindstrom *et al.*, 1969).

Mutagenicity and other short-term tests

2,4,5-Trimethylaniline was mutagenic to *Salmonella typhimurium* strains TA100 and TA98 in the presence of a liver preparation from Aroclor-induced rats and mice. In the same study, it did not induce breaks in the DNA of Chinese hamster V79 cells at concentrations of up to 3 mM for four hours. 2,4,6-Trimethylaniline had the opposite effects: it was not mutagenic to *S. typhimurium* strain TA100 (no doses given), but it induced breaks in the DNA of Chinese hamster V79 cells at concentrations of 3 and 10 mM for four hours (Zimmer *et al.*, 1980).

(b) Humans

No data were available to the Working Group.

3.3 Case reports and epidemiological studies of carcinogenicity in humans

No data were available to the Working Group.

4. Summary of Data Reported and Evaluation

4.1 Experimental data

2,4,5-Trimethylaniline and its hydrochloride were tested in two experiments in mice and in two experiments in rats by dietary administration. In one experiment in mice, 2,4,5-trimethylaniline produced an increased incidence of hepatocellular carcinomas in female mice. The other experiment in mice was considered inadequate for an evaluation. In one experiment in rats it produced an increased incidence of liver carcinomas and lung adenomas. In the other experiment in rats no significantly increased incidence of tumours was observed.

2,4,6-Trimethylaniline hydrochloride was tested in one experiment in mice and in one experiment in male rats. These experiments were considered inadequate for evaluation.

2,4,5-Trimethylaniline was mutagenic to *Salmonella typhimurium* with metabolic activation. The available data were inadequate to evaluate the mutagenicity of 2,4,6-trimethylaniline.

4.2 Human data

2,4,5-Trimethylaniline has been produced commercially in the past. 2,4,6-Trimethylaniline is believed to be produced in commercial quantities, but its uses and the extent of human exposure are unknown. The hydrochloride salts of these two compounds are not produced commercially.

No case report or epidemiological study was available to the Working Group.

4.3 Evaluation

There is *limited evidence* [1] for the carcinogenicity of 2,4,5-trimethylaniline in experimental animals. No evaluation of the carcinogenicity of 2,4,5-trimethylaniline to humans could be made.

The available data were inadequate for an evaluation of the carcinogenicity of 2,4,6-trimethylaniline in experimental animals to be made. No evaluation of the carcinogenicity of 2,4,6-trimethylaniline to humans could be made.

---

[1] See preamble, pp. 18-19.

5. References

Bowie, W.C., Arnault, L.T., Brouwer, E.A. & Lindstrom, H.V. (1965) Hematological manifestations by aromatic amines in rats (Abstract no. 1461). *Fed. Proc., 24,* 392

Egan, H., Fishbein, L., Castegnaro, M., O'Neill, I.K. & Bartsch, H., eds (1981) *Environmental Carcinogens - Selected Methods of Analysis,* Vol. 4, *Some Aromatic Amines and Azodyes in the General and Industrial Environment (IARC Scientific Publications No. 40),* Lyon, International Agency for Research on Cancer

IARC (1975) *IARC Monographs on the Evaluation of Carcinogenic Risk of Chemicals to Man,* Vol. 8, *Some aromatic azo compounds,* Lyon, pp. 199-206

Irvine, W.J. & Saxby, M.J. (1969) Steam volatile amines of Latakia tobacco leaf. *Phytochemistry, 8,* 473-476

Lindstrom, H.V., Bowie, W.C., Wallace, W.C., Nelson, A.A. & Fitzhugh, O.G. (1969) The toxicity and metabolism of mesidine and pseudocumidine in rats. *J. Pharmacol. exp. Ther., 167,* 223-234

National Cancer Institute (1979) *Bioassay of 2,4,5-Trimethylaniline for Possible Carcinogenicity (Tech. Rep. Ser. No. 160; DHEW Publ. No. (NIH) 79-1716),* Washington DC, US Government Printing Office

Reio, L. (1974) Paper chromatographic separation and identification of phenol derivatives and related compounds of biochemical interest using a reference system. Fifth supplement. *J. Chromatogr., 88,* 119-147

Sadtler Research Laboratories, Inc. (undated) *Sadtler Standard Spectra,* Philadelphia, PA

The Society of Dyers & Colourists (1971) *Colour Index,* 3rd ed., Vol. 4, Bradford, Yorkshire, Lund Humphries, pp. 4704, 4826

The Society of Dyers & Colourists (1975) *Colour Index,* revised 3rd ed., Vol. 6, Bradford, Yorkshire, Perkin House, pp. 6403, 6405

US Tariff Commission (1945) *Synthetic Organic Chemicals, US Production and Sales, 1941-43 (Report No. 153, Second Series),* Washington DC, US Government Printing Office, p. 79

US Tariff Commission (1970) *Synthetic Organic Chemicals, US Production and Sales, 1968 (TC Publication 327),* Washington DC, US Government Printing Office, p. 51

US Tariff Commission (1973) *Synthetic Organic Chemicals, US Production and Sales, 1971 (TC Publication 614),* Washington DC, US Government Printing Office, p. 69

Wade, A., ed. (1977) *Martindale, The Extra Pharmacopoeia,* 27th ed., London, The Pharmaceutical Press, p. 880

Weast, R.C., ed. (1979) *CRC Handbook of Chemistry and Physics,* 60th ed., Cleveland OH, Chemical Rubber Co., pp. C-148, D-212

Weisburger, E.K., Russfield, A.B., Homburger, F., Weisburger, J.H., Boger, E., Van Dongen, C.G. & Chu, K.C. (1978) Testing of twenty-one environmental aromatic amines or derivatives for long-term toxicity or carcinogenicity. *J. environ. Pathol. Toxicol., 2,* 325-356

Windholz, M., ed. (1976) *The Merck Index,* 9th ed., Rahway, NJ, Merck & Co., p. 1244

Zimmer, D., Mazurek, J., Petzold, G. & Bhuyan, B.K. (1980) Bacterial mutagenicity and mammalian cell DNA damage by several substituted anilines. *Mutat. Res., 77,* 317-326

ANTHRAQUINONES

2-AMINOANTHRAQUINONE

1. Chemical and Physical Data

1.1 Synonyms and trade names

Chem. Abstr. Services Reg. No.: 117-79-3
Chem. Abstr. Name: 9,10-Anthracenedione, 2-amino-
IUPAC Systematic Name: 2-Aminoanthraquinone
Synonyms: 2-Amino-9,10-anthraquinone; *beta*-aminoanthraquinone; aminoanthraquinone; *beta*-anthraquinonylamine

1.2 Structural and molecular formulae and molecular weight

$C_{14}H_9NO_2$ Mol. wt: 223.2

1.3 Chemical and physical properties of the pure substance

From Weast (1979), unless otherwise specified
(a) *Description*: Red needles
(b) *Boiling-point*: Sublimes
(c) *Melting-point*: 303-306°
(d) *Spectroscopic data*: Infra-red and ultra-violet spectral data have been published (Sadtler Research Laboratories, Inc., undated)
(e) *Solubility*: Insoluble in water and diethyl ether; slightly soluble in ethanol; soluble in acetone, benzene and chloroform
(f) *Reactivity*: Forms salts with mineral acids; can be acylated or alkylated on the nitrogen atom and nitrated or sulphonated in the ring (Chung, 1978)

1.4 Technical products and impurities

2-Aminoanthraquinone is available in western Europe as a powder containing 95% active ingredient; it is available in Japan as a yellowish-brown powder containing a minimum of 80% active ingredient.

2. Production, Use, Occurrence and Analysis

2.1 Production and use

(a) Production

2-Aminoanthraquinone was first used in 1901, as an intermediate in the manufacture of two vat dyes, blue vat indanthrone (Vat Blue 4) and yellow vat flavanthrone (Vat Yellow 1) (The Society of Dyers & Colourists, 1971). It can be prepared by: (a) heating aqueous ammonia and nitrobenzene with sodium anthraquinone-2-sulphonate, (b) ammonolysis of 2-chloroanthraquinone, (c) reaction of anthraquinone-2-carboxamide with hypochlorite or hypobromite (Hofmann reaction), or (d) reduction of 2-nitroanthraquinone (Chung, 1978). The first method is believed to be that used for commercial production in the US, and the second method is that used in Japan. It can also be produced as part of a crude aminoanthraquinone mixture when anthraquinone is nitrated, and the resulting mixture of nitro- and dinitroanthraquinones is reduced. One such mixture has been described as containing 65-80% 1-aminoanthraquinone, 20-35% 2-aminoanthraquinone and 1,5- and 1,8-diaminoanthraquinone (Chung & Farris, 1979).

2-Aminoanthraquinone has been produced commercially in the US since 1921 (US Tariff Commission, 1922). It is believed to be produced presently for captive consumption by one US company. It may also be produced as a component of a crude aminoanthraquinone mixture. Commercial production of an undisclosed amount (see preamble, p. 20) was reported by another US company in 1977 (US Environmental Protection Agency, 1980). Total US production of 2-aminoanthraquinone and its salt (probably the hydrochloride) by four companies in 1971 amounted to 91 thousand kg (US Tariff Commission, 1973). This was a dramatic decrease from the 520 thousand kg produced by five companies in 1965 (US Tariff Commission, 1967).

Imports of 2-aminoanthraquinone through the principal US customs districts in 1979 totalled 4.6 thousand kg (US International Trade Commission, 1980a).

2-Aminoanthraquinone is believed to be produced by one company in the Federal Republic of Germany, one in Italy and one in the UK. Although it has been produced commercially in Japan since before 1945, production in recent years has been negligible.

(b) Use

2-Aminoanthraquinone is believed to be used almost exclusively as a dye intermediate. The Society of Dyers & Colourists (1971) reported that 22 organic dyes and four organic pigments can be prepared from it. All of these products can be produced from intermediates other than 2-aminoanthraquinone; in 1979, only three were reported to be produced in commercial quantities in the US. Three US companies reported production of Vat Blue 6, 8.33% in 1979, but no production data were published (see preamble, p. 20) (US International

Trade Commission, 1980b). Total production by these same three companies in 1978 was 496 thousand kg (US International Trade Commission, 1979). In 1979, one US company reported commercial production of Pigment Blue 22 (Vat Blue 14), and another reported production of Pigment Blue 60 (US International Trade Commission, 1980b). In 1974, three US companies reported total sales of 36.8 thousand kg Vat Blue 14, 8.33% (US International Trade Commission, 1976). In 1970, three US companies reported total production of 30.4 thousand kg of Vat Blue 4, 10% (US Tariff Commission, 1972); Vat Blue 4 is also known as Pigment Blue 60 or, in the past, as indanthrone.

In 1976, one US company reported production of Vat Yellow 1, 12.5% (known in the past as flavanthrone); and another company reported production of Vat Blue 12, 6.5% (US International Trade Commission, 1977). Both can be prepared from 2-aminoanthraquinone.

Crude aminoanthraquinone mixtures (containing 65-80% 1-amino anthraquinone, 20-35% 2-aminoanthraquinone and 1,5- and 1,8-diamino anthraquinones) can be used to produce vat dyes with properties that are superior to those of dyes based on purer 1-aminoanthraquinone. Thus, the amounts of 2-aminoanthraquinone produced and used can be expected to increase if the crude mixture replaces 1-aminoanthraquinone, since the latter is reported to be a major starting material in the manufacture of many dyes (Chung & Farris, 1979).

2.2 Occurrence

2-Aminoanthraquinone has not been reported to occur as a natural product. No data on its occurrence in the environment were available to the Working Group.

2.3 Analysis

Amine mixtures can be analysed for their content of 2-aminoanthraquinone by thin-layer chromatography of the 1-fluoro-2,4-dinitrobenzene derivative (Franc & Koudelkova, 1979).

3. Biological Data Relevant to the Evaluation of Carcinogenic Risk to Humans

3.1 Carcinogenicity studies in animals

Oral administration

Mouse: Groups of 49-50 male and 50 female B6C3F$_1$ mice, six to nine weeks of age, were fed diets containing 5000 or 10000 mg/kg 2-aminoanthraquinone (technical grade of low purity with unspecified impurities). The doses were selected on the basis of a range-finding study [see section 3.2(a)]. Groups of 50 mice of each sex served as matched controls for each dose level. The low- and high-dose mice were treated for 78 and 80 weeks, respectively, and observed for an additional 15-16 weeks. By the end of the study, 82, 78, 94 and 86% of males and 78, 76, 88 and 76% of females were still alive in the low-dose control, high-dose control, low-dose and high-dose groups, respectively. An increased incidence of hepatocellular carcinomas was found in animals of each sex that received the high dose: in male mice, they occurred in 12/46 low-dose controls, 6/48 high-dose controls, 20/47 low-dose and 36/49 ($p < 0.001$) high-dose animals; in the females, the frequencies were 4/46, 1/50, 5/47 and 12/47 ($p = 0.001$), respectively (National Cancer Institute, 1978; Murthy et al., 1979). [The Working Group noted that insufficient data were given on the substance tested and on the impurities it contained.]

Rat: Groups of 50 male and 50 female Fischer 344 rats, six weeks of age, were fed 2-aminoanthraquinone in the diet. (Two batches from separate suppliers were used, and both were technical grades of low purity with unspecified impurities.) Males received 10000 or 20000 mg/kg of diet for the first 10 weeks, reduced to 2500 or 5000 mg/kg for the remaining 68 weeks; female rats received 2000 mg/kg of diet for 78 weeks. Animals were observed for an additional 28-32 weeks. A group of 50 male and 25 female animals served as matched controls and was observed for 107-109 weeks. Of the males, 54% of controls, 64% of the low-dose and 70% of the high-dose rats were still alive at the end of the study. The survival of female rats was not adequate for analysis of late-developing tumours. A compound-related increase in neoplastic nodules and hepatocellular carcinomas was observed in 18/41 low-dose and 18/45 high-dose males; the incidence in controls was 0/36 ($p < 0.001$). The incidence of hepatocellular carcinomas alone was also significantly different from that in controls (National Cancer Institute, 1978; Murthy et al., 1979). [The Working Group noted that insufficient data were given on the substances tested and on the impurities they contained.]

3.2 Other relevant biological data

(a) Experimental systems

Toxic effects

No LD_{50} could be established in rats with single oral administrations of up to 36 g/kg bw 2-aminoanthraquinone (Baker et al., 1975).

In a four-week subchronic toxicity feeding study in Fischer 344 rats and $B6C3F_1$ mice receiving 1-5% of the compound in the diet, reductions in weight gain were observed of up to 10% in males and 20% in females (National Cancer Institute, 1978).

Sprague-Dawley rats that received multiple gastric intubations and Fischer rats fed 2% 2-aminoanthraquinone in the diet showed cystic changes and enlargement of the kidneys (Griswold *et al.*, 1968; Gothoskar *et al.*, 1979a,b).

Effects on reproduction and prenatal toxicity

No data were available to the Working Group.

Absorption, distribution, excretion and metabolism

In rats fed 2% 2-aminoanthraquinone in the diet, the parent compound and *N*-acetyl-2-aminoanthraquinone, together with unspecified hydroxylated derivatives (free and glucuronate and sulphate conjugates) (Gothoskar *et al.*, 1979a) as well as *N*-formyl-2-aminoanthraquinone (Gothoskar *et al.*, 1979b) were detected in the urine. Feeding of 2% for 2-14 weeks induced both cytochrome P-450 dependent hydroxylations and UDP-glucuronosyl transferase activity (Ramanathan *et al.*, 1979).

Dietary administration of 2% of the compound to rats decreased the total concentration of vitamin A in the liver (Reddy & Weisburger, 1980).

Mutagenicity and other short-term tests

Although a crude preparation of 2-aminoanthraquinone induced reversions in *Salmonella typhimurium* TA1537, TA1538 and TA98, this activity was not seen after purification of the compound and was attributed to the presence of 1,2-diaminoanthraquinone (Brown & Brown, 1976).

(b) Humans

No data were available to the Working Group.

3.3 Case reports and epidemiological studies of carcinogenicity in humans

No data were available to the Working Group.

4. Summary of Data Reported and Evaluation

4.1 Experimental data

2-Aminoanthraquinone (technical grade of low purity) was tested in one experiment in mice and in one experiment in rats by dietary administration. It produced hepatocellular carcinomas in mice of both sexes and in male rats.

Purified 2-aminoanthraquinone was not mutagenic to *Salmonella typhimurium*.

4.2 Human data

2-Aminoanthraquinone has been produced commercially since at least 1921. Its use as an intermediate in the manufacture of dyes and pigments could result in occupational exposure.

No case report or epidemiological study was available to the Working Group.

4.3 Evaluation

There is *limited evidence*[1] for the carcinogenicity in experimental animals of the material tested, which was technical-grade 2-aminoanthraquinone of low purity.

In view of the uncertain purity of the compound tested and in the absence of data on humans, no evaluation of the carcinogenicity of 2-aminoanthraquinone could be made.

---

[1] See preamble, pp. 18-19.

5. References

Baker, J.R., Smith, E.R., Yoon, Y.H., Wade, G.G., Rosenkrantz, K. & Schmall, B. (1975) Nephrotoxic effect of 2-aminoanthraquinone in Fischer rats. *J. Toxicol. environ. Health*, *1*, 1-11

Brown, J.P. & Brown, R.J. (1976) Mutagenesis by 9,10-anthraquinone derivatives and related compounds in *Salmonella typhimurium*. *Mutat. Res.*, *40*, 203-224

Chung, R.H. (1978) *Anthraquinone derivatives*. In: Kirk, R.E. & Othmer, D.F., eds, *Encyclopedia of Chemical Technology*, 3rd ed., Vol. 2, New York, John Wiley & Sons, pp. 719-728, 749-757

Chung, R.H. & Farris, R.E. (1979) *Dyes, anthraquinone*. In: Kirk, R.E. & Othmer, D.F., eds, *Encyclopedia of Chemical Technology*, 3rd ed., Vol. 8, New York, John Wiley & Sons, pp. 254, 277-279

Franc, J. & Koudelkova, V. (1979) Thin-layer chromatography of aromatic amines and their derivatives after reactions with 1-fluoro-2,4-dinitrobenzene. *J. Chromatogr.*, *170*, 89-97

Gothoskar, S.V., Benjamin, T., Roller, P.P. & Weisburger, E.K. (1979a) Metabolic fate of 2-aminoanthraquinone, a probable hepatocarcinogen and a nephrotoxic agent in the Fischer rat. *Cancer Detect. Prev.*, *2*, 485-494

Gothoskar, S.V., Benjamin, T., Roller, P.P. & Weisburger, E.K. (1979b) *N*-Formylation of an aromatic amine as a metabolic pathway. *Xenobiotica*, *9*, 533-537

Griswold, D.P., Jr, Casey, A.E., Weisburger, E.K. & Weisburger, J.H. (1968) The carcinogenicity of multiple intragastric doses of aromatic and heterocyclic nitro or amino derivatives in young female Sprague-Dawley rats. *Cancer Res.*, *28*, 924-933

Murthy, A.S.K., Russfield, A.B., Hagopian, M., Mouson, R., Snell, J. & Weisburger, E.K. (1979) Carcinogenicity and nephrotoxicity of 2-amino-, 1-amino-2-methyl-, and 2-methyl-1-nitro-anthraquinone. *Toxicol. Lett.*, *4*, 71-78

National Cancer Institute (1978) *Bioassay of 2-Aminoanthraquinone for Possible Carcinogenicity (Tech. Rep. Ser. No. 144; DHEW Publ. No. (NIH) 78-1399)*, Washington DC, US Government Printing Office

Ramanathan, R., Reddy, T.V. & Weisburger, E.K. (1979) Drug metabolizing enzymes during feeding of the carcinogen 2-aminoanthraquinone (Abstract no. 2276). *Fed. Proc.*, *38*, 661

Reddy, T.V. & Weisburger, E.K. (1980) Hepatic vitamin A status of rats during feeding of the hepatocarcinogen 2-aminoanthraquinone. *Cancer Lett.*, *10*, 39-44

Sadtler Research Laboratories, Inc. (undated) *Sadtler Standard Spectra*, Philadelphia, PA

The Society of Dyers & Colourists (1971) *Colour Index*, 3rd ed., Vol. 4, Bradford, Yorkshire, Lund Humphries, pp. 4579, 4585, 4710

US Environmental Protection Agency (1980) *Chemicals in Commerce Information System (CICIS)*, Washington DC, Office of Pesticides & Toxic Substances, Chemical Information Division

US International Trade Commission (1976) *Synthetic Organic Chemicals, US Production and Sales, 1974 (USITC Publication 776)*, Washington DC, US Government Printing Office, pp. 58, 81

US International Trade Commission (1977) *Synthetic Organic Chemicals, US Production and Sales, 1976 (USITC Publication 833)*, Washington DC, US Government Printing Office, pp. 108-109

US International Trade Commission (1979) *Synthetic Organic Chemicals, US Production and Sales, 1978 (USITC Publication 1001)*, Washington DC, US Government Printing Office, pp. 87, 117, 139

US International Trade Commission (1980a) *Imports of Benzenoid Chemicals and Products, 1979 (USITC Publication 1083)*, Washington DC, US Government Printing Office, p. 8

US International Trade Commission (1980b) *Synthetic Organic Chemicals, US Production and Sales, 1979 (USITC Publication 1099)*, Washington DC, US Government Printing Office, pp. 98, 111

US Tariff Commission (1922) *Census of Dyes and Other Synthetic Organic Chemicals, 1921 (Tariff Information Series No. 26)*, Washington DC, US Government Printing Office, p. 21

US Tariff Commission (1967) *Synthetic Organic Chemicals, US Production and Sales, 1965 (TC Publication 206)*, Washington DC, US Government Printing Office, pp. 10, 64

US Tariff Commission (1972) *Synthetic Organic Chemicals, US Production and Sales, 1970 (TC Publication 479)*, Washington DC, US Government Printing Office, pp. 65, 87

US Tariff Commission (1973) *Synthetic Organic Chemicals, US Production and Sales, 1971 (TC Publication 614)*, Washington DC, US Government Printing Office, pp. 21, 24

Weast, R.C., ed. (1979) *CRC Handbook of Chemistry and Physics*, 60th ed., Cleveland, OH, Chemical Rubber Co., p. C-121

1-AMINO-2-METHYLANTHRAQUINONE

1. Chemical and Physical Data

1.1 Synonyms and trade names

Chem. Abstr. Services Reg. No.: 82-28-0
Chem. Abstr. Name: 9,10-Anthracenedione, 1-amino-2-methyl-
IUPAC Systematic Name: 1-Amino-2-methylanthraquinone
Synonyms: C.I. Disperse Orange 11; C.I. 60700; Disperse Orange; Disperse Orange (anthraquinone dye); 2-methyl-1-anthraquinonylamine
Trade names: Acetate Fast Orange R; Acetoquinone Light Orange JL; Artisil Orange 3RP; Celliton Orange R; Cilla Orange R; Duranol Orange G; Microsetile Orange RA; Nyloquinone Orange JR; Perliton Orange 3R; Serisol Orange YL; Supracet Orange R

1.2 Structural and molecular formulae and molecular weight

$C_{15}H_{11}NO_2$ Mol. wt: 237.3

1.3 Chemical and physical properties of the pure substance

From the Society of Dyers & Colourists (1971), unless otherwise specified

(a) Melting-point: 205°
(b) Spectroscopic data: Infra-red and ultra-violet spectral data have been published (Sadtler Research Laboratories, Inc., undated).
(c) Solubility: Soluble in acetone, benzene, ethanol, ethylene glycol monoethyl ether and linseed oil; slightly soluble in carbon tetrachloride
(d) Reactivity: Forms salts with mineral acids; can be acylated or alkylated on the nitrogen atom (Chung, 1978)

1.4 Technical products and impurities

No data were available to the Working Group.

2. Production, Use, Occurrence and Analysis

2.1 Production and use

(a) Production

1-Amino-2-methylanthraquinone was first synthesized by Romer and Link in 1883 (The Society of Dyers & Colourists, 1971). It can be prepared by the reduction of 2-methyl-1-nitroanthraquinone (see monograph, p. 205) with aqueous sodium sulphide (Chung & Farris, 1979). It is not known whether this is the method used for its commercial production.

1-Amino-2-methylanthraquinone is believed to be produced by three companies in the UK and by one in the Federal Republic of Germany.

It had been produced commercially in the US since 1948 (US Tariff Commission, 1949), but production was last reported by one company in 1970 (US Tariff Commission, 1972). Imports through the principal US customs districts were last reported in 1972, when only 120 kg were brought in (US Tariff Commission, 1973).

It is not produced in commercial quantities in Japan.

(b) Use

1-Amino-2-methylanthraquinone is believed to be used almost exclusively as a dye intermediate. The Society of Dyers & Colourists (1971) reported that it can be used as a dye, and that seven dyes can be prepared from it. No evidence was found that it has ever been produced for commercial use as a dye in the US, and none of the dyes that can be prepared from it are presently produced there in commercial quantities. Only two of them were produced earlier in the US: Solvent Blue 13, which was last produced by one company in 1974 (US International Trade Commission, 1976), and Acid Blue 47, which was last produced by one company in 1973 (US International Trade Commission, 1975).

2.2 Occurrence

1-Amino-2-methylanthraquinone has not been reported to occur as a natural product. No data on its occurrence in the environment were available to the Working Group.

2.3 Analysis

No information on quantitative methods of analysis for 1-amino-2-methylanthraquinone was available to the Working Group.

3. Biological Data Relevant to the Evaluation of Carcinogenic Risk to Humans

3.1 Carcinogenicity studies in animals

Oral administration

Mouse: Groups of 50 male and 50 female $B6C3F_1$ mice, six weeks of age, were fed diets containing technical-grade 1-amino-2-methylanthraquinone (of indeterminate purity, <68%, with unidentified impurities detected by nuclear magnetic resonance, ultra-violet spectrometry and thin-layer chromatography). The first group received an initial concentration of 300 mg/kg of diet; at week 17, this was raised to 1200 mg/kg but was lowered to the original level at week 42 because of deaths from toxicity. A second group received a dietary concentration of 600 mg/kg of diet for 73 weeks. Animals were observed for an additional 24-25 weeks. A group of 50 animals of each sex served as matched controls. By the end of week 43, 46-56% of animals in the first group had died, and there were insufficient numbers at risk from late-developing tumours; 74% of all animals in the second group and the controls lived until the end of the experiment. An increased incidence of hepatocellular tumours and neoplastic nodules was seen in high-dose females: controls, 4/44; low-dose, 2/16; high-dose, 12/43 ($P = 0.022$); but the incidence of hepatocellular carcinomas was not significantly increased. Adenocarcinomas of the kidney were found in two high-dose male mice, and many of the treated animals had compound-related non-neoplastic renal lesions (National Cancer Institute, 1978; Murthy *et al.*, 1979). [The Working Group noted the high mortality in high-dose mice.]

Rat: Groups of 50 male and 45-48 female Fischer 344 rats, six weeks of age, were fed diets containing technical-grade 1-amino-2-methylanthraquinone (same sample as used above) at initial concentrations of 300 and 600 mg/kg of diet. At week 17, these were increased to 1200 and 2400 mg/kg of diet, respectively, for the remaining 62 weeks. Animals were observed for 26-28 additional weeks. A group of 50 rats of each sex served as matched controls. By the end of the experiment, 68, 90 and 62% of males and 70, 78 and 56% of females were still alive in the control, low-dose and high-dose groups, respectively. Statistically significant increases in the incidence of hepatocellular carcinoma were observed in animals of both sexes: in males - 2/48 controls, 7/50 low-dose, 10/48 high-dose animals ($P = 0.014$); and in females - 1/49 controls, 3/45 low-dose and 10/44 high-dose animals ($P = 0.002$). Hepatocellular neoplastic nodules or hepatocellular carcinomas were found in a dose-related trend in males ($P < 0.001$): 3/48 controls, 25/50 low-dose ($P < 0.001$) and 24/48 high-dose ($P < 0.001$); and in females ($P = 0.005$): 2/49 controls, 11/45 low-dose ($P = 0.004$) and 11/44 high-dose ($P = 0.004$). Adenocarcinomas of the kidney occurred only in 4/48 high-dose males (not significant). When benign and malignant kidney tumour incidences were combined, that in males was as follows: 0/48 in controls, 6/50 in low-dose ($P = 0.015$) and 10/48 in high-dose ($P = 0.001$) animals (National Cancer Institute, 1978; Murthy *et al.*, 1979).

3.2 Other relevant biological data

(a) Experimental systems

Toxic effects

No LD_{50} values were available to the Working Group.

In a seven-week subchronic toxicity feeding study with up to 4.5% 1-amino-2-methylanthraquinone in the diet, all rats fed 1.5% or more, all female mice fed 0.5% or more and all male mice fed 0.24% or more in the diet died. Reductions in weight gain of 5-25% were recorded in rats and mice receiving 0.24% and 0.12% in the diet, respectively. Lesions observed in rats fed 0.24% in the diet and in mice fed 0.06% included enlarged kidneys and lymph nodes as well as discolouration of kidneys and adrenals (National Cancer Institute, 1978).

In a chronic feeding study in rats with 0.24% (time-weighted average concentration) 1-amino-2-methylanthraquinone in the diet, lesions observed included inflammatory changes in the lung, spleen, myocardium, bile duct, pancreas and kidneys. Similar lesions were reported in mice fed 0.06% (time-weighted average concentration), although to a lesser extent (National Cancer Institute, 1978).

Effects on reproduction and prenatal toxicity

No data were available to the Working Group.

Absorption, distribution, excretion and metabolism

No data were available to the Working Group.

Mutagenicity and other short-term tests

No data were available to the Working Group.

(b) Humans

No data were available to the Working Group.

3.3 Case reports and epidemiological studies of carcinogenicity in humans

No data were available to the Working Group.

4. Summary of Data Reported and Evaluation

4.1 Experimental data

1-Amino-2-methylanthraquinone (technical grade, maximum purity 68%) was tested in one experiment in mice and one experiment in rats by dietary administration. This material increased the incidence of hepatocellular carcinomas in rats of both sexes and of kidney adenomas and adenocarcinomas in males. The experiment in mice was inadequate for evaluation.

No data on the mutagenicity of this compound were available.

4.2 Human data

1-Amino-2-methylanthraquinone was first produced commercially in 1948. It has been used as a dye intermediate in the US and is still produced in commercial quantities in western Europe. Its continued production could lead to occupational exposure.

No case report or epidemiological study was available to the Working Group.

4.3 Evaluation

There is *limited evidence* [1] for the carcinogenicity in experimental animals of the material tested, which was technical-grade 1-amino-2- methylanthraquinone of low purity.

In view of the uncertain purity of the compound tested and in the absence of data on humans, no evaluation of the carcinogenicity of 1-amino-2-methylanthraquinone could be made.

---

[1] See preamble, pp. 18-19.

5. References

Chung, R.H. (1978) *Anthraquinone derivatives*. In: Kirk, R.E. & Othmer, D.F., eds, *Encyclopedia of Chemical Technology*, 3rd ed., Vol. 2, New York, John Wiley & Sons, pp. 719-725, 752-753, 757

Chung, R.H. & Farris, R.E. (1979) *Dyes, anthraquinone*. In: Kirk, R.E. & Othmer, D.F., eds, *Encyclopedia of Chemical Technology*, 3rd ed., Vol. 8, New York, John Wiley & Sons, pp. 214, 279

Murthy, A.S.K., Russfield, A.B., Hagopian, M., Mouson, R., Snell, J. & Weisburger, E.K. (1979) Carcinogenicity and nephrotoxicity of 2-amino, 1- amino-2-methyl-, and 2-methyl-1-nitro-anthraquinone. *Toxicol. Lett., 4*, 71-78

National Cancer Institute (1978) *Bioassay of 1-Amino-2-methylanthraquinone for Possible Carcinogenicity (Tech. Rep. Ser. No. 111; DHEW Publ. No. (NIH) 78-1366)*, Washington DC, US Government Printing Office

Sadtler Research Laboratories, Inc. (undated) *Sadtler Standard Spectra*, Philadelphia, PA

The Society of Dyers & Colourists (1971) *Colour Index*, 3rd ed., Vol. 4, Bradford, Yorkshire, Lund Humphries, pp. 4534, 4713

US International Trade Commission (1975) *Synthetic Organic Chemicals, US Production and Sales, 1973 (USITC Publication 728)*, Washington DC, US Government Printing Office, p. 64

US International Trade Commission (1976) *Synthetic Organic Chemicals, US Production and Sales, 1974 (USITC Publication 776)*, Washington DC, US Government Printing Office, p. 79

US Tariff Commission (1949) *Synthetic Organic Chemicals, US Production and Sales, 1948 (Report No. 164, Second Series)*, Washington DC, US Government Printing Office, p. 61

US Tariff Commission (1972) *Synthetic Organic Chemicals, US Production and Sales, 1970 (TC Publication 479)*, Washington DC, US Government Printing Office, p. 27

US Tariff Commission (1973) *Imports of Benzenoid Chemicals and Products, 1972 (TC Publication 601)*, Washington DC, US Government Printing Office, p. 9

2-METHYL-1-NITROANTHRAQUINONE

1. Chemical and Physical Data

1.1 Synonyms and trade names

Chem. Abstr. Services Reg. No.: 129-15-7
Chem. Abstr. Name: 9,10-Anthracenedione, 2-methyl-1-nitro-
IUPAC Systematic Name: 2-Methyl-1-nitroanthraquinone
Synonym: 1-Nitro-2-methylanthraquinone

1.2 Structural and molecular formulae and molecular weight

$C_{15}H_9NO_4$ Mol. wt: 267.2

1.3 Chemical and physical properties of the pure substance

From Weast (1979), unless otherwise specified
(a) *Description*: Pale-yellow needles
(b) *Melting-point*: 270-271°
(c) *Spectroscopic data*: Infra-red and ultra-violet spectral data have been published (Sadtler Research Laboratories, Inc., undated).
(d) *Solubility*: Insoluble in water and hot ethanol; slightly soluble in hot diethyl ether, hot benzene, hot acetic acid and chloroform; soluble in nitrobenzene
(e) *Reactivity*: The nitro group may be replaced by a variety of substituents (methoxy, phenoxy, sulphonic acid). Easily reduced to an amino group (Chung, 1978)

1.4 Technical products and impurities

No data were available to the Working Group.

2. Production, Use, Occurrence and Analysis

2.1 Production and use

(a) Production

2-Methyl-1-nitroanthraquinone can be prepared by the nitration of 2-methylanthraquinone with a mixture of nitric and sulphuric acids (Chung, 1978). This is believed to be the method used for commercial production.

2-Methyl-1-nitroanthraquinone is believed to be produced by one company in the UK. It was produced commercially in the US since 1946 (US Tariff Commission, 1948), but it is not believed to be produced in commercial quantities at present. Commercial production by two US companies was last reported in 1970 (US Tariff Commission, 1972); total US production by four US companies in 1968 amounted to 20.4 thousand kg (US Tariff Commission, 1970).

It is not produced in commercial quantities in Japan.

(b) Use

2-Methyl-1-nitroanthraquinone is believed to be used almost exclusively as a dye intermediate. The Society of Dyers & Colourists (1971) reported that three dyes can be prepared from it; however, no evidence was found that these dyes have ever been produced commercially in the US.

2-Methyl-1-nitroanthraquinone is also used commercially (probably without being isolated) to produce 1-amino-2-methylanthraquinone (see monograph, p. 199).

2-Methyl-1-nitroanthraquinone can be oxidized to 1-nitro-2-anthraquinonecarboxylic acid, which can be used to prepare five dyes (The Society of Dyers & Colourists, 1971); only two of these have been produced commercially in the US in recent years: Vat Brown 31, 28%, last produced by one company in 1974 (US International Trade Commission, 1976), and Vat Red 35, 12.5% (which can also be made from other intermediates), last produced by one company in 1970 (US Tariff Commission, 1972). Another dye intermediate, 1-nitro-2-anthraquinonecarbonyl chloride, can be produced from 1-nitro-2-anthraquinonecarboxylic acid, and two dyes can be prepared from this (The Society of Dyers & Colourists, 1971). One of these, Vat Red 10, 18%, was produced by one US company in 1979 (US International Trade Commission, 1980); and total sales of Vat Red 10 by three companies in 1966 were 43.1 thousand kg (US Tariff Commission, 1968). The other dye, Solubilized Vat Red 10, 31%, was last produced by one company in the US in 1967 (US Tariff Commission, 1969).

2.2 Occurrence

2-Methyl-1-nitro-anthraquinone has not been reported to occur as a natural product. No data on its occurrence in the environment were available to the Working Group.

2.3 Analysis

No information on quantitative methods of analysis for 2-methyl-1-nitroanthraquinone was available to the Working Group.

3. Biological Data Relevant to the Evaluation of Carcinogenic Risk to Humans

3.1 Carcinogenicity studies in animals

Oral administration

Mouse: Groups of 50 male and 50 female B6C3F$_1$ mice, six weeks of age, were fed diets containing 300 or 600 mg/kg 2-methyl-1-nitroanthraquinone (impurities unspecified). A group of 50 mice of each sex served as matched controls. None of the mice in either of the treated groups survived longer than 338 days. Subcutaneous haemangiosarcomas developed in the majority of treated mice at the two dose levels (88/90 males and 79/82 females); four of these tumours metastasized to the lung. Mesenteric haemangiosarcomas were observed in six male and eight female mice. One haemangiosarcoma of the spleen occurred among 49 male controls; no haemangiosarcoma at any site occurred in 48 female controls (Murthy *et al.*, 1977, 1979). [The Working Group noted that no attempt was made to measure the purity of the tested material or to characterize it.]

Rat: Groups of 50 male and 50 female Fischer 344 rats, six weeks of age were fed diets containing 600 or 1200 mg/kg 2-methyl-1-nitroanthraquinone (impurities unspecified) for 78 weeks and were observed for an additional 31 weeks. The doses were selected on the basis of a range-finding study [see section 3.2(a)]. A group of 50 rats of each sex served as matched controls. By the end of the study, 60, 70 and 54% of males and 44, 80 and 58% of females were still alive in the control, low-dose and high-dose groups, respectively. An increase in the incidence of hepatocellular carcinomas was observed only in males: 1/48 controls, 5/48 low-dose and 9/49 high-dose animals ($P = 0.008$). Fibromas of the subcutaneous tissue were observed in both male and female rats: in males - 3/48 controls, 10/49 low-dose ($P = 0.039$) and 34/49 high-dose animals ($P < 0.001$); in females - 1/50, 0/50 and 13/49 ($P < 0.001$), respectively. Subcutaneous haemangiosarcomas occurred in three high-dose male rats (National Cancer Institute, 1978). [The Working Group noted that no attempt was made to measure the purity of the tested material or to characterize it.]

3.2 Other relevant biological data

(a) Experimental systems

Toxic effects

No LD_{50} values were available to the Working Group.

In a six-week subchronic toxicity feeding study with up to 5% 2-methyl-1-nitroanthraquinone in the diet, male and female rats that received 0.15% showed reductions in weight gain of 14% and 3%, respectively. In a chronic feeding study in rats with 0.06 or 0.12% in the diet, hyperplasia of the lymphoid tissue and of basal cells of the stomach as well as inflammatory changes of the stomach were reported (National Cancer Institute, 1978).

Effects on reproduction and prenatal toxicity

No data were available to the Working Group.

Absorption, distribution, excretion and metabolism

No data were available to the Working Group.

Mutagenicity and other short-term tests

2-Methyl-1-nitroanthraquinone (purity unspecified) induced reverse mutations in *Salmonella typhimurium* strains TA1537, TA1538, TA98 and TA100, with or without a liver activation system from Aroclor 1254-induced rats (Brown & Brown, 1976).

(b) Humans

No data were available to the Working Group.

3.3 Case reports and epidemiological studies of carcinogenicity in humans

No data were available to the Working Group.

4. Summary of Data Reported and Evaluation

4.1 Experimental data

2-Methyl-1-nitroanthraquinone (of uncertain purity/impurity) was tested in one experiment in mice and one experiment in rats by dietary administration. It was carcinogenic in mice of both sexes, producing haemangiosarcomas, and in rats, producing hepatocellular carcinomas in males and subcutaneous fibromas in animals of both sexes.

2-Methyl-1-nitroanthraquinone was mutagenic to *Salmonella typhimurium*.

4.2 Human data

2-Methyl-1-nitroanthraquinone was first produced commercially in 1946. Its probable use as a dye intermediate could result in occupational exposure.

No case report or epidemiological study was available to the Working Group.

4.3 Evaluation

There is *sufficient evidence*[1] for the carcinogenicity in experimental animals of the material tested, which was 2-methyl-1-nitroanthraquinone of uncertain purity/impurity.

In view of the uncertain purity/impurity of the compound tested and in the absence of data on humans, no evaluation of the carcinogenicity of 2-methyl-1-nitroanthraquinone could be made.

---

[1] See preamble, p. 18.

5. References

Brown, J.P. & Brown, R.J. (1976) Mutagenesis by 9,10-anthraquinone derivatives and related compounds in *Salmonella typhimurium*. *Mutat. Res.*, *40*, 203-224

Chung, R.H. (1978) *Anthraquinone derivatives*. In: Kirk, R.E. & Othmer, D.F., eds, *Encyclopedia of Chemical Technology*, 3rd ed., Vol. 2, New York, John Wiley & Sons, pp. 716-719, 751-752, 757

Murthy, A.S.K., Baker, J.R., Smith, E.R. & Wade, G.G. (1977) Development of hemangiosarcomas in B6C3F$_1$ mice fed 2-methyl-1-nitroanthraquinone. *Int. J. Cancer*, *19*, 117-121

Murthy, A.S.K., Russfield, A.B., Hagopian, M., Mouson, R., Snell, J. & Weisburger, E.K. (1979) Carcinogenicity and nephrotoxicity of 2-amino-, 1-amino-2-methyl-, and 2-methyl-1-nitroanthraquinone. *Toxicol. Lett.*, *4*, 71-78

National Cancer Institute (1978) *Bioassay of 2-Methyl-1-nitroanthraquinone for Possible Carcinogenicity* (Tech. Rep. Ser. No. 29; DHEW Publ. No. (NIH) 78-829), Washington DC, US Government Printing Office

Sadtler Research Laboratories, Inc. (undated) *Sadtler Standard Spectra*, Philadelphia, PA

The Society of Dyers & Colourists (1971) *Colour Index*, 3rd ed., Vol. 4, Bradford, Yorkshire, Lund Humphries, pp. 4720, 4722, 4723

US International Trade Commission (1976) *Synthetic Organic Chemicals, US Production and Sales, 1974* (USITC Publication 776), Washington DC, US Government Printing Office, p. 82

US International Trade Commission (1980) *Synthetic Organic Chemicals, US Production and Sales, 1979* (USITC Publication 1099), Washington DC, US Government Printing Office, p. 98

US Tariff Commission (1948) *Synthetic Organic Chemicals, US Production and Sales, 1946* (Report No. 159, Second Series), Washington DC, US Government Printing Office, p. 73

US Tariff Commission (1968) *Synthetic Organic Chemicals, US Production and Sales, 1966* (TC Publication 248), Washington DC, US Government Printing Office, pp. 24, 111

US Tariff Commission (1969) *Synthetic Organic Chemicals, US Production and Sales, 1967* (TC Publication 295), Washington DC, US Government Printing Office, p. 108

US Tariff Commission (1970) *Synthetic Organic Chemicals, US Production Sales, 1968* (TC Publication 327), Washington DC, US Government Printing Office, pp. 25, 45

US Tariff Commission (1972) *Synthetic Organic Chemicals, US Production and Sales, 1970* (TC Publication 479), Washington DC, US Government Printing Office, pp. 44, 87

Weast, R.C., ed. (1979) *CRC Handbook of Chemistry and Physics*, 60th ed., Cleveland, OH, Chemical Rubber Co., p. C-124

NITROSO COMPOUNDS

N-NITROSODIPHENYLAMINE

1. Chemical and Physical Data

1.1 Synonyms and trade names

Chem. Abstr. Services Reg. No.: 86-30-6
Chem. Abstr. Name: Benzenamine, N-nitroso-N-phenyl-
IUPAC Systematic Name: N-Nitrosodiphenylamine
Synonyms: Diphenylnitrosamine; diphenyl N-nitrosoamine; N,N-diphenylnitrosamine; NDPA; NDPhA; nitrosodiphenylamine; N-nitroso-N-phenylaniline; nitrous diphenylamide
Trade names: Curetard A; Delac J; Naugard TJB; Redax; Retarder J; TJB; Vulcalent A; Vulcatard A; Vulkalent A; Vultrol

1.2 Structural and molecular formulae and molecular weight

$C_{12}H_{10}N_2O$ Mol. wt: 198.2

1.3 Chemical and physical properties of the pure substance

From Weast (1979), unless otherwise specified

(a) *Description*: Yellow plates
(b) *Melting-point*: 66.5°
(c) *Density*: 1.23 (Hawley, 1977)
(d) *Spectroscopic data*: Infra-red, nuclear magnetic resonance, ultra-violet (Sadtler Research Laboratories, Inc., undated) and mass spectral data (Mass Spectrometry Data Centre, 1974) have been published.
(e) *Solubility*: Insoluble in water; soluble in acetone, benzene, ethanol and ethylene dichloride (Hawley, 1977)
(f) *Stability*: Technical grades may decompose at above 85° to produce oxides of nitrogen (Uniroyal, Inc., 1979).
(g) *Reactivity*: May undergo trans-nitrosation reactions with secondary amines to convert them to N-nitrosamines (Uniroyal Inc., 1979)

1.4 Technical products and impurities

N-Nitrosodiphenylamine is available in the US as an orange-brown, amorphous solid typically containing 96% active ingredient, 0.1% ash, and with a melting range of 64-67° (Uniroyal, Inc., undated). It is also available in western Europe as a yellow to medium-brown, coarse powder containing a maximum of 0.1% matter that is insoluble in industrial methylated spirit, a maximum of 0.1% ash, a maximum of 20 mg/kg copper and with a maximal loss of 0.5% at 60° and a crystallizing-point of 64.8-65.8° (Vulnax International Ltd, undated). It is also available in Japan, as a yellowish-brown flake containing maxima of 0.3% moisture and 0.3% ash, and with a minimal melting-point of 62°.

2. Production, Use, Occurrence and Analysis

2.1 Production and use

(a) Production

N-Nitrosodiphenylamine can be prepared by the reaction of nitrous acid with diphenylamine (Layer & Kehe, 1978); this is believed to be the method used for its commercial production.

This compound has been produced commercially in the US since 1945 (US Tariff Commission, 1947). It is presently produced there by four companies, who reported a total production of 287 thousand kg in 1979 (US International Trade Commission, 1980). This volume represented a continuation of the decreasing production observed in recent years: in 1978, the same four companies reported a total production of 733 thousand kg (US International Trade Commission, 1979), down sharply from the 1587 thousand kg produced by six US companies in 1974 (US International Trade Commission, 1976).

Imports of N-nitrosodiphenylamine through the principal US customs districts were last reported separately in 1977, when they amounted to 23.6 thousand kg (US International Trade Commission, 1978).

N-Nitrosodiphenylamine is believed to be produced by two companies in France, one in the Federal Republic of Germany and one in the UK. It has been produced commercially in Japan since 1960; in 1979, two companies produced an estimated five million kg, and even larger quantities are believed to have been imported.

(b) Use

N-Nitrosodiphenylamine is believed to be used almost exclusively as an intermediate in the manufacture of *para*-nitrosodiphenylamine (see monograph, p. 227) and as a rubber-processing chemical. The Fischer-Hepp rearrangement of *N*-nitrosodiphenylamine can be used to produce *para*-nitrosodiphenylamine (Kehe, 1965), which can be reduced to *N*-phenyl-*para*-phenylenediamine, a rubber-processing chemical itself and an intermediate in the manufacture of other rubber-processing chemicals.

N-Nitrosodiphenylamine is an effective radical scavenger, and can be used to stabilize monomers, polymers and petroleum products (Uniroyal, Inc., undated). In rubber processing, its major use is believed to be as an anti-scorching agent, or vulcanization retarder, during rubber compounding. It works effectively at levels of 0.5-1.0% in a variety of rubbers at processing temperatures, and does not interfere with the action of the organic accelerators at vulcanizing temperatures.

N-Nitrosodiphenylamine has been reported to synergize the effects of halogen-containing flame retardants used in polymers (Uniroyal, Inc., undated), but no evidence was found that it is used commercially for that purpose.

In Japan, almost all the *N*-nitrosodiphenylamine used is for the manufacture of *N*-phenyl-*para*-phenylenediamine and its derivatives.

2.2 Occurrence

(a) Natural occurrence

N-Nitrosodiphenylamine has not been reported to occur as a natural product.

(b) Occupational exposure

Levels as high as 47 $\mu g/m^3$ *N*-nitrosodiphenylamine were found in the atmosphere of a factory manufacturing chemicals for tyre curing; a sample scraped from a staircase in the factory contained 15000 mg/kg. *N*-Nitrosodiphenylamine was not detected in the work place air of individual factories producing industrial rubber products, aircraft tyres and synthetic rubber and latex, or in the air of three tyre factories (Fajen *et al.*, 1979, 1980).

In one US tyre-manufacturing plant, work place air samples were found to contain levels of 6.7-13.1 $\mu g/m^3$ *N*-nitrosodiphenylamine (McGlothlin & Wilcox, 1980).

(c) Water and sediments

N-Nitrosodiphenylamine has been found in raw waste samples and secondary effluent samples from textile plants in concentrations of 2-20 $\mu g/l$ (Rawlings & DeAngelis, 1979; US Environmental Protection Agency, 1979a). It has also been found in effluents from ink-manufacturing facilities (US Environmental Protection Agency, 1980).

(d) Other

N-Nitrosodiphenylamine is formed during the aging of explosive products in which diphenylamine is a stabilizer for cellulose nitrate (Layer & Kehe, 1978).

Table 1. Methods for the analysis of *N*-nitrosodiphenylamine

Sample matrix	Sample preparation	Assay procedure[a]	Limit of detection	Reference
Amine mixtures	Dissolve and develop in hexane, diethyl ether and dichloromethane; detect with iodoplatinate reagent	TLC	not given	Gunatilaka (1976)
	Dissolve and elute with pentane and methanol	HPLC-UV	not given	Vodicka et al. (1978)
Propellants	Dissolve and elute with dichloromethane/hexane	HPLC-UV	not given	Doali & Juhasz (1974)
Effluents	Extract three times in dichloromethane; add hydrochloric acid; drain dichloromethane layer through sodium sulphate column; concentrate; remove diphenylamine interference by column chromatography (Florisil or alumina)	GC	not given	US Environmental Protection Agency (1979b)
Liquor	Extract with dichloromethane; dry organic layer over sodium sulphate; concentrate under vacuum; redissolve in dichloromethane	HPLC-TEA	10 mg/kg	Fine et al. (1976)

[a] Abbreviations: TLC - thin-layer chromatography; HPLC-UV - high-performance liquid chromatography with ultra-violet spectrometry; GC - gas chromatography; HPLC-TEA - high-performance liquid chromatography with thermal energy analysis

2.3 Analysis

Typical methods for the analysis of N-nitrosodiphenylamine in various matrices are summarized in Table 1.

3. Biological Data Relevant to the Evaluation of Carcinogenic Risk to Humans

3.1 Carcinogenicity studies in animals

(a) Oral administration

Mouse: Groups of 50 male and 50 female B6C3F$_1$ mice, six weeks of age, were fed diets containing N-nitrosodiphenylamine (98% pure by high-performance liquid chromatography, with two unspecified impurities) at two dose levels for each sex. Males received 10000 or 20000 mg/kg of diet for 101 weeks. Females initially received 5000 mg/kg of diet for 38 weeks, none for three weeks, and then 1000 mg/kg of diet for another 60 weeks; or they initially received 10000 mg/kg of diet for 38 weeks, none for three weeks, and then 4000 mg/kg of diet for another 60 weeks. The doses were selected on the basis of a range-finding study [see section 3.2(a)]. A group of 20 males and 20 females served as matched controls. By the end of the study, 90, 92 and 82% of males and 80, 84 and 62% of females were still alive in the control, low-dose and high-dose, respectively. Similar incidences and types of tumours were seen in control and treated mice, and the results of the statistical tests were not significant. Epithelial hyperplasia of urinary bladder mucosa occurred in 2/49 low-dose males, 7/46 high dose males, 3/47 low-dose females and 6/38 high-dose females. These lesions were not seen in the controls (Cardy *et al.*, 1979; National Cancer Institute, 1979).

Groups of 18 male and 18 female (C57BL/6XC3H/Anf)F$_1$ mice and (C57BL/6XAKR)F$_1$ mice received commercial N-nitrosodiphenylamine (impurities unspecified) according to the following schedule: 1000 mg/kg bw in dimethyl sulphoxide at seven days of age by stomach tube and the same amount (not adjusted for increasing body weight) daily up to four weeks of age; subsequently, the mice were fed 3769 mg/kg of diet until they reached 79 weeks of age. The dose was the maximum tolerated dose for infant and young mice. At the end of the observation period (18 months), 2, 15, 18 and 17 mice were still alive in the four groups, respectively. The total numbers of tumour-bearing animals were 7/15, 1/15, 3/18 and 3/18 in treated males and females of the two strains, compared with 22/79, 8/87, 16/90 and 7/82 in pooled controls. No statistically significant increases in the incidences of tumours were found (National Technical Information Service, 1968; Innes *et al.*, 1969). [The Working Group noted the small numbers of animals used.]

Rat: Groups of 50 male and 50 female Fischer 344 rats, six weeks of age, were fed diets containing 1000 or 4000 mg/kg N-nitrosodiphenylamine (98% pure by high-performance liquid chromatography, with two unspecified impurities) for 100 weeks. The doses were selected on the basis of a range-finding study [see section 3.2(a)]. A group of 20 rats of each sex served as matched controls. All animals under study received food and water *ad libitum*. No dose-related trend in mortality was seen in male rats: 80% of controls, 88% of the low-dose group and 86% of the high-dose group survived until the end of the period of observation, i.e., 100 weeks; however, in females the respective percentages were 90, 88 and 70 (P = 0.024). Significantly increased incidences were observed for the following neoplasms: (a) transitional-cell carcinomas of the urinary bladder: among males - 0/19 in controls, 0/46 in the low-dose group, 16/45 in the high-dose group (P = 0.001); among females - 0/18 in controls, 0/48 in the low-dose group and 40/49 in the high-dose group (P <0.001); (b) a dose-related trend (P = 0.003) in fibromas of the subcutis and skin among males: 1/20 in controls, 1/50 in the low-dose group and 10/50 in the high-dose group (Cardy *et al.*, 1979; National Cancer Institute, 1979).

A group of 25 male Wistar rats, with a mean body weight of 92 g at the beginning of the experiment (age not specified), were given 1070 µg *N*-nitrosodiphenylamine in 1 ml of 1% aqueous methylcellulose daily by gavage on five days a week for 45 weeks (total dose, 240 mg/rat). All 25 rats survived to 53 weeks, when the experiment was terminated; autopsies were done and various organs were examined histologically. No tumours were observed (Argus & Hoch-Ligeti, 1961). [The Working Group noted the short duration of the experiment and the low dose used.]

A group of 20 BD rats (sex unspecified) were given *N*-nitrosodiphenylamine in the drinking-water to give a daily dose of 120 mg/kg bw and a total dose of 65 g/kg bw. No tumours appeared within 700 days (Druckrey *et al.*, 1967). [The Working Group noted the inadequate reporting of this study.]

(b) Skin application

Mouse: Groups of 16 male and 24 female hairless hr/hr Oslo strain mice (age unspecified) were treated for 20 weeks by single weekly applications on the interscapular skin of *N*-nitrosodiphenylamine (purity unspecified), as 0.1 ml per animal of a 1% solution in acetone. After 80 weeks' observation, all surviving animals were killed and necropsied. Among 14 male and 21 female mice that survived until the end of the period of observation, three males had lung adenomas (Iversen, 1980). [The Working Group noted the lack of appropriate controls in this study.]

(c) Intraperitoneal administration

Rat: A group of 24 male CB stock rats, six to seven weeks of age, received i.p. injections once a week of 25 mg/rat *N*-nitrosodiphenylamine (purity unspecified) in polyethylene glycol 400 for six months (total dose, 325 mg/rat). A group of 24 rats injected with the vehicle only served as controls. The experiment was terminated after two years. By 18-24 months, 10 controls and five treated rats were still alive. Neoplasms were found in two treated rats: one was a hepatoma and one a pituitary adenoma; one control rat had a hepatoma (Boyland *et al.*, 1968). [The Working Group noted that only five dosed rats survived more than 18 months and that a low dose was used.]

3.2 Other relevant biological data

(a) Experimental systems

Toxic effects

The single oral LD_{50} of *N*-nitrosodiphenylamine administered by stomach tube to rats was 3000 mg/kg bw (Druckrey *et al.*, 1967).

In subchronic studies, Fischer rats and $B6C3F_1$ mice were fed diets containing up to 46000 mg/kg *N*-nitrosodiphenylamine for seven or 11 weeks. Female rats did not survive doses greater than 16000 mg/kg of diet; female mice survived higher doses. Male rats and male mice were not killed by the highest doses tested (10000 and 22000 mg/kg of diet, respectively). Reductions in weight gain ranged from 37% in female rats fed 16000 mg/kg to 14% in female mice fed 46000 mg/kg of diet. In carcinogenicity studies, inflammatory lesions were observed in the urinary bladders of mice (National Cancer Institute, 1979).

Effects on reproduction and prenatal toxicity

No data were available to the Working Group.

Absorption, distribution, excretion and metabolism

No data were available to the Working Group.

Mutagenicity and other short-term tests

N-Nitrosodiphenylamine at a dose of 250 μg/ml did not induce DNA repair in *Escherichia coli* in the presence of a rat liver metabolic activation system (Rosenkranz & Poirier, 1979).

It did not induce reverse mutations to streptomycin-independence in *E.coli*, either on plates without metabolic activation (Szybalski, 1958) or in liquid medium containing 0.05 mM *N*-nitrosodiphenylamine and a liver postmitochondrial fraction from uninduced rats (Nakajima *et al.*, 1974).

In several studies, doses of up to 2500 μg/plate *N*-nitrosodiphenylamine did not induce reversions in *Salmonella typhimurium* strains TA98, TA100, TA1535, TA1536, TA1537,

TA1538, G46, C3076 or D3052 in the presence of liver fractions derived from Aroclor-induced rats, Syrian hamsters or mice, or from polychlorinated biphenyl- or phenobarbital-induced rats, using both the plate incorporation and liquid preincubation assays (McCann et al., 1975; Purchase et al., 1976; Yahagi et al., 1977; Anderson & Styles, 1978; Dunkel, 1979; McMahon et al., 1979; Rosenkranz & Poirier, 1979; Simmon, 1979a; Bartsch et al., 1980).

In the host-mediated assay in mice, i.m. or i.g. doses of up to 5000 mg/kg bw N-nitrosodiphenylamine did not induce reverse mutations in S. typhimurium strains TA1530, TA1535 or TA1538 (Simmon et al., 1979).

A concentration of 5% w/v N-nitrosodiphenylamine did not induce mitotic recombination in Saccharomyces cerevisiae strain D3 in the presence of a metabolic activation system derived from the livers of Aroclor-induced rats (Simmon, 1979b).

No adenine reversions were induced in Neurospora crassa after incubation with 1 mM N-nitrosodiphenylamine for 30 min, without metabolic activation (Marquardt et al., 1963).

It did not induce TK-/- or HGPRT mutants in mouse lymphoma cells (strain L5178Y/TK+/-) at doses of up to 100 µg/ml for four hours, with or without a metabolic activation system from the livers of Aroclor-induced rats (Clive et al., 1979).

No 8-azaguinine-resistant mutations were induced by 0.5 or 0.1 mM N-nitrosodiphenylamine in V79 Chinese hamster cells, with or without a metabolic activation system derived from the livers of 3-methylcholanthrene- or phenobarbital-induced rats (Drevon et al., 1977; Kuroki et al., 1977a,b).

Fischer rat embryo cells exposed to up to 1 µg/ml N-nitrosodiphenylamine in the presence of a liver fraction from Aroclor-induced rats did not mutate to ouabain resistance (Mishra et al., 1978).

N-Nitrosodiphenylamine did not induce unscheduled DNA synthesis in human foreskin fibroblasts treated with up to 400 µg/ml (Lake et al., 1978) or in rat hepatocyte primary cultures treated with 10^{-3}M (Williams & Laspia, 1979).

N-Nitrosodiphenylamine did not morphologically transform Fischer rat embryo cells at doses of up to 1 µg/ml (Mishra et al., 1978) or Syrian hamster embryo cells at doses of up to 10 µg/ml (Pienta et al., 1977), and did not induce soft agar cloning ability in baby hamster kidney cells at doses of up to 250 mg/ml (Styles, 1977). No cellular or subcellular activation systems were provided in these studies.

No sex-linked recessive lethal mutations were induced by feeding 4 mM N-nitrosodiphenylamine to adult male Drosophila melanogaster (Vogel, 1976).

No depression of testicular DNA synthesis was observed in mice given an i.p. injection of 500 mg/kg bw N-nitrosodiphenylamine (Friedman & Staub, 1976).

(b) Humans

No data were available to the Working Group.

3.3 Case reports and epidemiological studies of carcinogenicity in humans

No data were available to the Working Group.

4. Summary of Data Reported and Evaluation

4.1 Experimental data

N-Nitrosodiphenylamine was adequately tested in one experiment in mice and in one experiment in rats by dietary administration. No carcinogenic effects were observed in mice. It was carcinogenic for male and female rats, producing transitional-cell carcinomas of the urinary bladder in animals given the high dose. In other experiments the results were not indicative of carcinogenic effects, but there were various inadequacies in the conducting and/or reporting of these experiments.

N-Nitrosodiphenylamine has been extensively tested in short-term assays in prokaryotes and eukaryotes for mutagenicity and other chromosomal effects. It was not mutagenic and did not induce unscheduled DNA synthesis or chromosomal damage under any test condition employed.

4.2 Human data

N-Nitrosodiphenylamine has been produced commercially since at least 1945. Its use as a chemical intermediate and as a rubber processing chemical could result in occupational exposure.

No case report or epidemiological study was available to the Working Group.

4.3 Evaluation

There is *limited evidence* [1] for the carcinogenicity of N-nitrosodiphenylamine in experimental animals.

No evaluation of the carcinogenicity of N-nitrosodiphenylamine to humans could be made.

---

[1] See preamble, pp. 18-19.

5. References

Anderson, D. & Styles, J.A. (1978) The bacterial mutation test. *Br. J. Cancer, 37*, 924-930

Argus, M.F. & Hoch-Ligeti, C. (1961) Comparative study of the carcinogenic activity of nitrosamines. *J. natl Cancer Inst., 27*, 695-709

Bartsch, H., Malaveille, C., Camus, A.-M., Martel-Planche, G., Brun, G., Hautefeuille, A., Sabadie, N., Barbin, A., Kuroki, T., Drevon, C., Piccoli, C. & Montesano, R. (1980) Validation and comparative studies on 180 chemicals with *S. typhimurium* strains and V79 Chinese hamster cells in the presence of various metabolizing systems. *Mutat. Res., 76*, 1-50

Boyland, E., Carted, R.L., Gorrod, J.W. & Roe, F.J.C. (1968) Carcinogenic properties of certain rubber additives. *Eur. J. Cancer, 4*, 233-239

Cardy, R.H., Lijinsky, W. & Hildebrandt, P.K. (1979) Neoplastic and nonneoplastic urinary bladder lesions induced in Fischer 344 rats and B6C3F1 hybrid mice by *N*-nitrosodiphenylamine. *Ecotoxicol. environ. Saf., 3*, 29-35

Clive, D., Johnson, K.O., Spector, J.F.S., Batson, A.G. & Brown, M.M.M. (1979) Validation and characterization of the L5178Y/TK$^{+/-}$ mouse lymphoma mutagen assay system. *Mutat. Res., 59*, 6l-108

Doali, J.O. & Juhasz, A.A. (1974) Application of high speed liquid chromatography to the qualitative analysis of compounds of propellant and explosives interest. *J. chromatogr. Sci., 12*, 51-56

Drevon, C., Kuroki, T. & Montesano, R. (1977) Microsome-mediated mutagenesis of a Chinese hamster cell line by various chemicals. *Dev. Toxicol. environ. Sci., 2*, 207-213

Druckrey, H., Preussmann, R., Ivankovic, S. & Schmaehl, D. (1967) Organotropic carcinogenic action of 65 different *N*-nitroso compounds on BD rats (Ger.). *Z. Krebsforsch., 69*, 103-201

Dunkel, V.C. (1979) Collaborative studies on the *Salmonella*/microsome mutagenicity assay. *J. Assoc. off. anal. Chem., 62*, 874-882

Fajen, J.M., Carson, G.A., Rounbehler, D.P., Fan, T.Y., Vita, R., Goff, U.E., Wolf, M.H., Edwards, G.S., Fine, D.H., Reinhold, V. & Biemann, K. (1979) *N*-Nitrosamines in the rubber and tire industry. *Science, 205*, 1262-1264

Fajen, J.M., Fine, D.H. & Rounbehler, D.P. (1980) N-*Nitrosamines in the factory environment*. In: Walker, E.A., Castegnaro, M., Griciute, L. & Borzsonyi, M., eds, N-*Nitroso Compounds: Analysis, Formation and Occurrence (IARC Scientific Publications No. 31)*, Lyon, International Agency for Research on Cancer, pp. 517-528

Fine, D.H., Ross, R., Rounbehler, D.P., Silvergleid, A. & Song, L. (1976) Analysis of nonionic nonvolatile *N*-nitroso compounds in foodstuffs. *J. agric. Food Chem., 24*, 1069-1071

Friedman, M.A. & Staub, J. (1976) Inhibition of mouse testicular DNA synthesis by mutagens and carcinogens as a potential simple mammalian assay for mutagenesis. *Mutat. Res., 37*, 67-76

Gunatilaka, A.A.L. (1976) Thin-layer chromatography of *N*-nitrosamines. *J. Chromatogr., 120*, 229-233

Hawley, G.G., ed. (1977) *The Condensed Chemical Dictionary*, 9th ed., New York, Van Nostrand-Reinhold, p. 619

Innes, J.R.M., Ulland, B.M., Valerio, M.G., Petrucelli, L., Fishbein, L., Hart, E.R., Pallotta, A.J., Bates, R.R., Falk, H.L., Gart, J.J., Klein, M., Mitchell, I. & Peters, J. (1969) Bioassay of pesticides and industrial chemicals for tumorigenicity in mice: a preliminary note. *J. natl Cancer Inst.*, *42*, 1101-1114

Iversen, O.H. (1980) Tumorigenicity of N-nitroso-diaethyl, -dimethyl and -diphenyl-amines in skin painting experiments. *Eur. J. Cancer*, *16*, 695-698

Kehe, H.J. (1965) *Diarylamines*. In: Kirk, R.E. & Othmer, D.F., eds, *Encyclopedia of Chemical Technology*, 2nd ed., Vol. 7, New York, John Wiley & Sons, pp. 42, 50, 53

Kuroki, T., Drevon, C. & Montesano, R. (1977a) Microsome-mediated mutagenesis in V79 Chinese hamster cells by various nitrosamines. *Cancer Res.*, *37*, 1044-1050

Kuroki, T., Drevon, C. & Montesano, R. (1977b) Microsome-mediated mutagenesis of a Chinese hamster cell line by nitrosamines. *Mutat. Res.*, *46*, 205-206

Lake, R.S., Kropko, M.L., Pezzutti, M.R., Shoemaker, R.H. & Igel, H.J. (1978) Chemical induction of unscheduled DNA synthesis in human skin epithelial cell cultures. *Cancer Res.*, *38*, 2091-2098

Layer, R.W. & Kehe, H.J. (1978) *Amines, aromatic (diarylamines)*. In: Kirk, R.E. & Othmer, D.F., eds, *Encyclopedia of Chemical Technology*, 3rd ed., Vol. 2, New York, John Wiley & Sons, pp. 331, 336, 338

Marquardt, H., Schwaier, R. & Zimmermann, F. (1963) Non-mutagenicity of nitrosamines in *Neurospora crassa* (Ger.). *Naturwissenschaften*, *50*, 135-136

Mass Spectrometry Data Centre (1974) *Eight Peak Index of Mass Spectra*, 2nd ed., Reading, UK, Atomic Weapons Research Establishment

McCann, J., Choi, E., Yamasaki, E. & Ames, B.N. (1975) Detection of carcinogens as mutagens in the *Salmonella*/microsome test: assay of 300 chemicals. *Proc. natl Acad. Sci. (USA)*, *72*, 5135-5139

McGlothlin, J.D. & Wilcox, T.G. (1980) *Interim Report No. 3, Health Hazard Evaluation Project No. HE 79-109, Kelly-Springfield Tire Company*, Cincinnati, OH, National Institute for Occupational Safety & Health

McMahon, R.E., Cline, J.C. & Thompson, C.Z. (1979) Assay of 855 test chemicals in ten tester strains using a new modification of the Ames test for bacterial mutagens. *Cancer Res.*, *39*, 682-693

Mishra, N.K., Wilson, C.M., Pant, K.J. & Thomas, F.O. (1978) Simultaneous determination of cellular mutagenesis and transformation by chemical carcinogens in Fischer rat embryo cells. *J. Toxicol. environ. Health*, *4*, 79-91

Nakajima, T., Tanaka, A. & Tojyo, K.-I. (1974) The effect of metabolic activation with rat liver preparations on the mutagenicity of several N-nitrosamines on a streptomycin-dependent strain of *E. coli*. *Mutat. Res.*, *26*, 361-366

National Cancer Institute (1979) *Bioassay of N-Nitrosidiphenylamine for Possible Carcinogenicity (Tech. Rep. Ser. No. 164; DHEW Publ. No. (NIH) 79-172*0), Washington DC, US Government Printing Office

National Technical Information Service (1968) *Evaluation of Carcinogenic, Teratogenic and Mutagenic Activities of Selected Pesticides and Industrial Chemicals*, Vol. 1, *Carcinogenic Study*, Washington DC, US Department of Commerce, p. 88

Pienta, R.J., Poiley, J.A. & Lebherz, W.B., III (1977) Morphological transformation of early passage golden Syrian hamster embryo cells derived from cryopreserved primary cultures as a reliable *in vitro* bioassay for identifying diverse carcinogens. *Int. J. Cancer, 19*, 642-655

Purchase, I.F.H., Longstaff, E., Ashby, J., Styles, J.A., Anderson, D., Lefevre, P.A. & Westwood, F.R. (1976) Evaluation of six short term tests for detecting organic chemical carcinogens and recommendations for their use. *Nature, 264*, 624-627

Rawlings, G.D. & DeAngelis, D.G. (1979) Toxicity removal in textile plant waste waters. *J. Am. Leather Chem. Assoc., 74*, 404-417

Rosenkranz, H. & Poirier, L.A. (1979) Evaluation of the mutagenicity and DNA modifying activity of carcinogens and non-carcinogens in microbial systems. *J. natl Cancer Inst., 62*, 873-892

Sadtler Research Laboratories, Inc. (undated) *Sadtler Standard Spectra*, Philadelphia, PA

Simmon, V.F. (1979a) *In vitro* mutagenicity assays of chemical carcinogens and related compounds with *Salmonella typhimurium*. *J. natl Cancer Inst., 62*, 893-899

Simmon, V.F. (1979b) *In vitro* assays for recombinogenic activity of chemical carcinogens and related compounds with *Saccharomyces cerevisiae* D3. *J. natl Cancer Inst., 62*, 901-909

Simmon, V.F., Rosenkranz, H.S., Zeiger, E. & Poirier, L.A. (1979) Mutagenic activity of chemical carcinogens and related compounds in the intraperitoneal host-mediated assay. *J. natl Cancer Inst., 62*, 911-918

Styles, J.A. (1977) A method for detecting carcinogenic organic chemicals using mammalian cells in culture. *Br. J. Cancer, 36*, 558-563

Szybalski, W. (1958) Special microbiological systems. II. Observations on chemical mutagenesis in microorganisms. *Ann. N.Y. Acad. Sci., 76*, 475-489

Uniroyal, Inc. (undated) *TJB - A Chemical Intermediate or Stabilizer*, Naugatuck, CT, Uniroyal Chemical Division

Uniroyal, Inc. (1979) *Product Safety Data Sheet - Retarder J®*, Naugatuck, CT, Uniroyal Chemical Division

US Environmental Protection Agency (1979a) Textile mills point source category; effluent limitations guidelines, pretreatment standards, and new source performance standards. *Fed. Regist., 44*(210), 62204-62229

US Environmental Protection Agency (1979b) Guidelines establishing test procedures for the analysis of pollutants. *US Code Fed. Regul., Title 40*, Part 136; *Fed. Regist., 44* (233), 69464-69470, 69496-69499

US Environmental Protection Agency (1980) Ink formulating - point source category; effluent limitations guidelines, pretreatment standards and new source performance standards. *US Code Fed. Regul., Title 40*, Part 447; *Fed. Regist., 45*(2), 928-939

US International Trade Commission (1976) *Synthetic Organic Chemicals, US Production and Sales, 1974 (USITC Publication 776)*, Washington DC, US Government Printing Office, pp. 136, 140

US International Trade Commission (1978) *Imports of Benzenoid Chemicals and Products, 1977 (USITC Publication 900)*, Washington DC, US Government Printing Office, p. 28

US International Trade Commission (1979) *Synthetic Organic Chemicals, US Production and Sales, 1978 (USITC Publication 1001)*, Washington DC, US Government Printing Office, pp. 213, 217

US International Trade Commission (1980) *Synthetic Organic Chemicals, US Production and Sales, 1979 (USITC Publication 1099)*, Washington DC, US Government Printing Office, pp. 175, 179

US Tariff Commission (1947) *Synthetic Organic Chemicals, US Production and Sales, 1945 (Report No. 157, Second Series)*, Washington DC, US Government Printing Office, p. 163.

Vodicka, L., Kriz, J., Burda, J. & Novak, P. (1978) High-performance liquid chromatography of compounds obtained during the production of N-nitrosodiphenylamine. *J. Chromatogr., 148,* 247-254

Vogel, E. (1976) *Mutagenicity of carcinogens in* Drosophila *as function of genotype-controlled metabolism.* In: de Serres, F.J., Fouts, J.R., Bend, J.R. & Philpot, R.M., eds, In vitro *Metabolic Activation in Mutagenesis Testing*, Amsterdam, Elsevier/North-Holland Biomedical Press, pp. 63-79

Vulnax International Ltd (undated) *Product: Vulcatard A*, Saint Cloud, France

Weast, R.C., ed. (1979) *CRC Handbook of Chemistry and Physics*, 60th ed., Cleveland, OH, Chemical Rubber Co., p. C-109

Williams, G.M. & Laspia, M.F. (1979) The detection of various nitrosamines in the hepatocyte primary culture/DNA repair test. *Cancer Lett., 6,* 199-206

Yahagi, T., Nagao, M., Seino, Y., Matsushima, T., Sugimura, T. & Okada, M. (1977) Mutagenicities of N-nitrosamines on *Salmonella. Mutat. Res., 48,* 121-130

para-NITROSODIPHENYLAMINE

1. Chemical and Physical Data

1.1 Synonyms and trade names

Chem. Abstr. Services Reg. No.: 156-10-5
Chem. Abstr. Name: Benzenamine, 4-nitroso-N-phenyl-
IUPAC Systematic Name: 4-Nitrosodiphenylamine
Synonyms: 4-Nitroso-N-phenylaniline; para-nitroso-N-phenylaniline; N-phenyl-para-nitrosoaniline
Trade names: Naugard TKB; TKB

1.2 Structural and molecular formulae and molecular weight

$C_{12}H_{10}N_2O$ Mol. wt: 198.2

1.3 Chemical and physical properties of the pure substance

From Windholz (1976), unless otherwise specified
(a) *Description*: Green to blue crystals
(b) *Melting-point*: 144-145°
(c) *Spectroscopic data*: Infra-red, nuclear magnetic resonance, ultra-violet (Sadtler Research Laboratories, Inc., undated) and mass spectral data (Mass Spectrometry Data Centre, 1974) have been published.
(d) *Solubility*: Slightly soluble in water and petroleum ether; soluble in benzene, chloroform, diethyl ether and ethanol
(e) *Stability*: An unusually stable nitroso compound (Uniroyal, Inc., undated)
(f) *Reactivity*: Undergoes reduction, alkylation and nitrosation. Forms salts with strong acids and bases; the basic salts rearrange to a quinoid structure (Uniroyal, Inc., undated)

1.4 Technical products and impurities

para-Nitrosodiphenylamine is available in the US as a purple solid, typically containing 95% active ingredient, 0.25% ash, and with a melting-range of 140-145° (Uniroyal, Inc., undated).

2. Production, Use, Occurrence and Analysis

2.1 Production and use

(a) Production

para-Nitrosodiphenylamine can be prepared by the Fischer-Hepp rearrangement of *N*-nitrosodiphenylamine (Kehe, 1965); this is believed to be the method used for its commercial production.

Commercial production of *para*-nitrosodiphenylamine in the US was first reported by one company in 1975 (US International Trade Commission, 1977), although it is believed to have been produced since at least 1970. Production of undisclosed commercial quantities (see preamble, p. 20) was reported by one company in 1979 (US International Trade Commission, 1980), and by two companies in 1978 (US International Trade Commission, 1979). One of the companies reported that its 1977 production was 45.4-454 thousand kg (US Environmental Protection Agency, 1980).

No evidence was found that *para*-nitrosodiphenylamine is produced in commercial quantities in western Europe. In Japan, it is produced (but probably not isolated) during the production of *N*-phenyl-*para*-phenylenediamine from *N*-nitrosodiphenylamine (see monograph, p. 213). Almost all of the estimated five million kg of *N*-nitrosodiphenylamine produced in Japan annually (and the larger quantity that is imported) is converted to *N*-phenyl-*para*-phenylenediamine *via para*-nitrosodiphenylamine.

(b) Use

para-Nitrosodiphenylamine is reported to be used as a chemical intermediate for dyes and pharmaceuticals and as a polymerization inhibitor during the manufacture of vinyl monomers such as styrene (Uniroyal, Inc., undated).

Use of *para*-nitrosodiphenylamine as a chemical intermediate is believed to be limited primarily to the production of *N*-phenyl-*para*-phenylenediamine and its derivatives. This diamine serves as an azoic diazo component, as an oxidation base, as a developer of diazotized direct dyes, and as an intermediate in the manufacture of other dyes. It is also believed to be used as an antioxidant in rubber and as an intermediate in the synthesis of other antioxidants. Although Hawley (1977) reported that *N*-phenyl-*para*-phenylenediamine has been employed as an intermediate for pharmaceuticals and photographic chemicals, no evidence was found that it is presently so used.

It has also been reported (Windholz, 1976) that *para*-nitrosodiphenylamine has been used as an accelerator in the vulcanization of rubber.

2.2 Occurrence

para-Nitrosodiphenylamine has not been reported to occur as a natural product. No data on its occurrence in the environment were available to the Working Group.

2.3 Analysis

Typical methods for the analysis of *para*-nitrosodiphenylamine in various matrices are summarized in Table 1.

Table 1. Methods for the analysis of para-nitrosodiphenylamine

Sample matrix	Sample preparation	Assay procedure[a]	Limit of detection	Reference
Amine mixtures	Dissolve and elute with pentane and methanol	HPLC-UV	not given	Vodicka et al. (1978)
Rubber	Hydrolyse nitrile rubber film	UV[b]	not given	Shilov et al. (1979)
	Extract with chloroform	UV[b]	not given	Fikhtengol'ts et al. (1979)

[a] Abbreviations: HPLC-UV - high-performance liquid chromatography with ultra-violet spectrometry; UV - ultra-violet spectrometry
[b] Not specific for *para*-nitrosodiphenylamine

3. Biological Data Relevant to the Evaluation of Carcinogenic Risk to Humans

3.1 Carcinogenicity studies in animals

Oral administration

Mouse: Two groups of 50 male and 50 female B6C3F$_1$ mice, six weeks of age, were fed diets containing *para*-nitrosodiphenylamine (technical grade, 73% active material, 25% water, with unspecified impurities). The low-dose group received 5000 mg/kg of diet for 40 weeks, then 2500 mg/kg for 17 weeks; the high-dose group received 10000 mg/kg of diet for 40 weeks, then, after a seven-week interval, 5000 mg/kg for 10 weeks. Both groups were observed for an additional 35 weeks. A group of 20 animals of each sex served as matched controls and were kept under observation for 92 weeks. Of the high-dose group, 19 males and 21 females died before 52 weeks due to toxicity. By the end of the study, 85, 88 and 60% of males and 90, 84 and 52% of females were still alive in the control, low-dose and high-dose groups, respectively. A statistically significant increase in the incidence of liver tumours was observed in treated male mice only: hepatocellular carcinoma or adenoma - 2/18 in controls, 22/42 in the low-dose animals ($P = 0.002$) and 12/13 in the high-dose animals ($P = 0.038$); incidences of hepatocellular carcinomas alone were 0/18, 10/42 ($P = 0.02$) and 1/31 in the three groups, respectively (National Cancer Institute, 1979). [The Working Group noted that the combined incidence of liver adenomas and carcinomas in the control group was lower than that observed in historical controls and that the incidence in the treated groups may be consistent with the normal range in historical controls.]

Rat: Two groups of 50 male and 50 female Fischer 344 rats, six weeks of age, were fed diets containing 2500 or 5000 mg/kg *para*-nitrosodiphenylamine (same sample as used above) for 78 weeks and were observed for an additional 27 weeks. The doses were selected on the basis of a range-finding study [see section 3.2(a)]. A group of 20 animals of each sex served as matched controls and were kept under observation for 105 weeks. By the end of the study, 90, 86 and 92% of males and 85, 84 and 92% of females were still alive in the control, low-dose and high-dose groups, respectively. A statistically significant increase in the incidence of hepatocellular carcinoma or neoplastic nodule was observed in male rats only: 0/20 in controls, 10/49 in the low-dose animals ($P = 0.024$) and 19/50 in the high-dose animals ($P < 0.001$); the incidence of hepatocellular carcinomas alone was not increased (National Cancer Institute, 1979).

3.2 Other relevant biological data

(a) Experimental systems

Toxic effects

The oral LD$_{50}$ of *para*-nitrodiphenylamine in rats was 2410 mg/kg bw (Lewis & Tatken, 1979).

In a four-week subchronic study, Fischer 344 rats and B6C3F$_1$ mice were fed diets containing up to 3.2% and 2.6% *para*-nitrosodiphenylamine, respectively. There was dose-related mortality in male and female mice, but not in rats. The rats showed a dose-dependent depression of mean body weight gain of up to 53% (National Cancer Institute, 1979).

Effects on reproduction and prenatal toxicity

No data were available to the Working Group.

Absorption, distribution, excretion and metabolism

No data were available to the Working Group.

Mutagenicity and other short-term tests

para-Nitrosodiphenylamine did not transform cells of embryos excised from pregnant hamsters treated intraperitoneally with up to 20 mg/kg bw (DiPaolo *et al.*, 1973).

(b) *Humans*

No data were available to the Working Group.

3.3 Case reports and epidemiological studies of carcinogenicity in humans

No data were available to the Working Group.

4. Summary of Data Reported and Evaluation

4.1 Experimental data

para-Nitrosodiphenylamine (technical grade) was tested in mice and rats by dietary administration. It produced lesions described as neoplastic nodules in the livers of male rats. The results of the study in mice were inconclusive.

No data on the mutagenicity of *para*-nitrosodiphenylamine were available.

4.2 Human data

para-Nitrosodiphenylamine has been produced commercially since at least 1970. Its use as a chemical intermediate could result in occupational exposure.

No case report or epidemiological study was available to the Working Group.

4.3 Evaluation

The available data are insufficient for an evaluation of the carcinogenicity of *para*-nitrosodiphenylamine in experimental animals.

In view of the few data in experimental animals and in the absence of data on humans, no evaluation of the carcinogenicity of *para*-nitrosodiphenylamine to humans could be made.

5. References

DiPaolo, J.A., Nelson, R.L., Donovan, P.J. & Evans, C.H. (1973) Host-mediated *in vivo-in vitro* assay for chemical carcinogenesis. *Arch. Pathol.*, *95*, 380-385

Fikhtengol'ts, V.S., Kogan, L.M., Krol, V.A., Monastyrskaya, N.B. & Khaikina, S.A. (1979) Spectrophotometric method for the determination of the functional groups in SKI-3-01 rubber (Russ.). *Prom-st. Sint. Kauch* (4), 6-9 [*Chem. Abstr.*, *91*, 5843t]

Hawley, G.G., ed. (1977) *The Condensed Chemical Dictionary*, 9th ed., New York, Van Nostrand-Reinhold, p. 42

Kehe, H.J. (1965) *Diarylamines*. In: Kirk, R.E. & Othmer, D.F., eds, *Encyclopedia of Chemical Technology*, 2nd ed., Vol. 7, New York, John Wiley & Sons, pp. 42, 50, 53

Lewis, R.J. & Tatken, R.L., eds (1979) *1978 Registry of Toxic Effects of Chemical Substances*, Cincinnati, OH, US Department of Health, Education, & Welfare, p. 497

Mass Spectrometry Data Centre (1974) *Eight Peak Index of Mass Spectra*, 2nd ed., Reading, UK, Atomic Weapons Research Establishment

National Cancer Institute (1979) *Bioassay of p-Nitrosodiphenylamine for Possible Carcinogenicity (Tech. Rep. Ser. No. 190; DHEW Publ. No. (NIH) 79-1746)*, Washington DC, US Government Printing Office

Sadtler Research Laboratories, Inc. (undated) *Sadtler Standard Spectra*, Philadelphia, PA

Shilov, A.D., Srednev, S.S., Basov, B.K. & Kogan, L.M. (1979) Determination of bound *p*-nitrosodiphenylamine in butadiene-nitrile rubber and latexes (Russ.). *Prom-st. Sint. Kauch* (10), 6-9 [*Chem. Abstr.*, *92*, 1303 18v]

Uniroyal, Inc. (undated) *TKB*, Naugatuck, CT, Uniroyal Chemical Division

US Environmental Protection Agency (1980) *Chemicals in Commerce Information System (CICIS)*, Washington DC, Office of Pesticides & Toxic Substances, Chemical Information Division

US International Trade Commission (1977) *Synthetic Organic Chemicals, US Production and Sales, 1975 (USITC Publication 804)*, Washington DC, US Government Printing Office, p. 39

US International Trade Commission (1979) *Synthetic Organic Chemicals, US Production and Sales, 1978 (USITC Publication 1001)*, Washington DC, US Government Printing Office, p. 71

US International Trade Commission (1980) *Synthetic Organic Chemicals, US Production and Sales, 1979 (USITC Publication 1099)*, Washington DC, US Government Printing Office, p. 52

Vodicka, L., Kriz, J., Burda, J. & Novak, P. (1978) High-performance liquid chromatography of compounds obtained during the production of *N*-nitrosodiphenylamine. *J. Chromatogr.*, *148*, 247-254

Windholz, M., ed. (1976) *The Merck Index*, 9th ed., Rahway, NJ, Merck & Co., p. 862

INORGANIC FLUORIDES USED IN DRINKING-WATER AND DENTAL PREPARATIONS

INORGANIC FLUORIDES USED IN DRINKING-WATER AND DENTAL PREPARATIONS

1. Chemical and Physical Data

SODIUM FLUORIDE

1.1 Synonyms and trade names

Chem. Abstr. Services Reg. No.: 7681-49-4

Chem. Abstr. Name and IUPAC Systematic Name: Sodium fluoride

Synonyms: Disodium difluoride; natrium fluoride; sodium fluoride cyclic dimer; sodium hydrofluoride; sodium monofluoride; trisodium trifluoride; villiaumite

Trade names: Antibulit; Cavi-Trol; Credo; FDA 0101; F1-Tabs; Flozenges; Fluoral; Fluorident; Fluorigard; Fluorineed; Fluorinse; Fluoritab; Fluorocid; Fluor-O-Kote; Fluorol; Flura; Flura-Drops; Flura-Gel; Flura-Loz; Flurcare; Flursol; Fungol B; Gel II; Gelution; Gleem; Iradicav; Karidium; Karigel; Kari-Rinse; Lemoflur; Lea-Cov; Luride; Luride Lozi-Tabs; Nafeen; NaFpak; Na Frinse; Nufluor; Ossalin; Ossin; Pediaflor; Pedident; Pennwhite; Pergantene; Phos-Flur; Point Two; Predent; Rafluor; Rescue Squad; So-Flo; Stay-Flo; Studafluor; Super-dent; T-Fluoride; Thera-Flur; Thera-Flur-N; Zymafluor

1.2 Structural and molecular formulae and molecular weight

NaF   Mol. wt: 42.0

1.3 Chemical and physical properties of the pure substance

From Weast (1979), unless otherwise specified

(a) *Description*: White, free-flowing crystalline powder (Wachter, 1980)

(b) *Boiling-point*: 1695°

(c) *Melting-point*: 993°

(d) *Density*: d_4^{1} 2.558

(e) *Refractive index*: 1.336 (conditions not given)

(f) *Solubility*: 42 g/l water at 10°; soluble in hydrogen fluoride; very slightly soluble in ethanol

1.4 Technical products and impurities

Sodium fluoride is available in the US, with the following specifications: minima of 97.0% active ingredient and 43.8% available fluorine; and maxima of 0.6% insoluble matter, 0.5% moisture and 400 mg/kg heavy metals (as lead) (Chemtech Industries, Inc., undated a). It is also available as a USP grade measured as containing 98-102% sodium fluoride (calculated on a dried basis), maxima of 0.012% chloride and 30 mg/kg heavy metals, and with a maximal loss of 1% on drying at 150° for four hours. It is also available as a USP oral solution measured as containing 95-105% of the labelled amount of sodium fluoride (US Pharmacopeial Convention, Inc., 1980).

Sodium fluoride is also available in Japan, where it must contain a minimum of 97% active ingredient.

FLUOSILICIC ACID

1.1 Synonyms and trade names

Chem. Abstr. Services Reg. No.: 16961-83-4
Chem. Abstr. Name: Silicate (2-), hexafluoro-, dihydrogen
IUPAC Systematic Name: Dihydrogen hexafluorosilicate (2-)
Synonyms: Dihydrogen hexafluorosilicate; fluorosilicic acid; hexafluorosilicic acid; hexafluosilicic acid; hydrofluorosilicic acid; hydrogen hexafluorosilicate; hydrosilicofluoric acid; sand acid; silicofluoric acid; silicon hexafluoride dihydride

Trade name: FKS

1.2 Structural and molecular formulae and molecular weight

H_2SiF_6 Mol. wt: 144.1

1.3 Chemical and physical properties of the pure substance

From Windholz (1976) and Weast (1979), unless otherwise specified
(a) Description: Colourless, fuming liquid
(b) Boiling-point: Decomposes
(c) Melting-point: 60-70% solution solidifies at about 19°, forming a crystalline dihydrate

(d) Density: d^{25} 1.4634 (60.97% solution); $d_{17.5}^{17.5}$ 1.2742 (30% solution)
(e) Refractive index: n_D^{25} 1.3465 (60.97% solution)
(f) Solubility: Soluble in hot and cold water
(g) Stability: Anhydrous liquid dissociates almost immediately into SiF_4 and HF
(h) Reactivity: Attacks glass and stoneware (Hawley, 1977)

1.4 Technical products and impurities

Fluosilicic acid is available in the US as commercial grade products containing 20-40% active ingredient (Essex Chemical Corp., 1980); a typical product is a clear, white liquid containing 30.3% active ingredient, 0.2% free silica and 0.1% sulphuric acid (The Harshaw Chemical Co., undated).

The American Water Works Association standard for fluosilicic acid used in water treatment specifies that it must contain 20-30% active ingredient, a maximum of 200 mg/kg heavy metals (as lead) and no soluble mineral or organic substance in quantities capable of inducing injurious health effects (American Water Works Association, 1971).

Fluosilicic acid is available in western Europe in commercial grades containing 20-24% active ingredient, a minimal fluorine content of 19%, a maximal chlorine content of 0.3%, a maximal phosphorus pentoxide content of 0.05% and 99.9% minimal weight loss at 600°.

SODIUM SILICOFLUORIDE

1.1 Synonyms and trade names
Chem. Abstr. Services Reg. No.: 16893-85-9
Chem. Abstr. Name: Silicate (2-), hexafluoro-, disodium
IUPAC Systematic Name: Disodium hexafluorosilicate (2-)
Synonyms: Disodium hexafluorosilicate; disodium silicofluoride; silicon sodium fluoride; sodium fluorosilicate; sodium fluosilicate; sodium hexafluorosilicate; sodium hexafluosilicate; sodium silicon fluoride
Trade names: Prodan; Salufer

1.2 Structural and molecular formulae and molecular weight

Na_2SiF_6 Mol. wt: 188.1

1.3 Chemical and physical properties of the pure substance

From Windholz (1976) and Weast (1979)
(a) *Description*: White, granular powder
(b) *Melting-point*: Decomposes at red heat
(c) *Density*: 2.679
(d) *Refractive index*: 1.312 (conditions not given)
(e) *Solubility*: 6.52 g/l water at 17°, 24.6 g/l water at 100°; insoluble in ethanol

1.4 Technical products and impurities

Sodium silicofluoride is available in the US as white crystals containing a minimum of 98% active ingredient and maxima of 0.5% moisture, 0.5% insoluble matter and 500 mg/kg heavy metals (as lead) (Chemtech Industries, Inc., undated b).

Sodium silicofluoride is also available in western Europe and Japan, where it must contain at least 98% active ingredient.

FLUORSPAR

1.1 Synonyms and trade names
Chem. Abstr. Services Reg. No.: 14542-23-5
Chem. Abstr. Name and IUPAC Systematic Name: Fluorite [CaF_2]
Synonyms: Acid-spar; calcium difluoride; calcium fluoride; liparite; met-spar
Trade Name: Irtran 3

1.2 Structural and molecular formulae and molecular weight

CaF_2 Mol. wt: 78.1

1.3 Chemical and physical properties of the pure substance

From Windholz (1976), unless otherwise specified
(a) *Description*: White powder or cubic crystals

(b) Boiling-point: 2500°
(c) Melting-point: 1403°
(d) Density: 3.18
(e) Refractive index: 1.433-1.435 (conditions not given) (Quan, 1978a)
(f) Solubility: Practically insoluble in water

1.4 Technical products and impurities

Three principal grades of fluorspar are available in the US. The specifications (calculated on an anhydrous basis) are as follows: acid-grade, minimum of 97% purity, 1.0-1.5% silica, 0.03-0.10% sulphide or free sulphur and a maximal moisture content of 0.1%; ceramic-grade, 85-97% pure, 2.5-3.0% silica, 1.0-1.5% calcite ($CaCO_3$), 0.12% ferric oxide and trace amounts of lead and zinc; metallurgical-grade, 60-85% pure, 0.3% sulphide or free sulphur, 2500-5000 mg/kg lead and minor amounts of phosphorus (Quan, 1978a). Optical grades for special glasses and for growing single crystals contain up to 99.99% calcium fluoride (Gall, 1980a). In countries other than the US, metallurgical-grade fluorspar contains a minimum of 80% calcium fluoride and a maximum of 15% silica (Quan, 1978a). In Japan, fluorspar must contain a minimum of 98% calcium fluoride.

STANNOUS FLUORIDE

1.1 Synonyms and trade names
Chem. Abstr. Services Reg. No.: 7783-47-3
Chem. Abstr. Name: Tin fluoride [SnF_2]
IUPAC Systematic Name: Tin fluoride
Synonyms: Fluoristan; tin bifluoride; tin difluoride
Trade names: Aim; Cap-Tin Mouthrinse; Crest; Gel-Kam; Iradicar SnF_2; Iradicar Stannous
  Fluoride; King's Gel-Tin; Stancare; Stanide

1.2 Structural and molecular formulae and molecular weight

SnF_2 Mol. wt: 156.7

1.3 Chemical and physical properties of the pure substance

From Lindahl & Meshri (1980a), unless otherwise specified
- (a) *Description*: White crystals
- (b) *Boiling-point*: 850°
- (c) *Melting-point*: 215°
- (d) *Density*: d^{25} 4.57 (Windholz, 1976)
- (e) *Solubility*: Readily soluble in water (30-39% at 20°) and anhydrous hydrogen fluoride (72-82% at 20°); practically insoluble in chloroform, diethyl ether and ethanol (Hawley, 1977)
- (f) *Stability*: Forms an oxyfluoride, $SnOF_2$, on exposure to air (Windholz, 1976); oxidizes and hydrolyses in aqueous solution on standing

1.4 Technical products and impurities

Stannous fluoride is available in the US as a USP grade, containing a minimum of 71.2% stannous tin and 22.3-25.5% fluoride on an anhydrous basis, and maxima of 0.5% weight loss on drying, 0.2% insoluble substances and 50 mg/kg antimony (US Pharmacopeial Convention, Inc., 1980).

SODIUM MONOFLUOROPHOSPHATE

1.1 Synonyms and trade names

Chem. Abstr. Services Reg. No.: 10163-15-2
Chem. Abstr. Name: Phosphorofluoridic acid, disodium salt
IUPAC Systematic Name: Disodium phosphorofluoridate
Synonyms: Disodium fluorophosphate; disodium monofluorophosphate; sodium fluorophosphate; sodium phosphorofluoridate
Trade names: Aquafresh; Colgate with MFP fluoride; Macleans Fluoride; MFP

1.2 Structural and molecular formulae and molecular weight

Na_2PO_3F Mol. wt: 144.0

1.3 Chemical and physical properties of the pure substance

From Pennwalt Corporation (undated)

(a) *Description*: White, finely divided solid
(b) *Melting-point*: Approximately 625°
(c) *Solubility*: 420 g/l water at 25°
(d) *Stability*: Dilute solutions are stable indefinitely, in the absence of acid or of metal ions which give insoluble fluorides.

1.4 Technical products and impurities

Sodium monofluorophosphate is available in the US as a National Formulary grade, with the following specifications: a minimal purity of 92%, a minimum of 12.7% total fluoride, a maximum of 1.2% fluoride ion, a minimum of 12.1% fluoride combined in PO_3F ion and maxima of 50 mg/kg heavy metals as lead, 3 mg/kg arsenic and 0.2% loss on drying (Pennwalt Corporation, undated).

Sodium monofluorophosphate is available in Japan with a minimum purity of 98%.

2. Production, Use, Occurrence and Analysis

SODIUM FLUORIDE

2.1 Production and use

(a) Production

Sodium fluoride can be prepared by fusing cryolite (Na_3AlF_6) with sodium hydroxide or by the reaction of 40% hydrofluoric acid with sodium hydroxide or sodium carbonate (Windholz, 1976). It is normally manufactured in commercial quantities by the latter method (Wachter, 1980); however, some is produced as a by-product of phosphoric acid production, and it is believed to be made in Japan by the reaction of sodium hydroxide with sodium silicofluoride.

Sodium fluoride is produced commercially in the US by four companies. Separate production data are not published, but it is estimated that 5-10 million kg are produced annually; two companies reported production in 1977 of 0.45-4.54 million kg each (US Environmental Protection Agency, 1980a). On the basis of the amount of hydrofluoric acid used for the production of fluoride salts, US production of sodium fluoride in 1976 is estimated to have been less than 6 million kg. Separate production data were last reported in 1972, when total production of sodium fluoride amounted to 5.6 million kg (US Bureau of the Census, 1976). Separate data on imports and exports of sodium fluoride are not published, but the quantities are believed to be negligible.

Sodium fluoride is believed to be produced by two companies each in France, the Federal Republic of Germany, Italy, Spain and the UK, and by one company in Belgium.

Commercial production of sodium fluoride in Japan was started in 1948; total production in 1979 by five companies is estimated to have been 4 million kg. Exports were approximately 1.6 million kg.

(b) Use

Sodium fluoride has been reported to be used as: a fluoridation agent in drinking-water; a flux in the manufacture of rimmed steel, aluminium and magnesium; a fungicide; a glass-frosting agent; a component of glues and adhesives; an insecticide; an agent in ore flotation; a pesticide; a stainless-steel pickling agent; a toothpaste ingredient; a component of vitreous enamels; and a component of wood preservatives (Chemtech Industries, Inc., undated a). Windholz (1976) reported that it has also been used: as a constituent of glass mixes; in electroplating; in heat-treating salt compositions; for disinfecting fermentation apparatuses in breweries and distilleries; in the manufacture of coated paper; in dental laboratories; in the removal of hydrogen fluoride from exhaust gases; and in veterinary medicine as an anthelmintic, pediculicide and acaricide. Hawley (1977) reported that it is also used in chemical cleaning and as the single crystal windows used in ultra-violet and infra-red radiation detection systems. It is also a component of laundry sours, used to remove soap prior to bleaching (Wachter, 1980).

It has been the object of clinical studies for the treatment of osteolytic lesions of bone in multiple myeloma (Gloor *et al.*, 1980) and primary osteoporosis (Riggs *et al.*, 1980).

Sodium fluoride was the first compound used when municipal drinking-water fluoridation was approved in the US in 1950; and this application is still believed to account for much of the total US consumption, although less expensive fluorides are now employed more widely in this way. Very small communities may still be using it because it is easy to handle. For this purpose, sufficient sodium fluoride is added to the drinking-water to bring the fluoride content to approximately 1 mg/l, which is considered optimal for the prevention of dental decay (Lindahl & Meshri, 1980a). Where fluoridation of drinking-water is not feasible or has not been implemented, sodium fluoride can be prescribed as a fluoride supplement to provide 1 mg/l fluoride in children's drinking-water or in water used to prepare their food.

Sodium fluoride (as an aqueous solution or as a solution or gel in combination with phosphoric acid) is also applied topically to children's teeth (American Dental Association, 1979). One major fluoride toothpaste sold in the US contains sodium fluoride.

The main use of fluosilicic acid in the US is believed to be in the manufacture of its sodium salt, sodium silicofluoride (discussed in detail below), which is used to produce synthetic cryolite. Fluosilicic acid is also used to make aluminium trifluoride by reaction with alumina trihydrate; this is one of its major uses in western Europe. Both cryolite and aluminium trifluoride are used as components of the electrolyte used in the reduction of alumina to aluminium; approximately 54% of the total US fluosilicic acid shipments in 1979 was used to produce aluminium chemicals (US Bureau of Mines, 1980a), and 30 million kg fluosilicic acid were used by the US aluminium industry in 1975 and 1976 (Coope, 1978).

Fluoridation of drinking-water is believed to be the second largest use of fluosilicic acid in the US and western Europe. In 1979, 46% of the total US fluosilicic acid shipments was used to produce water-fluoridation chemicals (US Bureau of Mines, 1980a). Fluosilicic acid is widely used in major US cities because of its low cost and its availability in large quantities as a by-product of the manufacture of phosphates. A 20-25% aqueous solution is metered into the water to provide the desired fluoride level.

The Commission of the European Communities (1978) requires that fluosilicic acid be labelled as causing burns and being toxic to the eyes.

The established permissible levels of inorganic fluorides in the working environment in various countries are summarized in Table 1.

SODIUM SILICOFLUORIDE

2.1 Production and use

(a) Production

Sodium silicofluoride can be prepared by the reaction of fluosilicic acid with sodium carbonate or sodium chloride. The principal method used for its commercial production is believed to be recovery of the precipitate which forms when sodium chloride is added to the aqueous fluosilicic acid produced from off-gases evolved during wet-process production of phosphoric acid or normal superphosphate.

Sodium silicofluoride is produced commercially in the US by at least six companies, and many of the US producers of fluosilicic acid can produce the sodium salt. Total US production of sodium silicofluoride in 1978 amounted to 54.2 million kg (US Bureau of the Census, 1979); imports of sodium silicofluoride in 1979 were about 3.33 million kg (US Bureau of the Census, 1980).

It is believed to be produced by two companies each in France, Italy, The Netherlands and Spain, and by one company each in Belgium, Denmark and Ireland.

Commercial production of sodium silicofluoride in Japan started in 1948; total production in 1979 by four companies was about 16.5 million kg, much of which is believed to have been exported.

(b) Use

The major uses of sodium silicofluoride have been reported to be as an intermediate in the production of synthetic cryolite and as a drinking-water additive. It is also used as a component of casting sands; to opaque vitreous enamels, such as china and porcelain; as a component of architectural, fibre, opal and window glass; as a gelling agent for natural rubber latex in the manufacture of foam rubber; to treat hides and skins before tanning (see IARC, 1981); in the production of zirconia pigments; in the extraction of beryllium; as an insecticide, fungicide, bactericide and rodenticide; as a pediculicide in veterinary medicine; as a rodent repellant in paperboard containers; as a slime-control agent in paper manufacture (see IARC, 1981); as a soil sterilant; as a preservative in glue, starch sizes, leather and wood (see IARC, 1981); as a mothproofing agent; and as a component of laundry sours (Olin Corporation, undated; The Tennessee Corporation, undated; Byrns, 1966; Bill Communications, Inc., 1975; Hawley, 1977; Windholz, 1976).

Sodium silicofluoride is widely used as a fluoridating agent for municipal drinking-water in both the US and western Europe, principally in medium-sized cities which do not wish to invest in the liquid metering equipment necessary for addition of fluosilicic acid.

Its use as an insecticide comprises application on lawns, shrubs, trees and ornamental plants; in houses, medical facilities, schools and commercial establishments (excluding those in which food is prepared or processed); and on specialized products such as carpeting, feathers, furs and woollen fabrics (US Environmental Protection Agency, 1973b, 1974). It is also a component of fumigant gas cartridges for killing rodents and other mammals in burrows or enclosed buildings (US Environmental Protection Agency, 1972).

In western Europe, sodium silicofluoride is used primarily for fluoridation of water and in the rubber industry; and in Japan, its main use is in the manufacture of synthetic cryolite.

The established permissible levels of inorganic fluorides in the working environment in various countries are summarized in Table 1. In Czechoslovakia, a separate regulation for sodium silicofluoride was established in 1976, limiting exposure to a time-weighted average of 1.5 mg/m^3 and to a ceiling level of 3 mg/m^3 (International Labour Office, 1977).

The Commission of the European Communities (1978) requires that sodium silicofluoride be labelled as toxic by inhalation, in contact with skin or if swallowed.

FLUORSPAR

2.1 Production and use

(a) Production

Fluorspar, the naturally occurring mineral form of calcium fluoride, was found in 1529 to lower the melting-point of minerals and to reduce the viscosity of slags. Significant mining of fluorspar started in England in about 1775 and in the US after 1820 (Gall, 1980a). It can be obtained by mining, by separation of heavy media and by froth flotation; it can also be prepared by the reaction of calcium carbonate with hydrofluoric acid (Windholz, 1976) or by reaction of a soluble calcium salt with sodium fluoride (Hawley, 1977). Very pure calcium fluoride is made by the former reaction. A method has been developed to produce synthetic fluorspar from fluorosilicic acid (Gall, 1980a); however, it is not believed to be in commercial use.

Total US domestic shipments of finished fluorspar by five companies amounted to 99.2 million kg in 1979; exports (96% to Canada) were 13.1 million kg, and imports were 927 million kg (more than 87% from Mexico and the Republic of South Africa) (US Bureau of Mines, 1980a).

Fluorspar is believed to be produced for sale by four companies each in France and the UK, by three companies in the Federal Republic of Germany, by two in Spain, and by an unknown number of companies in Italy and Sweden. Total production in the UK in 1978 was 189 million kg (Chemical Industries Association, Ltd, 1980).

Fluorspar is not produced commercially in Japan. It was first imported in 1948, and imports in 1979 are estimated to have been 467 million kg.

Total world production of fluorspar in 1979 is reported to have amounted to 4866 million kg (US Bureau of Mines, 1980a). Preliminary data on production (in 1977) and exportation (in 1976) in the major producing countries are given in Table 2.

(b) Use

In 1979, about 1030 million kg fluorspar were used in the US, as follows: 52% for production of hydrofluoric acid; 46% as a flux in steel manufacture (approximately 30% in basic oxygen furnaces and 8% each in open-hearth and electric furnaces), approximately 1% in the manufacture of glass (including fibreglass); 1% in iron and steel castings; and the remainder in primary aluminium and magnesium manufacture, enamel and pottery, welding-rod coatings, and other uses (US Bureau of Mines, 1980a).

Hydrofluoric acid is produced by the reaction of sulphuric acid with fluorspar It is used primarily in the manufacture of chlorofluorocarbons (principally fluorocarbons 12 and 22); in

Table 2. Production (1977) and exportation (1976) of fluorspar[a]

Country	Million kg Production	Exportation
China, People's Republic of	349[b]	141[b]
France	370	92
Italy	186	105
Mexico	955	576[b]
South Africa, Republic of	351	222[b]
Spain	399	226
Thailand	223	284
UK	200	35
USSR	501[b]	–
Mongolia	320[b]	

[a] From US Bureau of Mines (1980b)
[b] Estimate

the production of aluminium fluoride and synthetic cryolite; as a catalyst in the alkylation of isobutane with propylene or butylene to produce high-octane petrol components; and in the pickling of steel (Gall, 1980b).

In steel-making, fluorspar (principally metallurgical grade) is used to promote increased fluidity. Although the level added varies widely, the overall average in recent years has been 4-5 kg per 1000 kg of steel produced (Gall, 1980a).

It is also used in cements, dentifrices, phosphors, paint pigments and as a catalyst for wood preservatives (Hawley, 1977); in infra-red and ultra-violet transmission windows and lenses; in optical components of high-energy laser systems (Gall, 1980a); in water fluoridation; as a catalyst in dehydrations and dehydrogenations; and in high-temperature dry-film lubricants (Windholz, 1976).

Although methods of using fluorspar for water fluoridation by dissolving it in aluminium sulphate solutions have been developed, Maier (1970) reported that it was being used as such on a significant scale only in Brazil.

In Japan, fluorspar is used principally in the manufacture of hydrofluoric acid.

The established permissible levels of inorganic fluorides in the working environment in various countries are summarized in Table 1.

The US Environmental Protection Agency (1980c) has established a regulation requiring that there be no discharge to navigable waters in the US of wastewater pollutants from the processing of fluorspar by operations other than heavy media separation or flotation processes.

The US Food & Drug Administration (1979) proposed that nonprescription vitamin and mineral drug products must not contain fluorspar as a mineral supplement, on the basis that its use in such products is not justified since the small amounts of fluoride needed are provided by the diet.

Fluorspar is provisionally allowed for use in oral hygiene products in the European Communities, provided that, when mixed with other permitted fluorine compounds, the total fluoride concentration does not exceed 0.15% (Commission of the European Communities, 1976).

STANNOUS FLUORIDE

2.1 Production and use

(a) Production

Stannous fluoride was probably first prepared by Scheele in 1771; it was described by Guy Lussac and Thenard in 1809. It can be prepared by the reaction of aqueous hydrofluoric acid with stannous oxide or by the reaction of anhydrous hydrogen fluoride with metallic tin. Both methods are used for its commercial production (Lindahl & Meshri, 1980b).

Stannous fluoride is produced commercially in the US by two companies, in comparatively small amounts (Anon., 1978). It is believed to be produced by one company in France, one in the Federal Republic of Germany and one in The Netherlands. It is not produced commercially in Japan.

(b) Use

Stannous fluoride is used in the US exclusively as an anticaries ingredient in dentifrices and in topical applications for children's teeth (Anon., 1978). It was the anticaries agent in the first fluoride toothpaste marketed in the US in the 1950s, and today two major US fluoride toothpastes contain it. Available topical application products take the form of tablets, capsules and gels.

The US Food & Drug Administration (1980b) proposed that stannous fluoride be generally recognized as safe and effective in topical anticaries products when marketed at a level of 0.4% in dentifrices and anhydrous glycerine gels, and in a stable form for mixing with water to produce a dental rinse containing 0.1% stannous fluoride.

Stannous fluoride is provisionally allowed for use in oral hygiene products in the European Communities, provided that, when mixed with other permitted fluorine compounds, the total fluoride concentration does not exceed 0.15% (Commission of the European Communities, 1976).

SODIUM MONOFLUOROPHOSPHATE

2.1 Production and use

(a) Production

Sodium monofluorophosphate can be prepared by the reaction of monofluorophosphoric acid with sodium hydroxide; but it is produced commercially by the fusion of sodium metaphosphate and sodium fluoride (Lindahl, 1980).

It is produced in the US by only one company, so production data are not available (see preamble, p. 20); it is believed to be produced by one company in France and one in the UK. Commercial production of sodium monofluorophosphate in Japan was started in 1968; production in 1979 by the only manufacturer is estimated to have been 45 thousand kg, almost all of which was used in Japan.

(b) Use

Sodium monofluorophosphate is believed to be used exclusively as an anticaries ingredient in dentifrices for children's teeth. It is present in three major US fluoride toothpastes.

The US Food & Drug Administration (1980b) proposed that sodium monofluorophosphate be generally recognized as safe and effective in topical anticaries products when present at a level of 0.76% in dentifrices.

Sodium monofluorophosphate is provisionally allowed for use in oral hygiene products in the European Communities, provided that, when mixed with other permitted fluorine compounds, the total fluoride concentration does not exceed 0.15% (Commission of the European Communities, 1976).

2.2 Occurrence[1]

(a) Natural occurrence

Fluoride is widely distributed in the earth's crust and is believed to constitute 0.06-0.09% by weight of the upper lithosphere. Fluorspar is one of the chief fluorine-containing minerals, along with cryolite (Na_3AlF_6) and fluorapatite ($[Ca_3(PO_4)_2]_3CaF_2$) (National Research Council, 1971). Fluorspar deposits occur on every continent; world reserves have been estimated at 199.8-324 million tonnes (Quan, 1978b).

(b) Occupational exposure

It has been estimated that 350 thousand workers are potentially exposed to fluorides in the US (National Institute for Occupational Safety & Health, 1975).

Concentrations of inorganic fluoride found in various occupational environments and the corresponding concentrations found in the urine of workers in those environments are summarized in Table 3.

Table 3. Concentrations of fluoride in occupational environments and in the urine of workers[a]

Occupational environment	Concentration Air (mg/m^3)	Urine (mg/l)	Year of report
Aluminium plants	2.5-3.3	–	1944
Plant 1			
– furnace rooms	0.14-3.43	0.5-23.3 mg/24 h	1949
– other	–	1.83 mg/24 h (male) 1.58 mg/24 h (female)	1949
Plant 2			
– furnace rooms	0.34-0.91	–	1949
– other	0.015-0.141	–	1949

[1] In keeping with the approach taken in the review of the world literature on fluorides and their environmental effects by the National Research Council (1971), the term 'fluoride' is used in this section as a general term where exact differentiation between ionic and molecular forms or between gaseous and particulate forms is uncertain or unnecessary. Where specific forms are identified in the sources used, they are so designated.

Occupational environment	Concentration		Year of report
	Air (mg/m^3)	Urine (mg/l)	
Plant 3	0.6-2.35	-	1959
Plant 4	<0.34	2.0-2.8 (before shift) 3.2-6.5 (after shift)	1972
Plant 5	2.4-6.0^b	8.7-9.6	1972
Plant 6	-	4.63 (mean)^c	1972
Brazing			
Unspecified	0.24	3.1	1959
Silver - pipe shop	0.02-0.16	-	1971
- ships	0.28-0.8	-	1971
Chemical plant	0.063-8.2 (hydrogen fluoride)	0.33-4.48 (before shift) 0.95-26.6 (after shift)	1967
	1.03 (hydrogen fluoride)	0.33-3.52 (before shift) 1.55-8.8 (after shift)	1967
	0.18 (particulates)		
Cryolite processing	15-20^d	2.09-2.54 mg/24 h	1937^e
	-	2.41-43.41	1941
Fluorspar mines	0-1.92	-	1964
Inorganic fluoride manufacture	-	0.7-16.38	1951
Magnesium plants			
Foundry 1	6.0	2.23 (mean)	1944
Foundry 2	3.1-8.4	3.13 (mean)	
Foundry 3	-	6.39 (mean)	
Coke shop	0.143	0.9-4.1	1947
Foundry	0.286-6.37	0.5-7.5	
Furnace	0.314	1.0-3.4	
	0.48	3.9	1959
Phosphate processing	0.9-2.2	7.86	1954
	7.04	7.35	
	5.08-6.16	6.47	
	11.7	4.08	

Occupational environment	Concentration		Year of report
	Air (mg/m³)	Urine (mg/l)	
	4.7-20.0	3.67	
	4.7	2.9	
	15.8	2.89	
	1.2-24.7	2.86	
	1.6-3.2	3-4 (beginning of work week)	1960
		8-9 (end of work week)	
Fertilizer plant	0.15-0.62 (particulate)	1-9.6	1978[f]
	0.04-0.17 (gas)		
Pottery manufacture	≤3.5	–	1963
Silver soldering	0.17	2.1	1959
Steel plants			
Open-hearth furnaces	0.09-23	–	1948
Welding			
Unspecified	0.47	2.3	1959
Arc	0.1-10.0	0.5-4.8	1968
Automatic	1.4 (gas)	0.1-3.9	1969
	0.32 (particulate)		
Manual	2.29 (gas)	0.2-8.9	
	0.36 (particulate)		
Not specified	0-2.4	1.0 (before shift)	1952
		3.2 (after shift)	
	2.5-4.9	0.8-3.7 (before shift)	
		5.1-10.4 (after shift)	
	5.0-9.9	1.2-3.9 (before shift)	
		5.8-10.0 (after shift)	
	>10.0	3.5-5.6 (before shift)	
		10.0-19.0 (after shift)	

[a] From National Research Council (1971); National Institute for Occupational Safety & Health (1975)
[b] Samples taken in 1945-1946; time-weighted average fluoride exposures
[c] Samples taken in 1960-1962
[d] Based on 54.3% fluoride content in cryolite
[e] 2.50 mg/kg fluoride also found in teeth of workers
[f] From Fabbri et al. (1978)

(c) Air

Gaseous and solid fluorides are discharged into the atmosphere by active volcanoes and fumaroles. Fluoride emissions have been identified from the following volcanoes: Asama and Mihara (Japan), Oyama (Miyake Island), Satsuma (Iwo Jima), Klyuchevskoi (USSR) and the Valley of Ten Thousand Smokes (Alaska). Sodium silicofluoride and other fluorides have been determined in fumes from Vesuvius and Vulcano in Italy (National Research Council, 1971). More recently, samples of ash from the eruption of Mount St Helens, Washington, in May 1980 have been reported to contain up to 113 ppm (90 mg/m^3) of acid-labile fluoride (Taves, 1980).

Fluorides can be released into the atmosphere as a result of hydrogen fluoride manufacture, petroleum refining, magnesium and aluminium founding and metal brazing, from atomic energy installations using uranium hexafluoride and during test firing of rocket engines using fluorine or fluorine compounds as oxidizers (National Research Council, 1971). Major industrial sources of atmospheric fluoride emissions and their estimated magnitudes in the US in 1968 are summarized in Table 4.

Table 4. Estimated total fluoride emissions from major industrial sources in the US in 1968[a]

Source	Atmospheric emissions (thousand kg/yr)
Manufacture of:	
normal superphosphate fertilizer	8730
wet-process phosphoric acid	2700
triple superphosphate fertilizer	270
diammonium phosphate fertilizer	90
elemental phosphorus	4950
phosphate animal feed	90
aluminium	14 400
steel (open-hearth furnace)	15 120
steel (basic-oxygen furnace)	7560
steel (electric furnace)	13 410
brick and tile products	16 650
glass and frit	2430
Welding operations	2430
Nonferrous-metal foundries	3600
Combustion of coal	14 400
Total	106 830

[a] From National Research Council (1971)

The fluoride content of particulate matter collected from the air of US cities in 1959 was 0-0.08 $\mu g/m^3$ in nine cities, 0.16-0.23 $\mu g/m^3$ in three and 0.3-0.39 $\mu g/m^3$ in three. The concentration in Los Angeles was approximately 0.86 $\mu g/m^3$ (National Institute for Occupational Safety & Health, 1975).

Measurements taken in 1966 and 1967 in the US revealed that 97% of nonurban air samples contained no detectable fluoride (the highest concentration was 0.16 $\mu g/m^3$). In 87% of urban air samples, fluoride was present in concentrations of <0.05 $\mu g/m^3$, and concentrations in the remainder of the samples were in the range of 0.05-1.89 $\mu g/m^3$ (National Research Council, 1971).

Air samples taken in 1959 near a US phosphate mining and manufacturing complex contained 0-200 $\mu g/m^3$ fluoride; concentrations of 0-33.7 $\mu g/m^3$ were found in the same region during 1964-1969 (National Research Council, 1971). During 1949 and 1950, airborne fluoride concentrations of 0-280 $\mu g/m^3$ were found in an industrial area in which an aluminium reduction plant, an oil refinery and a smelter for magnesium and ferroalloys were situated. Air concentrations near a US aluminium factory were 0-12.9 $\mu g/m^3$ in 1963 and 0-2.5 $\mu g/m^3$ in 1964. The maximum fluoride concentrations found in the same area during April of 1965, 1966 and 1967 were 3.31, 4.77 and 4.65 $\mu g/m^3$, respectively (National Research Council, 1971).

Ninety percent of air samples taken in an industrial city in the Federal Republic of Germany during 1965 and 1966 contained concentrations of 0.5-3.8 $\mu g/m^3$ fluoride (National Research Council, 1971). In 1945, airborne fluoride concentrations of 0.02-0.22 $\mu g/m^3$ were measured outside an aluminium factory in Scotland (National Institute for Occupational Safety & Health, 1975).

(d) Water and sediments

Fluorides occur naturally in most waters, usually at levels of less than 1 mg/l. Certain spring waters contain elevated levels of fluoride derived from mineral deposits; and natural concentrations in some parts of the world may be as high at 10 mg/l and sometimes exceed 20 mg/l (National Research Council, 1971). The following are the highest levels found in water in a number of countries: Kenya, 2800 mg/l; Tanganyika, 95 mg/l; South Africa, 53 mg/l; Czechoslovakia, 28 mg/l; and Portugal, 22.8 mg/l (WHO, 1970).

US surface waters contain an average of 0.25 mg/l fluoride, although rivers have been reported to contain 0-6.5 mg/l and lakes up to 1627 mg/l. Various ground waters contain 0-35.1 mg/l, and US ground water contains an average of 0.4 mg/l (National Research Council, 1980). Pacific and Atlantic sea waters contain 1.2 mg/l fluoride. Water that has been trapped in sediments since their deposition (connate) and related waters have average fluoride concentrations of 2.7 mg/l, and thermal waters associated with volcanoes and epithermal mineral deposits have an average of 5.4 mg/l (Shawe, 1976)

In 1959, drinking-water from 49 US states was found to contain concentrations of 0-33.5 mg/l fluoride, although the highest values were rarely encountered (National Institute for Occupational Safety & Health, 1975). In 1962, the fluoride contents of the water supplies of the 100 largest US cities were measured: 92% contained <1 mg/l; the median level was 0.4 mg/l; and the maximum level was 7.0 mg/l (National Research Council, 1980). In 1970, the drinking-water supplies of 2630 US communities had natural fluoride concentrations greater than 0.7 mg/l. In 524 of those communities, fluoride concentrations were higher than 2 mg/l.

The water supplies of another 969 community contained 0.2-4.4 mg/l (National Research Council, 1977); most of the supplies that had not been fluoridated contained less than 0.3 mg/l (National Research Council, 1971).

In 1971, the following concentrations were reported in drinking-water in several countries: Canada and the US, 0.1 mg/l; UK and Newfoundland, trace amounts; Japan, 0.01-0.08 mg/l ingested daily from drinking-water; Norway, 0.01-0.07 mg/l; and the USSR, 0.2-0.3 mg/l (National Research Council, 1971). Drinking-water in an area of endemic fluorosis in India was reported to contain up to 16.2 mg/l fluoride (Singh *et al.*, 1963).

Ocean sediments and shales contain 700-900 mg/kg fluoride, and marine phosphates have an average of more than 30000 mg/kg (Shawe, 1976).

(e) Soil and plants

US soils have been reported to contain an average of less than 300 mg/kg fluoride (Shawe, 1976). Fluoride concentrations in soils increase with depth: in a study of 30 different US soils, 20-500 mg/kg were found at depths of 0-7.5 cm and levels of 20-1620 mg/kg at depths of 0-30 cm. Idaho and Tennessee soils had unusually high fluoride concentrations: 3870 mg/kg and 8300 mg/kg, respectively (National Research Council, 1971). In Tennessee soils, fluoride concentrations ranged from 80 mg/kg in a Fullerton silt loam to 338 mg/kg in a Montevallo silt loam (Brewer, 1966). Fluoride concentrations in soil in other countries have been reported as follows: USSR, 30-320 mg/kg; New Zealand, 68-540 mg/kg (WHO, 1970).

The following fluoride contents have been reported for micaceous clays: hydrous mica, 5800 mg/kg; muscovite, 400 mg/kg; Ordovician bentonite, 4500-7400 mg/kg; and sericite, 300-1800 mg/kg. Samples of 23 New Zealand soils had fluoride contents ranging from 68 mg/kg in Pinahi coarse sand to 540 mg/kg in Motatau clay (Brewer, 1966).

Plants can take up fluoride from soil, water and air; natural concentrations in the foliage of most plants range from 2-20 mg/kg (Brewer, 1966). Fluoride concentrations in 107 US alfalfa samples from areas assumed to be free of industrial pollution ranged from 0.8-36.5 mg/kg, with a mean of 3.6 mg/kg (National Research Council, 1971).

Although some plants, such as tea, accumulate fluoride from the soil and may contain as much as 400 mg/kg, there does not appear to be a relationship between the fluoride content of most plants and that of the soil; an association has, however, been established with increased concentrations of soluble fluoride in water. In addition, plants grown in acidic soil generally have higher fluoride concentrations (National Research Council, 1971).

Plants may contain fluorides as a result of industrial air pollution. The average fluoride contents of foliage from cherry and peach trees growing near an aluminium factory in the US were up to 13 mg/kg before the factory began operating, but increased to 65 mg/kg (cherry) and 76 mg/kg (peach) after the factory opened. At that time, alfalfa samples from the same area contained 6.5-166 mg/kg and pine needles 24-104 mg/kg. Concentrations of fluoride in air and foliage decreased with distance from the factory (National Research Council, 1971).

(f) Food, beverages and animal feeds

Water is the principal source of fluoride in the average US diet; however, it is also present in foods, and some foods cooked in fluoridated water may undergo an increase in fluoride content (National Research Council, 1971). Average daily fluoride intake from food in one US city was 1.6-1.9 mg when the water used for cooking was fluoridated and approximately 0.9 mg when it was not (National Research Council, 1980).

Table 5 summarizes the daily dietary intake of fluoride from food and water in several countries.

Table 5. Daily dietary fluoride[a]

Country	Fluoride ingested in food and water (mg)
Canada	0.18-0.3
Newfoundland	2.74[b]
Japan	0.47-2.66[c]
Norway	0.22-0.31
Sweden	0.9
Switzerland	0.5[d]

Country	Fluoride ingested in food and water (mg)
UK	0.3-0.5
US	0.2-0.3[d]
	0.34-0.80
	1.73-3.44[d]
	0.53-1.27[e]
USSR	0.6-1.2

[a] From National Research Council (1971, 1980)
[b] Tea accounted for 1 mg.
[c] Including 0.07-0.86 mg from green tea
[d] Exclusive of that in drinking-water
[e] From drinking-water only

A study of 16-19-year-old US males in 1971 reported a daily fluoride intake of 2.0-2.3 mg, while a study of children aged 4-18 years in 1973 reported an average fluoride intake of 0.3 mg/day. The fluoride content of hospital-prepared food in 16 US fluoridated communities in 1974 ranged from 1.6-3.4 mg/day, while in nonfluoridated communities the reported intake was 0.8-1.0 mg/day. A second study in 1974 reported an average fluoride intake over six years in a fluoridated city to be 2.0 mg/day (National Research Council, 1977).

The fluoride concentrations of major fresh foods in the US are summarized in Table 6.

Table 6. Fluoride concentrations in fresh foods[a]

Food	Concentration (mg/kg)	
	Data to 1959	Recent data
Meats	0.01-7.7	0.14-2
Eggs	0.00-2.05	—
Butter	0.04-1.5	—
Cheese	0.13-1.62	—

Food	Concentration (mg/kg)	
	Data to 1959	Recent data
Sugar	0.10–0.32	–
Fish	<0.10–24	≥1.0
Sardines	–	8–40
Shrimp	–	40 max
Cod, haddock, herring	–	50
Fish protein concentrate	–	2–5
		20–370
Citrus fruits	0.04–0.36	0.07–0.17
Noncitrus fruits	0.02–1.32	0.03–0.84
Cereals and cereal products	<0.10 – 20	0.18–2.8
Vegetables and tubers	0.10–3.0	0.02–0.9

[a] From National Research Council (1971, 1980)

Fluoride concentrations in beverages in the US are summarized in Table 7.

Table 7. Fluoride concentrations in beverages[a]

Beverage	Concentration (mg/l)
Wine	0.0 – 6.3
Beer	0.15 – 0.86
Tea infusion	0.1 – 2.0
Instant (solution)	0.2
Coffee beans	0.2 – 1.6
Instant (powder)	1.7

Beverage	Concentration (mg/l)
Milk	0.04 - 0.55
Coca-Cola	0.07
Orange juice	0.0 - 0.05

[a] From National Research Council (1971)

Most reviews indicate that fluoride concentrations in natural forage range from 5-10 mg/kg. However, a study of mixed forages in 1952 reported concentrations of 15-25 mg/kg. Fluoride concentrations in five pastures were reported to be 25-292, 33-174, 8-42, 12-117 and 7-89 mg/kg. In general, fluoride content is lower during early summer when the forage is growing rapidly, and rises during the autumn. In 1969, 90% of 168 US dairy feed samples contained less than 30 mg/kg fluoride, although some samples contained more than 200 mg/kg (National Research Council, 1971). Although most commercial feed-grade phosphates are defluorinated, they do contain some fluorides and can contribute from 3-20 mg/kg to a ration. Higher fluoride contents result from the use of un-defluorinated fertilizer-grade chemicals in animal feeds, which may contribute 81-710 mg/kg fluoride to the diet ration (National Research Council, 1974).

(g) Animals

Fluoride is concentrated in the calcified skeletal and dental tissues of animals as the mineral fluorapatite ($[Ca_3(PO_4)_2]_3 \cdot CaF_2$). It accumulates in those tissues over time, even when the levels of fluoride intake are low; in a number of studies, skeletal retention of fluoride has been found to be proportionate to the amount ingested (National Research Council, 1974).

It has been reported that bovine and rat sera contain 0.01-0.04 mg/l ionic (unbound) fluoride, representing 15-70% of the total plasma fluoride. Concentrations in the urine of cattle on a normal diet are usually less than 6 mg/kg (National Research Council, 1974). Honey-bees were found to contain 1 mg/kg fluoride (WHO, 1970).

(h) Human tissues and secretions

Fluoride is taken in *via* the diet, drinking-water, ambient air and fluoride-containing drugs or dentifrices.

Table 8 and Table 9 summarize fluoride concentrations found in human tissues after various exposures.

Table 8. Fluoride concentrations in human tissues after exposure to high levels in air[a]

Tissue	Fluoride concentration (mg/kg fresh weight)			
	Normal	Sodium fluoride-poisoning fatalities	Exposure in industrial areas	
			Long-term[b]	Short-term[c]
Blood	0.27 (4)[d]	9.2 (5)[d]	–	–
Brain	0.53 (4)	2.5 (2)	1.8	1.5
Lung	0.27 (5)	14.0 (2)	3.9	3.5
Heart	0.45 (3)	10.6 (1)	2.5	1.9
Spleen	0.28 (2)	11.8 (1)	1.7	1.8
Liver	0.54 (3)	9.3 (5)	1.6	1.4
Kidney	0.68 (2)	9.0 (3)	2.9	2.9
Thyroid	–	–	5.2	4.0
Aorta	–	–	29.4	28.2
Pancreas	–	–	1.8	1.7

[a] From Gettler & Ellerbrook (1939), WHO (1970) and National Research Council (1971)
[b] People over 15 years of age who had lived in industrial areas for at least 10 years
[c] People under 15 years of age who had lived in industrial areas for less than 10 years
[d] Numbers of subjects in parentheses

Table 9. Fluoride concentrations in human soft tissues after exposure to fluoridated drinking-water[a]

Tissue	Fluoride concentration (mg/kg dry weight) ± SD	
	0 - 1.0 mg/l in water	1.0 - 4.0 mg/l in water
Heart	2.29 ± 0.796	2.78 ± 1.364
Liver	2.34 ± 0.972	2.27 ± 1.109

Tissue	Fluoride concentration (mg/kg dry weight) ± SD	
	0 - 1.0 mg/l in water	1.0 - 4.0 mg/l in water
Lung	5.12 ± 3.104	6.18 ± 3.106
Kidney	3.26 ± 0.968	8.49 ± 6.63
Spleen	4.91 ± 4.143	3.53 ± 1.324
Aorta	41.0 ± 50.27	25.1 ± 19.54

[a] From National Research Council (1971)

Human placental tissue has also been found to contain fluorides: in a study of placental samples from two communities where the water supply contained 0.06 and 1-1.2 mg/l fluoride, mean concentrations of 0.74 and 2.09 mg/kg fluoride were found, respectively (National Institute for Occupational Safety & Health, 1975).

Fluoride is also found in human bone, in concentrations affected by age, sex, fluoride intake and specific bone type. The following ranges of fluoride were found in dried, fat-free bone from people living in three areas using different water supplies: with <0.5 mg/l in the water, <1000-2000 mg/kg of bone; with 0.8 mg/l, <2000-2500 mg/kg; and with 1.9 mg/l, <1000-4000 mg/kg (National Research Council, 1971). Samples of femura from people living in areas where the water supply contained <0.5 mg/l fluoride had levels of <500 mg/kg (age, 20 years) to almost 3000 mg/kg (age, 80-90 years) (WHO, 1970). A study carried out in 1976 reported bone fluoride contents of 1295-5745 mg/kg wet weight in people aged 17-82 years, most of whom lived in fluoridated communities (Cone et al., 1980).

The fluoride contents of the bones of people who were considered to have received exposure to 'excessive' amounts of fluoride (as cryolite dust, excessive fluoride in water or rock phosphate dust) are summarized in Table 10.

Table 10. Fluoride concentrations in the bones of people who had received exposure to 'excessive' amounts of fluoride[a]

Bone	Age of subject (years)	Fluoride (mg/kg in ash)	Normal fluoride content of bones of people of same age (mg/kg) ± SD
Vertebra	22	7500	450 ± 225
Sternum		6900	
Rib	37	20 830	690 ± 345
Vertebra		18 650	670 ± 335
Rib	48	15 900	850 ± 425
Vertebra		19 880	830 ± 415
Rib	52	9900	910 ± 455
Vertebra		9300	890 ± 445
Rib	72	11 200	1200 ± 600
Vertebra		13 100	1180 ± 590

[a] From National Research Council (1971)

Fluoride content of human teeth in relation to the fluoride content of drinking-water is shown in Table 11.

Table 11. Fluoride concentrations in dentine and enamel in relation to level of fluoride ingestion throughout life[a]

Ages of subjects	Fluoride in water (mg/l)	Fluoride concentration in teeth (mg/kg in ash)			
		Enamel surface	Interior	Whole	Dentine (whole)
Adult	0.1	–	–	86	332
	7.6	–	–	658	1968
20-49 years	<0.5	–	–	108	508
20-35 years	1.2	–	–	180	922
20-35 years	1.9	–	–	320	1290
Adult	<0.25	590	80	–	–
	1.4	960	110	–	–
	2.0	1310	270	–	–
20-29 years	0.1	571	48	–	–
	1.0	889	129	–	–
	3.0	1930	152	–	–
	5.0	3370	570	–	–

[a] National Research Council (1971)

The levels of fluoride in urine are directly related to the amount of fluoride in drinking-water. In the US, people drinking water that contains little or no fluoride usually excrete 0.2-0.5 mg/kg in their urine. With a water supply containing 1 mg/l, normal urinary concentrations are 0.5-1.5 mg/l. Younger people, who are actively laying down bone mineral, excrete less fluoride in their urine than do adults: children 5-6 years old were found to excrete 0.16 mg fluoride per 24 h, whereas 10-12-year olds excreted 0.35 mg per 24 h. In another study, fluoride concentrations in adult urine increased from 0.3 mg/l to 0.9-1.0 mg/l about a week after the water supply was fluoridated, whereas the urine of children contained 0.9 mg/l fluoride only three years after fluoridation had been initiated (National Research Council, 1971).

Fluoride is also excreted in human faeces. A person ingesting an average US diet and drinking non-fluoridated water excretes less than 0.2 mg/day in the faeces (National Research Council, 1971).

The fluoride content of total human body ash was found to be 0.74-4.76 g (National Research Council, 1971).

2.3 Analysis

Methods for analysing fluorides in environmental samples, such as plants, animals (soft tissues and body fluids), water, soils, minerals and air, using gravimetric, enzymatic, titrimetric, colorimetric and electrochemical methods have been summarized (National Research Council, 1971; National Institute for Occupational Safety & Health, 1975).

Typical methods for the analysis of total fluoride, sodium fluoride, fluosilicic acid, sodium silicofluoride and fluorspar in various matrices are summarized in Table 12.

Table 12. Methods for the analysis of inorganic fluorides

Sample matrix	Sample preparation	Assay procedure[a]	Limit of detection	Reference
Ores, slags, mould dressings, fluorspars	Fuse in boric acid-sodium carbonate mixture; extract with aluminium chloride-hydrochloric acid solution; add sodium 5-sulphosalicylate buffer	ISE	not given	Boniface & Jenkins (1977)
Steel-making slags	Irradiate	NAA	not given	Chiba (1978)
Toothpaste	Dissolve in acid medium; react sodium fluoride with trimethylchlorosilane to form trimethylfluorosilane; extract with benzene	GC/FID	not given	Todorovic et al. (1977)

Sample matrix	Sample preparation	Assay procedure[a]	Limit of detection	Reference
Sodium silico-fluoride	React with thorium nitrate in presence of sodium alizarine sulphonate	T	not given	Koicheva (1974)
Air	Draw through filter; place filter in micro-diffusion vessel with alcoholic sodium hydroxide and distilled water-silver perchlorate-perchloric acid mixture; bake; wash with water; add acetic acid buffer	ISE[b]	0.016 mg/m^3	National Institute for Occupational Safety & Health (1975)
Water	Add to acetic acid buffer	ISE	not given	Harwood (1969)
Tooth enamel	Embed tooth in resin; bombard with γ-ray emissions	CPA	not given	Baijot-Stroobants & Vreven (1979)
	Polish; embed in resin	EMA	not given	Lyon & Hefferren (1970)
	Immerse in perchloric acid; add sodium 2-(*para*-sulphophenyl-azo)-1,8-dihydroxy-naphthalene	C	not given	Lyon & Hefferren (1970)
Urine	Dilute with acetic acid buffer	ISE[b]	0.02 mg/l	National Institute for Occupational Safety & Health (1975)

[a] Abbreviations: ISE, ion-specific electrode; NAA, neutron activation analysis; GC/FID, gas chromatography with flame ionization detection; T, titration; CPA, charged-particle activation; EMA, electron microprobe analysis; C, colorimetry

[b] Specific to fluoride ion

3. Biological Data Relevant to the Evaluation of Carcinogenic Risk to Humans

3.1 Carcinogenicity studies in animals[1]

Oral administration

Mouse: Groups of 54 male and 54 weanling female Swiss CD1 mice were given 10 mg/l sodium fluoride in doubly deionized drinking-water for life, to give a dose of about 70 μg/day fluorine. An equal number of animals served as matched controls. No fluorine was detected in the diet of the animals. Dead animals were weighed and necropsied, gross lesions were recorded, and visible tumours and tissues were examined histologically. The body weight of males was not affected, but that of females was somewhat increased when compared with the corresponding controls. Males given sodium fluoride survived one to two months longer than controls; the life spans of treated and control female mice were similar. Tumours were observed in 24/71 control and 22/72 treated mice, in similar locations and of similar types (Kanisawa & Schroeder, 1969). [The sexes of the animals in which the tumours occurred were not given.]

A group of 50 female DBA mice, seven to ten weeks of age, were fed 900 mg/kg of diet sodium fluoride until the surviving animals were 97-100 weeks of age. An equal number of mice fed a diet without addition of sodium fluoride served as matched controls and were observed for the same period. The treatment resulted in drastic reductions in body weight. Mammary gland carcinomas occurred in 37/47 controls and in 20/40 treated animals. The mean age at appearance of the first tumour was 71 ± 2.3 weeks in controls and 76 ± 3.5 weeks in treated mice (Tannenbaum & Silverstone, 1949).

Groups of 94 C3H and 46 DBA female mice, four to twelve months of age, were given 0.4, 1.0 or 4.0 mg/l sodium fluoride in distilled drinking-water for seven to twelve months. Groups of 96 C3H and 45 DBA female mice were given pure distilled water and served as matched controls. All animals in this series (both treated and control) were also fed a diet containing 20-38 mg/kg fluorine. Other groups, of 65 and 36 C3H mice and 66 and 66 DBA mice, two to nine months of age, received 1.0 and 10.0 mg/l, respectively, sodium fluoride in distilled water for 10-17 months. Groups of 64 C3H and 66 DBA mice served as matched controls. All animals in this second series were fed a mixed-grain diet containing a negligible amount of fluorine. The percentages of deaths due to mammary gland carcinomas were 54% in the controls and 59% in all the experimental groups combined. Among the mice that received 10.0 mg/l fluoride, 63% died of mammary gland carcinomas, compared with 50% of controls (Taylor, 1954). [The Working Group noted that no distinction with regard to tumour incidence was made between the two strains, and that the studies were, in general, inadequately reported.]

---

[1] The Working Group was aware of a study in progress to assess the carcinogenicity of sodium fluorides in mice and rats by oral and inhalation administration, by injection and by skin painting (Toxicology Information Subcommittee, 1981).

3.2 Other relevant biological data

(a) Experimental systems

Toxic effects

A number of reviews are available on the toxic effects of inorganic fluorides (Eager, 1969; WHO, 1970; Marier & Rose, 1971; Shupe, 1971; Shupe *et al.*, 1972; Taves, 1979).

Data on the acute toxic doses of inorganic fluorides are summarized in Table 13. Acute toxic effects in animals include severe damage to the kidney, gut and liver (Segreto *et al.*, 1961; Taylor *et al.*, 1961a; Shupe, 1971; Lim *et al.*, 1975).

The sustained ingestion of excessive quantities of soluble and insoluble inorganic fluorides induces progressive changes in the teeth and skeleton in all species studied. These changes include dental fluorosis, osteoporosis, osteosclerosis, hyperostoses, osteophytosis and osteomalacia (Shupe *et al.*, 1972). Such effects were seen in cattle whose bones contained more than 5000 mg/kg dry weight fluoride (Marier & Rose, 1971).

The 30-day lethal dose of fluoride in drinking-water of rats was 205 mg/l (Taylor *et al.*, 1961b). Chronic administration to rats of >3.8 mg/kg fluoride (as sodium or stannous fluoride) daily in food and drinking-water resulted in changes in the teeth and liver and structural and functional changes in the kidneys (Jankauskas, 1974; Lim *et al.*, 1975; Roman *et al.*, 1977).

Other organs that may be affected by high levels of fluoride ion (e.g., i.p. administration of 406 mg sodium fluoride over 15 days) are the thyroid, adrenals and pancreas (Ogilvie, 1953; Makhni *et al.*, 1979). A daily intake of fluoride ion too small to give rise to dental changes does not appear to interfere with growth or to lead to pathological changes in the kidneys or other organs (Heyroth, 1952; Lim *et al.*, 1975; Schlatter, 1978).

Effects on reproduction and prenatal toxicity

Fetal serum concentrations of fluoride ion are below or similar to maternal serum concentrations in farm animals (Ericsson & Malmnas, 1962; Bawden *et al.*, 1964; Zipkin & Babeaux, 1965; Shupe *et al.*, 1972).

High i.p. doses of stannous fluoride ($\geqslant$10 mg/kg bw) were reported to cause varying degrees of embryolethality and teratogenicity in groups of five to seven mice (Stratmann, 1979). [The Working Group noted the small sizes of the groups.]

According to Messer *et al.* (1973), fluoride is essential for reproduction; when the intake of female mice was restricted to 0.1-0.3 mg/l in the drinking-water, infertility increased.

Table 13. LD_{50} values in experimental animals for some inorganic fluorides

Compound	Species (strain)	Route of administration	LD_{50} (mg/kg bw F)	Reference
Sodium fluoride	Mouse (Swiss white)[b]	oral[c]	44.3	Lim et al. (1978)
	Mouse (Swiss white)	i.p.	17.2	Lim et al. (1978)
	Mouse	oral	46.0	Leone et al. (1956)
	Mouse	i.v.	23.0	Leone et al. (1956)
	Rat (Rochester)[b]	oral[c]	51.6	Lim et al. (1978)
	Rat	oral	32.0	Shourie et al. (1950)
	Rat (Rochester)	i.v.[d]	11.8[e]	Taylor et al. (1961a)
	Rat (Rochester)	i.p.	24	Stokinger (1949)
Stannous fluoride	Mouse (Swiss white)	oral[c]	25.5	Segreto et al. (1961)
	Mouse (Swiss white)[b]	oral	31.2	Lim et al. (1978)
	Rat (Rochester)[b]	oral[c,f]	45.7	Lim et al. (1978)
Sodium monofluoro-phosphate	Mouse (Swiss white)[b]	oral[c]	93.9	Lim et al. (1978)
	Rat	oral	75.0	Shourie et al. (1950)

[a] Calculated from mortality data at 24 h. Values have been rounded to the nearest decimal.
[b] Animals fasted but given normal access to water for 24 h before treatment
[c] Administered via stomach tube under light ether anaesthesia
[d] The toxicity of intraperitoneally administered sodium fluoride was also studied, but the LD_{50} was not computed. It was, however, reported to be similar to that with i.v. administration.
[e] Mortality at 30 days
[f] Administered in an aqueous glycerol (5.0 M)-tartaric acid (0.01 M) solution to prevent hydrolysis and precipitation of tin salts

However, impaired reproductive performance has also been reported in mice, rats and cattle after intake of large amounts of fluoride, e.g., 100 mg/l in the drinking-water (Cass, 1961; Hodge & Smith, 1965; Messer et al., 1973).

Absorption, distribution, excretion and metabolism

Fluoride ion is rapidly and extensively absorbed from the gut (see, e.g., Cremer & Buttner, 1970). In rats, absorption of sodium fluoride, sodium silicofluoride, sodium monofluorophosphate and stannous fluoride is similar (Wallace-Durbin, 1954; Perkinson et al., 1955; Ericsson & Ullberg, 1958; Shupe et al., 1972; Andreas et al., 1978). Formation of complexes with cations such Ca, Fe, Mg and Al decreases absorption of fluoride ion (Largent, 1954; Hodge, 1961; Deshpande & Bestor, 1964; Shupe et al., 1972).

Fluoride is transported in the blood in the free rather than the protein-bound form (Chen et al., 1956); it is distributed rapidly throughout all soft tissues but is not accumulated (Wallace-Durbin, 1954; Armstrong & Singer, 1966). It is generally accepted that the fluoride ion is of principal importance in the toxicology of inorganic fluorides (Hodge, 1961).

Excretion occurs mainly *via* the kidneys; renal clearance is greater than that of chloride (Smith et al., 1955; Chen et al., 1956; Carlson et al., 1960a) and is pH-dependent. Thus, tubular reabsorption is increased and clearance decreased when the urine is acidic (Whitford et al., 1976; Whitford & Pashley, 1979). Little fluoride seems to be excreted into the milk (e.g., 0.2% of an i.v. dose in a cow) (Perkinson et al., 1955).

Fluoride accumulates in bones and teeth; and in a three-month experiment in rats, 25-35% of ingested fluoride was retained in the calcified tissues, independent of the concentration of fluoride in the drinking-water (1-100 mg/l) (Taylor et al., 1961b). When fluoride was withdrawn from the drinking-water of rats, the fluoride content of the bones decreased by 44% within the first eight weeks and more slowly thereafter (Miller & Phillips, 1953).

Effects on intermediary metabolism

Fluoride can inhibit a number of enzymes *in vitro*, including cholinesterase (Eager, 1969), DNA polymerase (Hellung-Larsen & Klenow, 1969), and enzymes (such as rat liver microsomal esterases) that are involved in the metabolism of carcinogens (Irving, 1966) and of glycogen (Zebrowski et al., 1964). In intact cells, fluoride is known to inhibit the glycolysis pathway and protein synthesis (Vesco & Colombo, 1970).

Administration of fluoride to animals can lead to reductions in the levels of some enzymes, such as serum alkaline phosphatases and esterases (Riekstniece et al., 1965; Ferguson, 1971).

Mutagenicity and other short-term tests

Sodium fluoride did not induce reverse mutations in *Salmonella typhimurium* strains TA1535, TA1537, TA1538, TA98 or TA100 when tested at up to 500 µg/plate in the absence, or at up to 2000 µg/plate in the presence, of a liver activation system from Aroclor 1254-induced rats. It did not induce gene conversion in *Saccharomyces cerevisiae* strain D4 in the same study (Martin *et al.*, 1979).

No sex-linked recessive lethals were induced in *Drosophila melanogaster* when sodium fluoride was administered by injection of a 10^{-3}M solution (Mukherjee & Sobels, 1968) or when adults were fed 0.025% in a honey solution (Mendelson, 1976). However, Mitchell & Gerdes (1973) reported a significant enhancement in the frequency of sex-linked recessive lethals after treatment with concentrations of 6% sodium fluoride or 25% stannous fluoride. [The Working Group questioned whether sodium fluoride is soluble at the levels used, and noted the unusually low spontaneous mutation frequencies reported.]

Sodium monofluorophosphate did not induce dominant lethal mutations in mature sperm or oocytes of *Drosophila* fed concentrations of up to 60 mM for as long as 36 h (Buchi, 1977).

Feeding of sodium fluoride to mice at concentrations of up to 50 mg/kg of diet for seven generations did not induce chromosomal aberrations or increase the frequency of sister chromatid exchanges in bone marrow (Kram *et al.*, 1978). In two other experiments with mice, Martin *et al.* (1979) found no evidence of chromosomal damage in either bone-marrow or testicular cells. In the first of these experiments, Swiss-Webster mice were fed for two years on a low-fluoride diet (0.5 mg/kg) and given drinking-water containing either 0 or 50 mg/l sodium fluoride. In the second experiment, BALB/c male mice were given the same low-fluoride diet and up to 100 mg/l sodium fluoride in their drinking-water for six weeks. Earlier, it had been claimed that chromosomal changes were induced in testicular and bone-marrow cells of BALB/c mice with doses as low as 1 mg/l sodium fluoride in drinking-water for three to six weeks (Anon., 1976; Mohamed & Chandler, 1977). The validity of this finding has been questioned by Martin *et al.* (1979).

Cytological changes have been observed in the chromosomes of cow and ewe oocytes when cultured in the presence of up to 0.1 and 0.2 mg/ml sodium fluoride, respectively and in cultures of mouse oocytes at concentrations below 0.4 mg/ml. The effects were not dose-related. No cytogenetic effects were induced in oocytes of mice exposed to sodium fluoride as a single, acute dose (500 µg intravenously) or chronically (250 µg subcutaneously daily for 16 days) (Jagiello & Liu, 1974).

No increase in chromosomal aberrations was observed in human lymphocytes treated with up to 3×10^{-3}M (Voroshilin *et al.*, 1973) or up to 10 mg/l sodium fluoride (Kralisz & Szymaniak, 1978).

Genetic effects, including anaphase lagging, bridges, tetraploidy, multipolar anaphases and increase in the frequency of abnormal mitotic figures, have been induced in several plant species by sodium fluoride (Mohamed *et al.*, 1966; Hakeem & Shehab, 1970; Mouftah & Smith, 1971; Bale & Hart, 1973a,b; Galal & Abd-Alla, 1976; Temple & Weinstein, 1978).

(b) Humans

Toxic effects

(i) *Acute toxicity*

Immediate effects of the ingestion of a toxic dose of a soluble inorganic fluoride occur in the gut; these include vomiting, abdominal pain and diarrhoea. The severity of the symptoms is dose-related (Princi, 1960; Hodge & Smith, 1972; Hoffman *et al.*, 1980; Spoerke *et al.*, 1980). These effects have also been reported to occur with lower doses (1-1.2 mg/day in tablet form) in some particularly sensitive individuals (Feltman & Kosel, 1961; Shea *et al.*, 1967).

Convulsions have been observed following absorption of 0.2-27.5 g fluoride (Roholm, 1936; Princi, 1960), and repeated ventricular fibrillation after ingestion of 120 g sodium fluoride (Abukurah *et al.*, 1972). Lethal doses of inorganic fluoride lie between 50 and 225 mg/kg bw (Smith & Hodge, 1959; Schlatter, 1978; Spoerke *et al.*, 1980). In fatal cases of fluoride poisoning, death is usually due to respiratory paralysis (Spoerke *et al.*, 1980). Autopsy shows acute congestion of the abdominal viscera, swelling of the liver and kidneys, tubular necrosis, haemorrhages in the lungs and dilatation of the right chambers of the heart (Lidbeck *et al.*, 1943).

Allergic responses to fluoride have been reported, although the role of fluoride in these responses has been questioned (Shea *et al.*, 1967; Zanfagna, 1976).

(ii) *Chronic toxicity*

With chronic exposure, toxic effects are seen in teeth, bones, kidneys, the reproductive system and blood (Schlatter, 1978). Smith & Hodge (1959) related fluoride intake to toxic effects as follows (Table 14):

Table 14. *Effects of various fluoride concentrations in water and food, or daily intake*

Fluoride level	Effect
≥2 ppm	Mottled enamal
8 ppm	Osteosclerosis
≥20-80 mg/day	Crippling fluorosis
>50 ppm	Thyroid changes
100 ppm	Growth retardation
>125 ppm	Kidney changes

Continuous or intermittent exposure to inorganic fluorides can lead to appreciable accumulation of fluoride in bone and to the development of osteosclerosis and other bone changes (WHO, 1970). However, bone changes consistent with skeletal fluorosis were detected in only 23 cases in a study of 170000 X-ray records (spine and pelvis) of people from communities where the fluoride concentrations in the drinking-water exceeded 0.7 mg/l (Stevenson & Watson, 1957).

In areas of endemic fluorosis and high poverty, the combination of excessive fluoride intake and nutritional insufficiency may lead not only to crippling skeletal malformations (see, e.g., WHO, 1970; Maiya et al., 1977) but also to neurological disorders (Siddiqui, 1955; Singh et al., 1963; Singh & Jolly, 1970) and may be associated with haematological abnormalities (Schlatter, 1978).

In areas of fluorosis in which the fluoride concentration in the drinking-water ranged from 5.0-16.2 mg/l, decreased urea clearance and glomerular filtration rate, together with increased blood urea have been observed (Siddiqui, 1955; Singh et al., 1963). In addition to skeletal changes, an increased incidence of renal calculi has also been reported (Herman, 1955).

Adverse effects of water-borne fluoride (5-13 mg/l) have been reported on the heart (Okushi, 1954; Takamori et al., 1956; Jansen & Thomson, 1974; see also Waldbott, 1961). Earlier reports of dysfunction of the thyroid (Wilson, 1941; Spira, 1944, 1946; Murray et al., 1948) have not been confirmed (e.g., Singh et al., 1963; Latham & Grech, 1967).

Effects on reproduction and prenatal toxicity

After injection of $^{18}F^-$, fetal serum concentrations did not exceed 25% of that in the maternal blood (Ericsson & Malmnas, 1962). Fluoride that crosses the placenta is deposited in fetal bones and teeth in amounts that increase with the age of the fetus (Brzezinski *et al.*, 1960) and with the fluoride intake of the mother (Gedalia *et al.*, 1964). Mottled dental enamel has been found in the teeth of children whose mothers drank water containing 12-18 mg/l fluoride (Smith & Smith, 1935). The fluoride content of maternal milk ranges from <0.1 to 0.2 mg/l. It increases by about 15-40% with daily supplements of 5 mg fluoride in the diet or drinking-water (WHO, 1970).

Early reports of an association between the presence of naturally occurring fluorides in water supplies and cases of Down's syndrome, ascertained from records of specialized institutions (Rapaport, 1956) and from birth and death certificates (Rapaport, 1959), have not been confirmed in later studies in which there was more complete ascertainment and/or consideration of mother's age (Berry, 1958; Needleman *et al.*, 1974; Erickson *et al.*, 1976; Erickson, 1980). No association with other congenital malformations has been observed (Erickson *et al.*, 1976).

Absorption, distribution, excretion and metabolism

This field has been covered in a number of reviews (WHO, 1970; National Institute for Occupational Safety & Health, 1975; National Research Council, 1977). Fluoride ion is absorbed rapidly: 96-97% from solutions of sodium and calcium fluoride, and 62% from calcium fluoride in solid form (Largent, 1961). Absorption begins in the stomach and continues in the duodenum and jejunum (Schlatter, 1978). It may be affected by levels of both dietary organic material (protein and fat) and ions (Ca^{++}, Mg^{++}, $PO_4^{=}$) (Waldbott, 1961).

Fluoride is excreted mainly in the urine; however, excretion in sweat may account for 25% of total excretion, and for up to 50% when sweating is excessive (McClure *et al.*, 1945; Crosby & Shepherd, 1957).

Fluoride levels found in human tissues are given in section 2.2; 96-97% of fluoride in human plasma is freely diffusible (Carlson *et al.*, 1960b). The rate of skeletal accumulation (retention) of fluoride is influenced by the amount ingested and the amount already accumulated (Largent, 1961). It has been estimated (Myers *et al.*, 1980) that if a person were to move from a high-fluoride (8 mg/l in the drinking-water) to a low-fluoride (<0.3 mg/l) locality, the half-life of the fluoride in his bones would be approximately eight years.

Effects on intermediary metabolism

As in animals (see p. 274), certain enzymes in human tissues are inhibited by fluoride (Frajola, 1960; Abukurah *et al.*, 1972).

Mutagenicity and chromosomal effects

No data were available to the Working Group.

3.3 Case reports and epidemiological studies of carcinogenicity in humans

Although the Working Group was aware of a number of epidemiological studies on the relation between cancer and occupational exposures to inorganic fluorides concurrently with a variety of other chemicals (e.g., in the aluminium production industry), the only studies reviewed in this section are those which relate to inorganic fluorides in drinking-water. The epidemiological studies considered are summarized in Table 16.

All are descriptive (sometimes called correlational or ecological) studies, in which both exposure and disease are measured in population aggregates. For the purpose of inferring causal relationships, they are generally less satisfactory and contain more sources of error than analytical studies, in which ascertainment of exposure and disease status is made separately for individual people rather than cities or counties (see preamble, p. 16). For example, because of the constant migration in and out of large population centres, the exposures of aggregates that are ascertained at one moment in time refer to a different set of people than do the measures of disease made after a suitable latent period. Variations in disease frequency between the population units may be due to differences in their demographic composition, degree of industrialization, exposure to environmental pollutants, ethnic distribution or life style, to mention only a few possibilities. Since the age-sex-race compositions of populations are often known, it is possible to eliminate or to reduce substantially the bias from this source by the use of appropriately standardized rates. The unexplained variations in the adjusted rates between population units are the appropriate background against which to evaluate the significance of differences among groups of units classified by their exposure. However, because of the small number of such units typically available for study, it is difficult to control adequately for confounding effects. Particular weaknesses which apply to several of the studies considered below are failure to consider each cancer site separately, and failure to account for the variability in rates between population units in the analysis and presentation of results.

Hagan *et al.* (1954) identified 32 US cities with a population in 1950 of 10000 or more, in which the water supplies contained natural fluoride in concentrations of 0.7 mg/l or greater, and which were near 32 cities of similar size with natural fluoride levels of less than 0.25 mg/l which could serve as controls. The total populations were 0.9 million in the high-fluoride cities and 1.3 million in the low. Cancer death rates in 1950, indirectly standardized for age, race and sex by reference to the rates prevailing in the US in 1950, averaged 135.4 per 100000 for high-level cities and 139.1 for low, a difference which is not statistically significant. Equally, there were no differences in mortality from heart disease, intracranial vascular lesions, nephritis or liver cirrhosis. [No attempt was made in this study to consider other characteristics of the sampled cities which might be related to cancer mortality; however, the comparability of rates for other causes of death provides some assurance that no large bias was present.]

Nixon & Carpenter (1974) identified 17 towns in England and Wales, each served by a single water supply containing natural fluoride [0.5-6 mg/l]. They selected as controls another 17 towns of similar size and geographic distribution in which the water supplies were virtually fluoride-free [0-0.5 mg/l]. Using age-specific cancer death rates (in five-year intervals) for all of England and Wales, they computed standardized mortality ratios for each town, sex and year from 1950 through 1965. Averages of these ratios for cancer and other selected causes over the 16-year period were related in a linear regression analysis to fluoride level, total water hardness in mg/l, latitude, social class index, domestic air pollution and rainfall. Crude correlation coefficients between fluoride level and cancer deaths were -0.29 and -0.18 for men and women, respectively. Regression adjustment for the other independent variables yielded partial correlations of -0.15 and -0.12. None were statistically significant, nor were the crude or partial coefficients computed for infectious diseases, diseases of the stomach, cardiovascular diseases, accidents or for all causes of death combined. [Due to the severe selection criteria, the fluoridated and non-fluoridated towns included in this analysis were situated in only two parts of the country. The consistently negative findings for several causes of death support the comparability of the towns with regard to factors which may influence mortality in general.]

For each of 18 local authority districts in England and Wales in which the natural fluoride level in water was 1 mg/l or more, Kinlen (1975) selected a nearby control district of similar size where the water contained a level of 0.2 mg/l or less. He matched 31 districts with levels of 0.5-0.99 mg/l to those with 0.1 mg/l or less, aggregating control districts, where necessary, to achieve populations of similar size. Urban areas were matched with urban, and rural with rural. Incidence data were obtained for cancers occurring at nine specific sites (thyroid, kidney, stomach, oesophagus, colon, rectum, bladder, bone and breast) during all or parts of the period 1961-1968. Using cancer rates for the entire sample, stratified by sex and age in 10-year intervals, he calculated expected numbers of site-specific cancer cases for each of the four groups: high ($\geqslant$1.0 mg/l fluoride) with a population of 482000; medium (0.5-0.99) with 779000; low ($\leqslant$0.2) with 510000; and very low ($\leqslant$0.1) with 897000 people. For no site was there a trend in the ratio of observed to expected cases with increasing fluoride level. In a separate study of artificial fluoridation (1 mg/l), he found no significant differences in the ratios of observed to expected cases in three fluoridated areas involving 11 local authority districts with an aggregate population of 1.3 million, compared with matched non-fluoridated control districts where the natural fluoride levels were less than 0.15 mg/l. [No account was taken in the data analysis of the variation within fluoride groups in rates between areas. Consideration of such variation would not alter the negative findings.]

Burk & Yiamouyiannis (1975) compared the trends in crude cancer mortality rates over the period 1950-1970 for two groups of US cities: (i) the 10 largest cities (1960 census) in which the water supply was fluoridated before 1960 and in all of which the cancer death rate per 100000 was greater than 155 in 1953; and (ii) the 10 largest cities where the water supply had not been fluoridated as of 1969. In subsequent analyses, they replaced four of the control cities with the next largest non-fluoridated cities, so as to ensure that cancer death rates were in excess of 155 in 1953 in all 10 (Yiamouyiannis & Burk, 1975). Whereas the 1950 rates were approximately 180 for both groups, by 1970 they had risen to nearly 200 for the non-fluoridated group and to over 220 for the fluoridated group.[1] A later analysis of data for the same cities, obtained from local health departments (Yiamouyiannis & Burk, 1977), considered trends in annual cancer death rates separately by age category: 0-24, 25-44, 45-64 and 65+ years. There were slight decreases in the cancer death rates among the two lowest age groups in both fluoridated and non-fluoridated cities. The rates for the 45-64-year-old group increased faster in the fluoridated cities (from 338.3 in 1952 to 375.1 in 1969) than in the non-fluoridated cities (323.5 to 347.2). For the highest age group, the rates rose from 1032.8 to 1069.9 and from 974.4 to 977.6 in the fluoridated and non-fluoridated cities, respectively. The age-specific rates were combined by the direct method to yield an age-adjusted rate for each city, using a reference population with an age distribution intermediate between those for the fluoridated and non-fluoridated populations in 1952 and 1969. These adjusted rates were 187.7 and 179.1 in 1952 and 218.3 and 200.5 in 1969 for the fluoridated and non-fluoridated cities, respectively. A regression analysis showed no significant correlation between the increases in either the age-adjusted or the age-specific (45-64 years) rates and the increase in each city's non-white population. [Standard epidemiological practice in the comparison of cancer rates is to make simultaneous adjustments for sex, race and age in 5- or 10-year intervals. In these analyses, sex was ignored, and the 20-year age groups were so broad as probably to provide insufficient control. The apparently negative conclusion from the regression analyses relating cancer to race, which relies on considerations of statistical significance, does not exclude the possibility that changes in race distribution, especially when taken in conjunction with changes in age and sex, might account for the larger increase observed in the fluoridated cities.]

In separate analyses of published data for the same 20 cities, both Oldham & Newell (1977) and Doll & Kinlen (1977, 1978) pointed out that much of the secular increase in cancer mortality could be explained by changes in the age-sex-race composition of the cities over time. Since the percentage of the population which was older and non-white grew faster in the fluoridated cities, one would expect the rates to increase faster for this group, as was indeed observed. Oldham & Newell used 1950 US population rates specific for sex, race (white *versus* non-white) and age (in 10-year intervals from five to 85+ years) to calculate expected rates for both 1950 and 1970, whereas Doll & Kinlen derived the 1970 expectations from 1970 US rates.

Smith (1980) repeated the Doll-Kinlen analyses, correcting for an error in the original data.[2] According to his calculations, the expected cancer death rates for the fluoridated and non-fluoridated cities in 1950 were 146.9 and 155.5, respectively, whereas rates of 180.8 and 179.0 were observed. The corresponding excess death rates (difference of observed minus expected) were 33.9 and 23.5, whereas the relative rates (ratio of observed to expected) were 1.23 and 1.15. By 1970, the expected rates had increased to 184.9 and 168.9, whereas

---

[1] Data from Figure 1 in Yiamouyiannis & Burk (1977); see also Taves (1977), Figure 1.

[2] As noted by Kinlen & Doll (1977), the actual number of cancer deaths in the non-fluoridated cities in 1970 was 14 272 rather than 14 487.

the observed rates were 217.4 and 194.2; these figures correspond to excess rates of 32.5 and 25.3 in the fluoridated and non-fluoridated cities, and to relative rates of 1.18 and 1.15. Thus, both the excess and relative death rates decreased between 1950 and 1970 in the fluoridated cities, whereas in the non-fluoridated cities the excess rate increased and the relative rate remained unchanged.

Taves (1977) extended these calculations by incorporating data for the 20 largest fluoridated and 15 largest non-fluoridated cities in the US. Using the 1950 US population rates specific for age, race and sex, he calculated expected numbers of cancer deaths for each group of cities. Ratios of observed to expected deaths (SMRs) for the 20 fluoridated cities were 1.20, 1.20 and 1.21 for the time periods 1949-1951, 1959-1961 and 1969-1971, respectively; for the non-fluoridated cities the SMRs were 1.13, 1.09 and 1.14. [While these results all indicate a stability over time in the excess or relative death rate with reference to the US population for both fluoridated and non-fluoridated cities, a more convincing analysis would take into consideration trends in directly adjusted rates.]

With regard to the above comment of the Working Group, Kinlen & Doll (1981) analysed data obtained from the US National Center for Health Statistics for the numbers of cancer deaths in 20 cities in 1970, classified by sex, ethnic group and age in 10-year intervals. They combined the age-sex-race-specific rates by direct standardization to the US 1970 population. Ratios of directly standardized rates for the fluoridated and non-fluoridated cities were identical to ratios of indirectly standardized rates, calculated according to methods used in their earlier publications. They also used the pooled cancer mortality rates for the two sets of cities in 1970 as a basis for indirect standardization. It was thus determined that the relative rates from 1950 to 1970 increased by 1.4% for the fluoridated cities and by 2.6% for the non-fluoridated. Similar results were obtained using the original group of 10 non-fluoridated cities selected only on the basis of size (Burk & Yiamouyiannis, 1975).

[None of the preceding analyses of the 20 US cities attempted to control for other differences between the two sets of cities which could confound an association.]

Strassburg & Greenland (1979) noted that the 10 largest fluoridated cities tended generally to be more heavily industrialized than the 10 largest non-fluoridated cities.

Hoover *et al.* (1976) carried out a regression analysis of sex- and site-specific age-adjusted cancer mortality rates for US counties on the basis of independent variables representing fluoridation status, population density, education, race/ethnic group, industrialization and geographic region. Whereas fluoridation status was strongly associated with mortality from cancer at several sites when used as the only independent variable in the regression equation, adjustment for the other factors reduced the association to statistically non-significant levels for all cancers except that of the stomach in men. [Since 62 sex-site combinations were examined, three associations significant at the $P = 0.05$ level would be expected by chance alone.]

Hoover *et al.* (1976) also classified all US counties into three groups according to the year in which the 1960 population of communities served by fluoridated water supplies first reached two-thirds of that of the whole county, i.e., in the periods 1950-1954, 1955-1959 or 1960-1964. A control group consisted of counties with two-thirds or more urban population (a criterion met by all but one of the fluoridated counties), in none of which communities fluoridated their water before 1970. Age- and calendar year-specific rates (five-year intervals) for the control counties were multiplied by the appropriate person-years to yield expected numbers of cancer deaths by sex and five-year calendar period for each group of fluoridated counties. Analyses were based on data for whites only. Summary ratios of observed to expected deaths showed that the cancer mortality for the counties first fluoridated in 1950-1954 was about 10% higher than that for the control counties at the time of fluoridation, and remained at this increased level for the three succeeding five-year periods. A similar stability in time was observed for counties in which two-thirds of the population first had fluoridated water in 1955-1959 and 1960-1964. When the deaths were examined by specific cancer site and sex, some sites showed a slight upward trend and others a downward trend. However, in no instance was a trend consistently observed for all three groups of fluoridated counties. [Since a large number of different sites were examined, some would be expected by chance alone to show departures from the overall finding for all sites combined.]

Hoover *et al.* (1976) also compared the changes in cancer incidence rates between 1947-1948 and 1969-1971 for two large US cities: Birmingham, Alabama, which remained largely unfluoridated before 1970; and Denver, Colorado, which was 66% fluoridated by 1955. Age-adjusted rate ratios (skin cancers excluded) for Denver to Birmingham were 1.04 and 1.02 for men and women, respectively, in 1947-1948, and 0.97 and 1.07 in 1969-1971. The ratio increased over this time period for 19 individual site/sex combinations, decreased for 19 and was unchanged for six. [This comparison of temporal changes for only two population units contributes little relevant information to a possible relationship between exposure to fluoride and cancer.]

Hoover *et al.* (1976) also identified 37 Texas counties with a 50% or greater urban population and then classified them according to whether the natural fluoride concentration in their community water supplies was intermediate (0.7-1.2 mg/l), high (1.3-1.9 mg/l) or very high (2.0+ mg/1). An additional 16 urban counties not listed in the fluoridation census served as controls. SMRs for white men and women during 1950-1969 were calculated for each of the four exposure categories by reference to the rates in the pooled population, with stratification by age, urbanization (counties >71% urban *versus* <71% urban) and education (counties in which the population had had a median length of schooling of >10.2 years *versus* <10.2 years). The SMRs for all cancer sites combined showed no indication of a trend with fluoride level: they were equal to 1.0 in the control, intermediate and high exposure groups and to 0.9 in the very high exposure group, for both men and women. Mortality from cancer at several specific sites in the alimentary and digestive tracts showed decreasing trends with increasing levels of fluoride, as did that from skin cancer; however, for Hodgkin's disease and for tumours of the testis, 'other endocrine' and connective tissues, the highest SMRs were found among those exposed to very high levels of fluoride. [Although no comment was made regarding the statistical significance of the observed trend, such variations in results for different sites could well be due to random statistical fluctuations.] Linear regression analyses were made of mortality rates for each county, directly standardized for age to the 1960 US population, on the basis of their fluoride levels and of potentially confounding variables, including percent urban, median years of schooling, percent non-white and percent foreign-born. Whereas three such analyses from among the 64 sex-site combinations would be expected to yield significant results at the $P = 0.05$ level, four were observed: for rectal cancer in men, ovarian cancer in women and brain cancer in people of both sexes. In all four cases, the association tended towards less cancer in people in the more heavily fluoridated counties.

Raman et al. (1977) selected the 400 largest municipalities in Canada in 1971, combined them to form 100 municipal areas, and then grouped them according to the year in which their water supplies were first fluoridated (Table 15):

Table 15. Grouping of Canadian municipalities by Raman et al. (1977)

Group	Year fluoridated	No. of cities	Aggregate population (in millions) in 1971
I	1945-1958	14	1.6
II	1959-1963	6	2.4
III	1964-1968	10	1.3
IV	1969-1973	7	0.5
Non-fluoridated		42	6.6
Total		79	12.4

Twenty-one municipal areas were omitted because the data required to calculate age- and sex-specific cancer rates for the four time periods shown above were not available. Cancer death rates by sex, age-adjusted by direct standardization to the 1971 male or female Canadian population, were calculated for each of the five groups and for each of the five-year periods surrounding the census years 1956, 1961, 1966 and 1971. In general, and especially for the non-fluoridated cities, the observed death rates exceeded those for all Canada. There was, however, no systematic trend towards an increase in relative cancer mortality following the period of fluoridation. Such changes as there were in fluoridated areas were accompanied by similar changes in non-fluoridated areas. Regression analyses of both age-specific and age-adjusted rates for individual cities and time periods likewise failed to detect a shift following fluoridation. Separate analyses for four broad site categories (respiratory system, large bowel, stomach and the leukaemias) were similarly negative. [No attempt was made in this study to control for other factors, such as ethnic group, industrialization or social class, which may have contributed to the regional variation in cancer mortality.]

Erickson (1978) determined that of all 57 US cities with a 1970 population in excess of 250000 persons, 24 with a total population of 16 million had had fluoridated water supplies since before 1960, while 22 with 11 million population had not, until at least 1971. For each city he computed death rates adjusted for age, sex and race (black or white only) in 34 cause-of-death categories for the period 1969-1971, using indirect standardization based on the pooled rates for all cities studied. The adjusted rates were used as dependent variables in linear regression analyses, which included fluoridation status, population density and median duration of education as independent variables. Two other variables, median income and percentage of the work force engaged in manufacture, were also considered for adjustment purposes in preliminary analyses but were discarded when it became clear they were not important independent predictors of mortality for most causes. On average, the fluoridated cities had a higher population density, were more heavily industrialized and had a population with fewer years of schooling. Whereas the mean crude death rate from malignant neoplasms was 206.2 per 100000 person-years for the fluoridated compared with 183.0 for the non-fluoridated cities, the mean adjusted rates were 199.9 and 192.3. After further correction for population density and duration of education in the regression analysis, these rates became 195.3 and 196.9; the differences were not statistically significant. Similar negative results were observed for seven subgroupings of cancer deaths by site (digestive, respiratory, breast, genital, urinary, leukaemia and other) as well as for most other causes. [This analysis of a relatively large number of population units shows clearly how the crude differences in mortality between the fluoridated and non-fluoridated cities can be explained by variations in their demographic and social composition.]

Rogot *et al.* (1978) found that of US cities with a 1950 population of 25000 or more, 227 had had continuously fluoridated water supplies between 1950 and 1970 whereas 187 had not but had low levels (<0.7 mg/l) of natural fluoride. For each city, they calculated SMRs for all causes of death, and for cancer and heart disease separately, for the three periods 1949-1950, 1959-1961 and 1969-1971, with reference to the 1960 US population rates by age, sex and race. The average change in the SMRs, after direct adjustment for differences in 1950 SMRs between the fluoridated and non-fluoridated cities, showed an increase in cancer mortality of 8% over the period 1950-1970 but a decrease of 13% in heart disease mortality in both groups of cities. While the cancer increase was greatest in small cities, in those for which a low SMR was calculated for 1950 and in those in southern and north-central states, there was no indication that it was affected by fluoridation status. [This study is of interest because of the large number of population units. No consideration was given to variations in the SMR measures between cities.]

Glattre & Wiese (1979) computed average levels of fluoride in water supplies registered with the Directorate of Health in southern Norway. Municipalities in which 60% or more of the inhabitants received water from such supplies were divided into three groups according to the computed levels. There were 71 municipalities with 1.2 million people receiving water with a concentration of 0.0-0.05 mg/l, 28 municipalities with 0.9 million people at 0.06-0.10 mg/l, and 22 municipalities with 0.3 million people at 0.11-0.50 mg/l. The average age-adjusted mortality rates from oral and pharyngeal cancer for 1971-1975 were negatively related to fluoride level: they were 4.7 (per 100000 person-years) ($\pm$ 0.7 standard error of the mean), 3.9 ($\pm$ 0.7) and 3.1 ($\pm$ 0.8) for men in the low, medium and high fluoride groups, respectively, and 1.5 ($\pm$ 0.3), 1.3 ($\pm$ 0.3) and 0.5 ($\pm$ 0.2) for women. When restricted to the 70 municipalities where 80% or more of the inhabitants obtained water from registered supplies, the average rates were 5.3, 4.7 and 3.4 for men in the three fluoride groups and 1.5, 1.1 and 0.6 for women. A regression analysis of the mortality rates for the 121 municipalities on the basis of actual fluoride levels showed a significant negative slope for women; however, the marked variation in the rates from one municipality to another obscured any association in men. The authors remarked that rates of smoking and drinking were known to be slightly higher in the municipalities in the high-fluoride group. [No formal analysis was made of the effect of variations in alcohol/tobacco consumption, nor of other potentially confounding variables, on the results.]

Richards & Ford (1979) identified 10 localities in New South Wales, Australia, excluding major urban areas, with an aggregate population in 1971 of 163000 and in which the water supplies had been fluoridated between 1956 and 1967. For comparison they selected an additional 10 cities, with 192000 total population where the water had a low natural fluoride content (< 0.7 mg/l). SMRs for malignant neoplasms during the period 1970-1972, using age- and sex-specific rates for the 1971 New South Wales population as standard, were 0.92 for the fluoridated cities and 0.95 for the non-fluoridated. Only two of the SMRs for individual cities, one for a fluoridated city and one not, were statistically significantly different from 1.0. The city-to-city variation was unrelated to fluoridation status. [The basis on which the cities were selected is not entirely clear from the report.]

Goodall & Foster (1980) compared the changes in cancer mortality rates from 1961 to 1976 in two sets of New Zealand population groups, one consisting of six groups whose water supplies were fluoridated prior to 1968 and the other of four groups with non-fluoridated water. The total cancer mortality for men aged 45 years and over changed from 629.5 per 100000 to 691.1 in the groups with fluoridated water, an increase of 10%; whereas in those with non-fluoridated water there was a 29% increase, from 567.7 to 733.5. For women aged 45 years and older there was a 5% decrease, from 484.7 to 463.2 with fluoridation, compared with a 2% increase of 501.4 to 511.9 without fluoridation. Similar trends were evident when the rates were examined separately within each of the 10-year age groups from 45-54 to 85+ years. [The rationale used to select the two sets of groups was not adequately explained. No attempt was made to correct for differences in ethnic composition, industrialization or other factors which could account for the observed differences.]

Cook-Mozaffari et al. (1981) calculated age-standardized cancer mortality ratios (SMRs), relative to the average rates for all of England and Wales for the periods 1959-1963, 1969-1973 and 1974-1978, for Birmingham, where the water supply was fluoridated in 1964, and for six other cities in which the populations were of comparable size. The SMR for Birmingham increased by 6.4% between the first two periods; this was almost identical to the average increase for the other cities (6.5%; range, 2.7-9.5%). Between the last two periods, the SMR for Birmingham increased by 1.0%, while it decreased by an average of -0.9% in the other cities (range, -6.4-3.4%). However, the increase in Birmingham was not statistically significant for either men or women, and it differed significantly from only one of the changes observed in other cities. Consideration of changes in the crude rates over time (Cook-Mozaffari & Doll, 1981) likewise showed that Birmingham's experience was typical of those of other cities. [These analyses by Cook-Mozaffari et al. (1981) and Cook-Mozaffari & Doll (1981) were performed as extensions of analyses made by Burk on the basis of which he stated that cancer mortality in Birmingham had risen as a result of fluoridation. Unfortunately, Burk's analyses have not been published in a readily available form and therefore cannot be cited here.]

Table 16. Summary of epidemiological studies considered

Type of population unit (country)	Time period(s)	Fluoridated/High level		
		No. of units	Total population (millions)	Range of fluoride concentrations (mg/l)
Cities (US)	1950	32	0.9	≥0.7
Towns (England & Wales)	1950–1965	17	NS[b]	0.5 – 6
Local authority districts (England & Wales)	1961–1968	18	0.5	≥1.0
Local authority districts (England and Wales)	1961–1968	31	0.8	0.5–0.99
Local authority districts (England & Wales)	1961–1968	11	1.3	Artificial
Cities (US)	1950–1970	10	NS	Artificial
Cities (US)	1949–1951 1959–1961 1969–1971	20	NS	Artificial
Counties (US)	NS	NS	NS	Artificial
Counties (US)	1950–1954 1955–1959 1960–1964	61	NS	Artificial

Non-fluoridated/Low level			Cause of death categories	Confounding factors considered[a]	Reference
No of units	Total population (millions)	Range of fluoride concentrations (mg/l)			
32	1.3	<0.25	All cancers	Race	Hagan et al. (1954)
17	NS	≤0.5	All cancers and 5 other categories	Water hardeners, latitude, social class, air pollution, rainfall	Nixon & Carpenter (1974)
18	0.5	≤0.2	9 cancer sites	Geography, type of area	Kinlen (1975)
35	0.9	≤0.1	9 cancer sites	Geography, type of area	Kinlen (1975)
11	1.3	<0.15	9 cancer sites	Geography, type of area	Kinlen (1975)
10	NS	NS	All cancers	Race	Burk & Yiamouyiannis (1975) (see text for re-analyses of data)
15	NS	NS	All cancers	Race	Taves (1977)
NS	NS	NS	Cancer at 62 sex/site combinations	Population density, education, race/ethnic group, industrialization, geography, race	Hoover et al. (1976)
156	NS	NS	Cancer at 62 sex/site combinations	Urbanization, socio-economic status	Hoover et al. (1976)

Type of population unit (country)	Time period(s)	Fluoridated/High level		
		No. of units	Total population (millions)	Range of fluoride concentrations (mg/l)
Cities (US)	1947-1948 1969-1971	1	NS	Artificial
Counties (Texas, US)	1950-1969	37	NS	0.7 - >2.0
Municipal areas (Canada)	1945-1958 1959-1963 1964-1968 1969-1973	37	5.8	Artificial
Cities (US)	1969-1971	24	16	Artificial
Cities (US)	1949-1950 1959-1961 1969-1971	227	40.5	Artificial
Municipal areas (Norway)	1971-1975	50	1.2	0.06 - 0.5
Cities (Australia)	1970-1972	10	0.16	Artificial
Areas (New Zealand)	1961-1976	6	NS	Artificial
Cities (England & Wales)	1959-1963 1969-1973 1974-1978	1	>0.4	Artificial

[a] Other than age and sex
[b] NS = Not specified

Non-fluoridated/Low level			Cause of death categories	Confounding factors considered[a]	Reference
No. of units	Total population (millions)	Range of fluoride concentrations (mg/l)			
1	NS	NS	All cancers (except skin) Cancer at 19 sex/site combinations		Hoover et al. (1976)
16	NS	NS	Cancer at 62 sex/site combinations	Race/ethnic group, urbanization, education	Hoover et al. (1976)
42	6.6	NS	All cancers in 4 broad sub-groups		Raman et al. (1977)
22	11	NS	34 causes of death, including 7 cancer site groups	Race, population density, education, income, industrialization	Erickson (1978)
187	22.4	<0.7	All causes of death, cancer, heart disease	Race	Rogot et al. (1978)
71	1.2	0-0.05	Oral and pharyngeal cancer		Glattre & Wiese (1979)
10	0.19	<0.7	All cancers		Richards & Ford (1979)
4	NS	NS	All cancers		Goodall & Foster (1980)
6	>2.4	NS	All cancers		Cook-Mozzafari et al. (1981)

4. Summary of Data Reported and Evaluation

4.1 Experimental data

Sodium fluoride was tested in three experiments in three different strains of mice by oral administration. The available data are insufficient to allow an evaluation to be made.

Sodium fluoride was not mutagenic to *Salmonella typhimurium* or *Drosophila melanogaster* and did not induce gene conversion in *Saccharomyces cerevisiae*.

4.2 Human data

Significant mining of fluorspar (calcium fluoride) started in about 1775. The natural occurrence of some inorganic fluorides and their use in water fluoridation and anti-caries dental products results in widespread exposure of the general population. In addition, the numerous industrial applications of these chemicals result in significant occupational exposure and emissions to the environment.

Only studies on water fluoridation and cancer were reviewed. The relationship between cancer mortality or incidence and both natural and artificial fluoride in drinking-water has been investigated in a large number of descriptive epidemiological studies of population aggregates, carried out in Australia, Canada, New Zealand, Norway, the United Kingdom and the United States. Because of the uneven distribution of natural fluoride in the earth's crust, and the fact that local communities make independent decisions with regard to fluoridation, some of these studies could be viewed roughly as natural experiments. When proper account was taken of the differences among population units, in demographic composition, and in some cases also in their degree of industrialization and other social factors, none of the studies provided any evidence that an increased level of fluoride in water was associated with an increase in cancer mortality.

4.3 Evaluation

The available data are inadequate for an evaluation of the carcinogenicity of sodium fluoride, the only inorganic fluoride tested, in experimental animals.

Variations geographically and in time in the fluoride content of water supplies provide no evidence of an association between fluoride ingestion and mortality from cancer in humans.

5. References

Abukurah, A.R., Moser, A.M., Jr, Baird, C.L., Randall, R.E., Jr, Setter, J.G. & Blanke, R.V. (1972) Acute sodium fluoride poisoning. *J. Am. med. Assoc., 222*, 816-817

American Dental Association (1979) *Accepted Dental Therapeutics*, 39th ed., Chicago, IL, pp. 316-338

American Water Works Association (1971) *AWWA Standard for Fluosilicic Acid* (AWWA B703-71), Denver, CO

American Wood-Preservers' Association (1979) *Wood Preservation Statistics, 1978*, Washington DC, p. 257

Andreas, M., Drauschke, M., Knoch, M., Gobel, H. & Heintzschel, D. (1978) Fluorine metabolism in the rat. III. Distribution in the gastrointestinal tract after oral application (Ger.). *Stomatol. Deutsch. Democr. Repub., 28*, 552-556

Anon. (1976) Fluorides in water may be genetic hazard. *Chemical Engineering News*, 20 September, p. 30

Anon. (1978) Stannous fluoride. *Chemical Marketing Reporter*, 10 July, p. 34

Armstrong, W.D. & Singer, L. (1966) *Distribution of fluoride between compartments of body water*. In: Gaillard, P.J., Van den Hooff, A. & Steendijk, R., eds, *Fourth European Symposium on Calcified Tissues*, Amsterdam, Excerpta Medica, pp. 1-2

Baijot-Stroobants, J. & Vreven, J. (1979) Determination by charged particles activation of fluoride uptake in human dental enamel. *Caries Res., 13*, 211-217

Bale, S.S. & Hart, G.E. (1973a) Studies on the cytogenetic and genetic effects of fluoride on barley. I. A comparative study of the effects of sodium fluoride and hydrofluoric acid on seedling root tips. *Can. J. Genet. Cytol., 15*, 695-702

Bale, S.S. & Hart, G.E. (1973b) Studies on the cytogenetic and genetic effects of fluoride on barley. II. The effect of treatments of seedling coleoptiles with sodium fluoride. *Can. J. Genet. Cytol., 15*, 703-712

Bawden, J.W., Wolkoff, A.S. & Flowers, C.E., Jr (1964) Placental transfer of F^{18} in sheep. *J. dent. Res., 43*, 678-683

Berry, W.T.C. (1958) A study of the incidence of mongolism in relation to the fluoride content of water. *Am. J. ment. Defic., 62*, 634-636

Bill Communications, Inc. (1975) *Materials and Compounding Ingredients for Rubber*, New York, p. 375

Boniface, H.J. & Jenkins, R.H. (1977) Determination of fluoride in steelworks materials with a fluoride ion-selective electrode. *Analyst, 102*, 739-744

Brewer, R.F. (1966) *Fluorine*. In: Chapman, H.D., ed., *Diagnostic Criteria for Plants and Soils*, Riverside, CA, University of California, Division of Agricultural Sciences, pp. 180-196

Brzezinski, A., Bercovici, B. & Gedalia, J. (1960) Fluorine in the human fetus. *Obstet. Gynecol., 15*, 329-331

Buchi, R. (1977) Induced dominant lethal mutations and cytotoxic effects in germ cells of *Drosophilla melanogaster* with Trenimon, PDMT and sodium monofluorophosphate. *Genetics, 87*, 67-81

Burk, D. & Yiamouyiannis, J. (1975) Letter (July 21) to Hon. J.J. Delaney; fluoridation and cancer. *Congr. Rec., 591-508-41226*, Washington DC, pp. 23730-23733

Byrns, A.C. (1966) *Fluorine compounds, inorganic, silicon.* In: Kirk, R.E. & Othmer, D.F., eds, *Encyclopedia of Chemical Technology*, 2nd ed., Vol. 9, New York, John Wiley & Sons, pp. 650-661

Carlson, C.H., Armstrong, W.D., Singer, L. & Hinshaw, L.B. (1960a) Renal excretion of radiofluoride in the dog. *Am. J. Physiol.*, *198*, 829-832

Carlson, C.H., Armstrong, W.D. & Singer, L. (1960b) Distribution and excretion of radiofluoride in the human. *Proc. Soc. exp. Biol. Med.*, *104*, 235-239

Cass, J.S. (1961) Fluorides: a critical review. IV. Response of livestock and poultry to absorption of inorganic fluorides - cont'd. *J. occup. Med.*, *3*, 527-543

Chemical Industries Association, Ltd (1980) *UK Chemical Industry Statistics Handbook*, London, The Chameleon Press Ltd, p. 81

Chemtech Industries, Inc. (undated a) *Product Data: Sodium Fluoride* (No. D0579-000510), St Louis, MO

Chemtech Industries, Inc. (undated b) *Product Data: Sodium Silicofluoride* (No. D047980), St Louis, MO

Chen, P.S., Jr, Smith, F.A., Gardner, D.E., O'Brien, J.A. & Hodge, H.C. (1956) Renal clearance of fluoride. *Proc. Soc. exp. Biol. Med.*, *92*, 879-883

Chiba, M. (1978) Determination of fluorine in steel making slags by 14 MeV neutron activation analysis (Jpn.). *Nippon Kinzoku Gakkaishi*, *42*, 1161-1166 [*Chem. Abstr.*, *90*, 80338n]

Commission of the European Communities (1975) Council Directive of 16 June 1975 concerning the quality required of surface water intended for the abstraction of drinking water in the Member States. *Off. J. Eur. Communities*, *L 194*, 30

Commission of the European Communities (1976) Council Directive of 27 July 1976 on the approximation of the laws of the Member States relating to cosmetic products. *Off. J. Eur. Communities*, *L 262*, 193-194

Commission of the European Communities (1978) *Directives Concerning the Elimination of Technical Barriers to Trade in Industrial Products Adopted by the Council and the Commission between the 1st January 1976 and 1st September 1978*, Vol. II, Brussels, pp. 62, 66

Commission of the European Communities (1980a) Council Directive of 15 July 1980 relating to the quality of water intended for human consumption. *Off. J. Eur. Communities*, *L 229*, 20

Commission of the European Communities (1980b) Council Directive of 15 July 1980 on the approximation of the laws of the Member States relating to the exploitation and marketing of natural mineral waters. *Off. J. Eur. Communities*, *L 229*, 10

Cone, M.V., Baldauf, M.F., Martin, F.M. & Ensminger, J.T. (1980) *Chemicals Identified in Human Biological Media, a Data Base (First Annual Report, October 1979)*, Vol. 1, Part 2, Oak Ridge National Laboratory. Prepared for US Environmental Protection Agency, Springfield, VA, National Technical Information Service, pp. 310-312

Cook-Mozaffari, P. & Doll, R. (1981) Trends in mortality from cancer following fluoridation of water supplies. *J. Epidemiol. Community Health* (in press)

Cook-Mozaffari, P., Bulusu, L. & Doll, R. (1981) A search for an effect of the fluoridation of water supplies on the risk of dying from cancer. *J. Epidemiol. Community Health* (in press)

Coope, B., ed. (1978) Fluorspar - down but not out. *Industrial Minerals*, February, Worcester Park, Surrey, Metal Bulletin, Ltd, pp. 39, 41, 43

Cremer, H.-D. & Buttner, W. (1970) *Absorption of fluorides*. In: *Fluorides and Human Health*, Geneva, World Health Organization, pp. 75-91

Crosby, N.D. & Shepherd, P.A. (1957) Studies on patterns of fluid intake, water balance and fluoride retention. *Med. J. Aust.*, 2, 341-346

Deshpande, S.S. & Bestor, J.F. (1964) Absorption and retention of fluoride from ingested stannous fluoride dentifrice. *J. pharm. Sci.*, 53, 803-807

Doll, R. & Kinlen, L. (1977) Fluoridation of water and cancer mortality in the USA. *Lancet, i*, 1300-1302

Doll, R. & Kinlen, L. (1978) Cancer mortality and fluoridation [Letter]. *Lancet, i*, 150

Eager, R.Y. (1969) *Toxic Properties of Inorganic Fluorine Compounds*, Amsterdam, Elsevier

Erickson, J.D. (1978) Mortality in selected cities with fluoridated and non-fluoridated water supplies. *New Engl. J. Med.*, 298, 1112-1116

Erickson, J.D. (1980) Down syndrome, water fluoridation, and maternal age. *Teratology*, 21, 177-180

Erickson, J.D., Oakley, G.P., Jr, Flynt, J.W., Jr & Hay, S. (1976) Water fluoridation and congenital malformations: no association. *J. Am. dent. Assoc.*, 93, 981-984

Ericsson, Y. & Malmnas, C. (1962) Placental transfer of fluorine investigated with F^{18} in man and rabbit. *Acta obstet. gynecol. scand.*, 41, 144-158

Ericsson, Y. & Ullberg, S. (1958) Autoradiographic investigations of the distribution of F^{18} in mice and rats. *Acta odontol. scand.*, 16, 363-381

Essex Chemical Corp. (1980) *Material Safety Data Sheet - Hydrofluosilicic Acid*, Clifton, NJ

Fabbri, L., De Rosa, E., Potenza, I., Mapp, C., Rossi, A., Brighenti, F. & Forin, F. (1978) Fluorosis hazard in the production of phosphate fertilizers (Ital.). *Med. Lav.*, 69, 594-604 [*Chem. Abstr.*, 91, 43887r]

Feltman, R. & Kosel, G. (1961) Prenatal and postnatal ingestion of fluorides - fourteen years of investigation - final report. *J. dent. Med.*, 16, 190-199

Ferguson, D.B. (1971) Effects of low doses of fluoride on serum proteins and a serum enzyme in man. *Nature - New Biol.*, 231, 159-160

Frajola, W.J. (1960) *Fluoride and enzyme inhibition*. In: Muhler, J.C. & Hine, M.K., eds, *Fluorine and Dental Health*, Bloomington, IN, Indiana University Press, pp. 60-69

Galal, H.E. & Abd-Alla, S.A. (1976) Chromosomal aberrations and mitotic inhibition induced by sodium fluoride and diethyl amine in root-tip cells of *Allium cepa, Allium sativum* and *Vicia faba. Egypt. J. Genet. Cytol.*, 5, 262-280

Gall, J.F. (1980a) *Fluorine compounds, inorganic, calcium*. In: Kirk, R.E. & Othmer, D.F., eds, *Encyclopedia of Chemical Technology*, 3rd ed., Vol. 10, New York, John Wiley & Sons, pp. 707-717

Gall, J.F. (1980b) *Fluorine compounds, inorganic, hydrogen*. In: Kirk, R.E. & Othmer, D.F., eds, *Encyclopedia of Chemical Technology*, 3rd ed., Vol. 10, New York, John Wiley & Sons, pp. 733, 742-746, 752-753

Gedalia, I., Brzezinski, A., Portuguese, N. & Bercovici, B. (1964) The fluoride content of teeth and bones of human foetuses. *Arch. oral Biol.*, 9, 331-340

Gettler, A.O. & Ellerbrook, L. (1939) Toxicology of fluorides. *Am. J. med. Sci.*, 197, 625-638

Glattre, E. & Wiese, H. (1979) Inverse relationship between fluoride and cancer in mouth and throat? *Acta odontol. scand.*, *37*, 9-14

Gloor, H.J., Wetterwald, O. & Truniger, B. (1980) Treatment of myelomas by fluorides (Ger.). *Schweiz. med. Wochenschr.*, *110*, 807-812

Goodall, C.M. & Foster, F.H. (1980) Fluoridation and cancer mortality in New Zealand. *N. Z. med. J.*, *92*, 164-167

Hagan, T.L., Pasternack, M. & Scholz, G.C. (1954) Waterborne fluorides and mortality. *Public Health Rep.*, *69*, 450-454

Hakeem, H. & Shehab, A. (1970) Morpho-cytological studies on the effect of sodium fluoride solutions on *Vicia faba*. *U.A.R. J. Bot.*, *13*, 9-27

The Harshaw Chemical Company (undated) *Harshaw Fluorides for Industry, Product Data/Prices: Hydrofluosilicic Acid, 30%, Bulletin 21980*, Cleveland, OH

Harwood, J.E. (1969) The use of an ion-selective electrode for routine fluoride analyses on water samples. *Water Res.*, *3*, 273-280

Hawley, G., ed. (1977) *The Condensed Chemical Dictionary*, 9th ed., New York, Van Nostrand-Reinhold, pp. 151, 392, 791, 812-813

Hellung-Larsen, P. & Klenow, H. (1969) On the mechanism of inhibition by fluoride ions of the DNA polymerase reaction. *Biochim. biophys. Acta*, *190*, 434-444

Herman, J.R. (1955) Fluorine in urinary tract calculi. *Proc. Soc. exp. Biol. Med.*, *91*, 189-191

Heyroth, F.F. (1952) Toxicological evidence for the safety of the fluoridation of public water supplies. *Am. J. public Health*, *42*, 1568-1575

Hodge, H.C. (1961) Metabolism of fluorides. *J. Am. med. Assoc.*, *177*, 313-316

Hodge, H.C. & Smith, F.A. (1965) *Biological properties of inorganic fluorides*. In: Simons, J.H., ed., *Fluorine Chemistry*, Vol. 4, New York, Academic Press, pp. 113-119

Hodge, H.C. & Smith, F.A. (1972) *Fluorides*. In Lee, D.H.K., ed., *Metallic Contaminants and Human Health*, New York, Academic Press, pp. 163-187

Hoffman, R., Mann, J., Calderone, J., Trumbull, J. & Burkhart, M. (1980) Acute fluoride poisoning in a New Mexico elementary school. *Pediatrics*, *65*, 897-900

Hoover, R.N., McKay, F.W. & Fraumeni, J.F., Jr (1976) Fluoridated drinking water and the occurrence of cancer. *J. natl Cancer Inst.*, *57*, 757-768

IARC (1981) *IARC Monographs on the Evaluation of the Carcinogenic Risk of Chemicals to Humans*, Vol. 25, *Wood, Leather and Some Associated Industries*, Lyon, International Agency for Research on Cancer

International Labour Office (1977) *Occupational Exposure Limits for Airborne Toxic Substances (Occupational Safety & Health Series No. 37)*, Geneva, pp. 118-119, 190-191

Irving, C.C. (1966) Enzymatic deacetylation of *N*-hydroxy-2-acetylaminofluorene by liver microsomes. *Cancer Res.*, *26*, 1390-1396

Jagiello, G. & Liu, J.-S. (1974) Sodium fluoride as potential mutagen in mammalian eggs. *Arch. environ. Health*, *29*, 230-235

Jankauskas, J. (1974) Effects of fluoride on the kidney (A review). *Fluoride*, *7*, 93-105

Jansen, I. & Thomson, H.M. (1974) Heart deaths and fluoridation. *Fluoride*, *7*, 52-57

Kanisawa, M. & Schroeder, H.A. (1969) Life term studies on the effect of trace elements on spontaneous tumors in mice and rats. *Cancer Res.*, *29*, 892-895

Kinlen, L. (1975) Cancer incidence in relation to fluoride level in water supplies. *Br. dent. J.*, *138*, 221-224

Kinlen, L. & Doll, R. (1977) Cancer and fluoride [Letter]. *Lancet, ii*, 1039

Kinlen, L.J. & Doll, R. (1981) Fluoridation of water and cancer mortality in US cities: a re-examination. *J. Epidemiol. Community Health* (in press)

Koicheva, G. (1974) Method for determining the fluorine ion content in working solutions of sodium fluosilicate during fluoridation of drinking water (Bulg.). *Tr. Nauchnoizsled. Inst. Vodosnabdyavane, Kanaliz. Sanit. Tekh.*, *10*, 61-72 [*Chem. Abstr.*, *86*, 95750g]

Kralisz, W. & Szymaniak, E. (1978) Evaluation of the cytogenetic activity of sodium fluoride *in vitro* (Pol.). *Czas. Stomat.*, *31*, 1109-1113

Kram, D., Schneider, E.L., Singer, L. & Martin, G.R. (1978) The effects of high and low fluoride diets on the frequencies of sister chromatid exchanges. *Mutat. Res.*, *57*, 51-55

Largent, E.J. (1954) *Metabolism of inorganic fluorides*. In: Shaw, J.H., ed., *Fluoridation as a Public Health Measure*, Washington DC, American Association for the Advancement of Science, pp. 49-78

Largent, E.J. (1961) *Fluorosis. The Health Aspects of Fluorine Compounds*, Columbus, OH, Ohio State University Press, p. 24

Latham, M.C. & Grech, P. (1967) The effects of excessive fluoride intake. *Am. J. publ. Health*, *57*, 651-660

Leone, N.C., Geever, E.F. & Moran, N.C. (1956) Acute and subacute toxicity studies of sodium fluoride in animals. *Publ. Health Rep. (Wash.)*, *71*, 459-467

Lidbeck, W.L., Hill, I.B. & Beeman, J.A. (1943) Acute sodium fluoride poisoning. *J. Am. med. Assoc.*, *121*, 826-827

Lim, J.K.J., Jensen, G.K. & King, O.H., Jr (1975) Some toxicological aspects of stannous fluoride after ingestion as a clear, precipitatefree solution compared to sodium fluoride. *J. dent. Res.*, *54*, 615-625

Lim, J.K., Renaldo, G.J. & Chapman, P. (1978) LD_{50} of SnF_2, NaF, and Na_2PO_3F in the mouse compared to the rat. *Caries Res.*, *12*, 177-179

Lindahl, C.B. (1980) *Fluorine compounds, inorganic, phosphorus*. In: Kirk, R.E. & Othmer, D.F., eds, *Encyclopedia of Chemical Technology*, 3rd ed., Vol. 10, New York, John Wiley & Sons, pp. 783-784, 786-788

Lindahl, C.B. & Meshri, D.T. (1980a) *Fluorine compounds, inorganic, introduction*. In: Kirk, R.E. & Othmer, D.F., eds, *Encyclopedia of Chemical Technology*, 3rd ed., Vol. 10, New York, John Wiley & Sons, pp. 655-659

Lindahl, C.B. & Meshri, D.T. (1980b) *Fluorine compounds, inorganic, tin*. In: Kirk, R.E. & Othmer, D.F., eds, *Encyclopedia of Chemical Technology*, 3rd ed., Vol. 10, New York, John Wiley & Sons, pp. 819-820

Lyon, H.W. & Hefferren, J.J. (1970) *Demonstration of Topically Applied Cariostatic Agents in Human Enamel and Dentin (Final Report, October 30)*, American Dental Association Research Institute. Prepared for Office of Naval Research, Department of the Navy, Springfield, VA, National Technical Information Service

Maier, F.J. (1970) 25 years of fluoridation. *J. Am. Water Works Assoc.*, *62*, 3-8

Maiya, M., Ratnakar, K.S., Jituri, K.H. & Hande, H.S. (1977) Endemic fluorosis in the Karnataka state. An epidemiological study. *J. Assoc. Phys. India*, *25*, 43-48

Makhni, S.S., Sidhu, S.S., Singh, P. & Grover, A.S. (1979) Long-term effects of fluoride administration in rabbits, an experimental study. iii. Histological changes in the parathyroid and correlation with changes in the bone. *Fluoride, 12*, 124-128

Marier, J.R. & Rose, D. (1971) *Environmental Fluorides (NRC Canada Publ. No. 12,226)*, Ottawa, National Research Council of Canada

Martin, G.R., Brown, K.S., Matheson, D.W., Lebowitz, H., Singer, L. & Ophaug, R. (1979) Lack of cytogenetic effects in mice or mutations in *Salmonella* receiving sodium fluoride. *Mutat. Res., 66*, 159-167

McClure, F.J., Mitchell, H.H., Hamilton, T.S. & Kinser, C.A. (1945) Balances of fluorine ingested from various sources in food and water by five young men. *J. ind. Hyg. Toxicol., 27*, 159-170

Mendelson, D. (1976) Lack of effect of sodium fluoride on a maternal repair system in *Drosophila* oocytes. *Mutat. Res., 34*, 245-250

Messer, H.H., Armstrong, W.D. & Singer, L. (1973) Influence of fluoride intake on reproduction in mice. *J. Nutr., 103*, 1319-1327

Miller, R.F. & Phillips, P.H. (1953) The metabolism of fluorine in the bones of the fluoride-poisoned rat. *J. Nutr., 51*, 273-281

Mitchell, B. & Gerdes, R.A. (1973) Mutagenic effects of sodium and stannous fluoride upon *Drosophila melanogaster*. *Fluoride, 6*, 113-117

Mohamed, A.H. & Chandler, M.E. (1977) *Cytological Effects of Sodium Fluoride on Mice*. In: *Hearings before a Subcommittee on the Committee on Government Operations House of Representatives, 95th Congress, 1st Session, Sept. 21 and Oct. 12*, Washington DC, pp. 42-60

Mohamed, A.H., Applegate, H.G. & Smith, J.D. (1966) Cytological reactions induced by sodium fluoride in *Allium cepa* root tip chromosomes. *Can. J. Genet. Cytol., 8*, 241-244

Mouftah, S.P. & Smith, J.D. (1971) Mitotic aberrations in bean chromosomes induced by sodium fluoride (Abstract). *Texas J. Sci., 22*, 296

Mukherjee, R.N. & Sobels, F.H. (1968) The effects of sodium fluoride and iodo acetamide on mutation induction by X-irradiation in mature spermatozoa of *Drosophila*. *Mutat. Res., 6*, 217-225

Murray, M.M., Ryle, J.A., Simpson, B.W. & Wilson, D.C. (1948) *Thyroid Enlargement and Other Changes Related to the Mineral Content of Drinking Water (with a Note on Goitre Prophylaxis)*, Memo 18, London, Medical Research Council

Myers, D.M., Plueckhahn, V.D. & Rees, A.L.G. (1980) *Report of the Committee of Inquiry into the Fluoridation of Victorian Water Supplies for 1979-80*, Melbourne, F.D. Atkinson, p. 41

National Institute for Occupational Safety & Health (1975) *Criteria for a Recommended Standard: Occupational Exposure to Inorganic Fluorides*, Cincinnati, OH, US Department of Health, Education, & Welfare

National Research Council (1971) *Biologic Effects of Atmospheric Pollutants: Fluorides*, Washington DC, National Academy of Sciences, pp. 5-65, 77-78, 94-100, 133-288

National Research Council (1974) *Effects of Fluorides in Animals*, Washington DC, National Academy of Sciences, pp. 1-19, 40-46, 62-70

National Research Council (1977) *Drinking Water and Health*, Vol. 1, Washington DC, National Academy of Sciences, pp. 369-400, 433-435, 474-487

National Research Council (1980) *Drinking Water and Health*, Vol. 3, Washington DC, National Academy Press, pp. 279-283, 376-379

Needleman, H.L., Pueschel, S.M. & Rothman, K.J. (1974) Fluoridation and the occurrence of Down's syndrome. *New Engl. J. Med., 291*, 821-823

Nixon, J.M. & Carpenter, R.G. (1974) Mortality in areas containing natural fluoride in their water supplies, taking account of socioenvironmental factors and water hardness. *Lancet, ii*, 1068-1071

Ogilvie, A.L. (1953) Histologic findings in the kidney, liver, pancreas, adrenal, and thyroid glands of the rat following sodium fluoride administration. *J. dent. Res., 32*, 386-397

Okushi, I. (1954) Changes of the heart muscle due to chronic fluorosis. I. Electrocardiogram and cardiac X-ray picture made in inhabitants of high-fluoride zone. *Shikoku Acta med., 5*, 159-165

Oldham, P.D. & Newell, D.J. (1977) Fluoridation of water supplies and cancer - a possible association? *Appl. Stat., 26*, 125-135

Olin Corporation (undated) *Chemicals for Industry*, New Haven, CT, Olin Chemicals Division

Pennwalt Corporation (undated) *Product Information: Sodium Monofluorophosphate, NF*, Tulsa, OK, Ozark-Mahoning Special Chemicals Division

Perkinson, J.D., Whitney, I.B., Monroe, R.A., Lotz, W.E. & Comar, C.L. (1955) Metabolism of fluorine 18 in domestic animals. *Am. J. Physiol., 182*, 383-389

Princi, F. (1960) Fluorides: a critical review. III. The effects on man of the absorption of fluoride. *J. occup. Med., 2*, 92-99

Quan, C.K. (1978a) *Fluorine*. In: *Mineral Commodity Profiles, MCP-20, August 1978*, Washington DC, Bureau of Mines, US Government Printing Office, pp. 5-6

Quan, C.K. (1978b) *Fluorspar*. In: *Minerals Yearbook, 1976*, Vol. 1, *Metals, Minerals and Fuels*, Washington DC, Bureau of Mines, US Government Printing Office, pp. 551-575

Raman, S., Becking, G., Grimard, M., Hickman, J.R., McCullough, R.S. & Tate, R.A. (1977) *Fluoridation and Cancer: An Analysis of Canadian Drinking Water Fluoridation and Cancer Mortality Data*, Ottawa, Information Directorate, Department of National Health and Welfare

Rapaport, I. (1956) Study on mongolism. Pathogenic role of fluorine (Fr.). *Bull. Acad. natl. Med. (Paris), 140*, 529-531

Rapaport, I. (1959) New studies on mongolism. On the pathogenic role of fluorine (Fr.). *Bull. Acad. natl Med. (Paris), 143*, 366-370

Richards, G.A. & Ford, J.M. (1979) Cancer mortality in selected New South Wales localities with fluoridated and non-fluoridated water supplies. *Med. J. Aust., 2*, 521-523

Riekstniece, E., Myers, H.M. & Glass, L.E. (1965) *In vivo* effects of sodium fluoride on serum proteins and enzymes as studies with starch gel electrophoresis. *Arch. oral Biol., 10*, 107-117

Riggs, B.L., Hodgson, S.F., Hoffman, D.L., Kelly, P.J., Johnson, K.A. & Taves, D. (1980) Treatment of primary osteoporosis with fluoride and calcium. Clinical tolerance and fracture occurrence. *J. Am. med. Assoc., 243*, 446-449

Rogot, E., Sharrett, A.R., Feinleib, M. & Fabsitz, R.R. (1978) Trends in urban mortality in relation to fluoridation status. *J. Epidemiol., 107*, 104-112

Roholm, K. (1936) Acute fluorine poisoning (Ger.). *Dtsch. Z. ges. gerichtl. Med., 27*, 174-188

Roman, R.J., Carter, J.R., North, W.C. & Kauker, M.L. (1977) Renal tubular site of action of fluoride in Fischer 344 rats. *Anesthesiology, 46*, 260-264

Schlatter, C.H. (1978) *Metabolism and toxicology of fluorides.* In: Courvoisier, B., Donath, A. & Baud, C.A., eds, *Symposium Centre d'Etudes des Maladies Osteo-articulaires de Geneve, II. Fluoride and Bone*, Geneva, Editions Medecine et Hygiene, pp. 1-21

Segreto, V.A., Yeary, R.A., Broks, R. & Harris, N.O. (1961) Toxicity study of stannous fluoride in Swiss strain mice. *J. dent. Res., 40*, 623

Shawe, D.R., ed. (1976) *Geology and Resources of Fluorine in the United States* (*Geological Survey Professional Paper 933*), Washington DC, US Government Printing Office, pp. 1-4

Shea, J.J., Gillespie, S.M. & Waldbott, G.L. (1967) Allergy to fluoride. *Ann. Allergy, 25*, 388-391

Shourie, K.L., Hein, J.W. & Hodge, H.C. (1950) Preliminary studies of the caries inhibiting potential and acute toxicity of sodium monofluorophosphate. *J. dent. Res., 29*, 529-533

Shupe, J.L. (1971) *Clinical and pathological effects of fluoride toxicity in animals.* In: Ciba Foundation, *Carbon - Fluorine Compounds; Chemistry, Biochemistry and Biological Activities*, Amsterdam, Elsevier, pp. 357-388

Shupe, J.L., Olson, A.E. & Sharma, R.P. (1972) Fluoride toxicity in domestic and wild animals. *Clin. Toxicol., 5*, 195-213

Siddiqui, A.H. (1955) Fluorosis in Nalgonda district, Hyderabad-Deccan. *Br. med. J., iv*, 1408-1413

Singh, A. & Jolly, S.S. (1970) *Chronic toxic effects on the skeletal system.* In: *Fluoride and Human Health*, Geneva, World Health Organization, pp. 238-249

Singh, A., Jolly, S.S., Bansal, B.C. & Mathur, C.C. (1963) Endemic fluorosis. Epidemiological, clinical and biochemical study of chronic fluorine intoxication in Panjab (India). *Medicine, 42*, 229-246

Smith, A.H. (1980) An examination of the relationship between fluoridation of water and cancer mortality in 20 large US cities. *N.Z. med. J., 661*, 413-416

Smith, F.A. & Hodge, H.C. (1959) *Fluoride toxicity.* In: Muhler, J.C. & Hine, M.K., eds, *Fluorine and Dental Health*, Bloomington, IN, Indiana University Press, pp. 11-37

Smith, F.A., Gardner, D.E. & Hodge, H.C. (1955) Investigations on the metabolism of fluoride. III. Effect of acute renal tubular injury on urinary excretion of fluoride by the rabbit. *Arch. ind. Health, 11*, 2-10

Smith, M.C. & Smith, H.V. (1935) The occurrence of mottled enamel on the temporary teeth. *J. Am. dent. Assoc., 22*, 814-817

Spira, L. (1944) Incidence of dystrophies caused by fluorine in organs regulated by the parathyroid glands. *J. Hyg., 43*, 402-408

Spira, L. (1946) Disturbance of pigmentation in fluorosis. *Acta med. scand., 126*, 65-84

Spoerke, D.G., Bennett, D.L. & Gullekson, D.J.-K. (1980) Toxicity related to acute low dose sodium fluoride ingestions. *J. Fam. Pract., 10*, 139-140

Stevenson, C.A. & Watson, A.R. (1957) Fluoride oestosclerosis. *Ann. J. Roentgenol., 78*, 13-18

Stokinger, H.E. (1949) *Toxicity following inhalation of fluorine and hydrogen fluoride.* In: Voegtlin, C. & Hodger, H.C., eds, *Pharmacology and Toxicology of Uranium Compounds*, Ch. 17, New York, McGraw-Hill, pp. 1021-1057

Strassburg, M.A. & Greenland, S. (1979) Methodologic problems in evaluating the carcinogenic risk of environmental agents. *J. environ. Health, 41,* 214-217

Stratmann, K.-R. (1979) Comparison of the embryotoxic action of inorganic fluorides (Ger.) *Dtsch. zahnaerztl. Z., 34,* 484-486

Takamori, T., Miyanaga, S., Kawahara, H., Okushi, I., Hirao, M. & Wakatsuki, H. (1956) Electrocardiographical studies of the inhabitants in high fluorine districts. *Tokushima J. exp. Med., 3,* 50-53

Tannenbaum, A. & Silverstone, H. (1949) Effect of low environmental temperature, dinitrophenol, or sodium fluoride on the formation of tumors in mice. *Cancer Res., 9,* 403-410

Taves, D.R. (1977) *Fluoridation and cancer mortality.* In: Hiatt, H.H., Watson, J.D. & Winsten, J.A., eds, *Origins of Human Cancer,* Vol. 4, Cold Spring Harbor, NY, Cold Spring Harbor Laboratory, pp. 357-366

Taves, D.R. (1979) *Claims of harm from fluoridation.* In: Johansen, E., Taves, D.R. & Olsen, T.E., eds, *Continuing Evaluation of the Use of Fluorides (American Association for the Advancement of Science Selected Symposium 11),* Boulder, CO, Westview Press, pp. 295-321

Taves, D.R. (1980) Fluoride distribution and biological availability in the fallout from Mount St Helens, 18 to 21 May 1980. *Science, 210,* 1352-1354

Taylor, A. (1954) Sodium fluoride in the drinking water of mice. *Dent. Dig., 60,* 170-172

Taylor, J.M., Scott, J.K., Maynard, E.A., Smith, F.A. & Hodge, H.C. (1961a) Toxic effects of fluoride on the rat kidney. I. Acute injury from single large doses. *Toxicol. appl. Pharmacol., 3,* 278-289

Taylor, J.M., Gardner, D.E., Scott, J.K., Maynard, E.A., Downs, W.L., Smith, F.A. & Hodge, H.C. (1961b) Toxic effects of fluoride on the rat kidney. II. Chronic effects. *Toxicol. appl. Pharmacol., 3,* 290-314

Temple, P.J. & Weinstein, L.H. (1978) Is hydrogen fluoride mutagenic in plants? *J. Air Poll. Control Assoc., 28,* 151-152

The Tennessee Corporation (undated) *Sodium Silicofluoride,* Washington DC

Todorovic, P., Laban-Bozic, O. & Kapor, S. (1977) Determination of sodium fluoride and chlorophyll in medicinal toothpaste (Serbo-Croat). *Farm. Vestn. (Ljubljana), 28,* 309-313 [*Chem. Abstr., 89,* 117934m]

Toxicology Information Subcommittee (1981) Carcinogenesis bioassay of sodium fluoride. *Tox-Tips,* January, pp. 56-19 - 56-20

US Bureau of the Census (1976) *Current Industrial Reports, Inorganic Chemicals, 1975 (Series M28A(75)-14),* Washington DC, US Department of Commerce, p. 8

US Bureau of the Census (1979) *Current Industrial Reports, Inorganic Chemicals, 1978 (Series M28A(78)-13),* Washington DC, US Department of Commerce, p. 8

US Bureau of the Census (1980) *US General Imports (IM146/December 1979),* Washington DC, US Department of Commerce, p. 1367

US Bureau of Mines (1980a) *Mineral Industry Surveys, Fluorspar in 1979,* US Department of the Interior, Washington DC, US Government Printing Office, pp. 1, 3, 5

US Bureau of Mines (1980b) *Minerals Yearbook, 1977,* Vol. 1, *Metals and Minerals,* US Department of the Interior, Washington DC, US Government Printing Office, pp. 406-411

US Environmental Protection Agency (1972) *EPA Compendium of Registered Pesticides*, Vol. 4, Washington DC, US Government Printing Office, p. G-1

US Environmental Protection Agency (1973a) *EPA Compendium of Registered Pesticides*, Vol. 2, Part 1, Washington DC, US Government Printing Office, pp. S-59-00.01 - S-59-00.03

US Environmental Protection Agency (1973b) *EPA Compendium of Registered Pesticides*, Vol. 3, Washington DC, US Government Printing Office, pp. S-10.1 - S-10.6, S-11.2 - S-11.3

US Environmental Protection Agency (1974) *EPA Compendium of Registered Pesticides*, Vol. 3, Washington DC, US Government Printing Office, pp. S-10.7 - S-10.8, S-11.1

US Environmental Protection Agency (1978) Pesticides programs. Notice of rebuttable presumption against registration and continued registration of pesticide products containing inorganic arsenic. *Fed. Regist.*, *43*(202), 48267-48269, 48324-48342

US Environmental Protection Agency (1979) Water programs; determination of reportable quantities for hazardous substances. *US Code Fed. Regul.*, *Title 40*, Part 117; *Fed. Regist.*, *44*(169), 50766-50779

US Environmental Protection Agency (1980a) *Chemicals in Commerce Information System (CICIS)*, Washington DC, Office of Pesticides and Toxic Substances, Chemical Information Division

US Environmental Protection Agency (1980b) Inorganic chemicals manufacturing point source category effluent limitations guidelines, pretreatment standards, and new source performance standards. *US Code Fed. Regul.*, *Title 40*, Part 415; *Fed. Regist.*, *45*(144), 49450, 49471, 49474, 49499

US Environmental Protection Agency (1980c) Mineral mining and processing point source category. *US Code Fed. Regul.*, *Title 40*, Part 436, pp. 78-79

US Food & Drug Administration (1979) Vitamin and mineral drug products for over-the-counter human use. *US Code Fed. Regul.*, *Title 21*, Part 345; *Fed. Regist.*, *44*(53), 16126-16130, 16174-16175, 16179-16181, 16197-16201

US Food & Drug Administration (1980a) Quality standards for foods with no identity standards. Bottled water. *US Code Fed. Regul.*, *Title 21*, Part 103.35, pp. 49-52

US Food & Drug Administration (1980b) Anticaries drug products for over-the-counter human use. Establishment of a monograph; notice of proposed rulemaking. *US Code Fed. Regul.*, *Title 21*, Part 355; *Fed. Regist.*, *45*(62), 20666-20691

US Occupational Safety & Health Administration (1980) Air contaminants. *US Code Fed. Regul.*, *Title 29*, Part 1910.1000, p. 73

US Pharmacopeial Convention, Inc. (1980) *The US Pharmacopeia*, Rockville, MD, pp. 730, 740-741

Vesco, C. & Colombo, B. (1970) Effect of sodium fluoride on protein synthesis in HeLa cells: inhibition of ribosome dissociation. *J. mol. Biol.*, *47*, 335-352

Voroshilin, S.I., Plotko, E.G., Gatiyatullina, E.Z. & Giliova, E.A. (1973) Cytogenetic effect of inorganic fluorine compounds on human and animal cells *in vivo* and *in vitro* (Russ.). *Genetica*, *9*, 115-120

Wachter, K. (1980) *Fluorine compounds, inorganic, sodium*. In: Kirk, R.E. & Othmer, D.F., eds, *Encyclopedia of Chemical Technology*, 3rd ed., Vol. 10, New York, John Wiley & Sons, pp. 797-798

Waldbott, G.L. (1961) The physiologic and hygienic aspects of the absorption of inorganic fluorides. *Arch. environ. Health, 2*, 155-167

Wallace-Durbin, P. (1954) The metabolism of fluorine in the rat using F^{18} as a tracer. *J. dent. Res., 33*, 789-800

Weast, R.C., ed. (1979) *CRC Handbook of Chemistry and Physics*, 60th ed., Boca Raton, FL, Chemical Rubber Co., pp. B79, B126

Whitford, G.M. & Pashley, D.H. (1979) *The effect of body fluid pH on fluoride distribution, toxicity and renal clearance*. In: Johansen, G., Taves, D.R. & Olsen, T.O., eds, *Continuing Evaluation of the Use of Fluorides* (*American Association for the Advancement of Science Selected Symposium 11*), Boulder, CO, Westview Press, pp. 187-221

Whitford, G.M., Pashley, D.H. & Stringer, G.I. (1976) Fluoride renal clearance: a pH-dependent event. *Am. J. Physiol., 230*, 527-532

WHO (1970) *Fluorides and Human Health*, Geneva

Wilson, D.C. (1941) Fluorine in the aetiology of endemic goitre. *Lancet, i*, 211-212

Windholz, M., ed. (1976) *The Merck Index*, 9th ed., Rahway, NJ, Merck & Co., pp. 211-212, 541, 1114-1115, 1135

Yiamouyiannis, J. & Burk, D. (1975) Letter (December 15) to Hon. J.J. Delaney; cancer in our drinking water? *Congr. Rec., H 12731-34*, 40773-40775

Yiamouyiannis, J. & Burk, D. (1977) Fluoridation and cancer. Age-dependence of cancer mortality related to artificial fluoridation. *Fluoride, 10*, 102-125

Zanfagna, P.E. (1976) Allergy to fluoride. *Fluoride, 9*, 36-41

Zebrowski, E.J., Suttie, J.W. & Phillips, P.H. (1964) Metabolic studies in fluoride fed rats (Abstract no. 502). *Fed. Proc., 23*, 184

Zipkin, I. & Babeaux, W.L. (1965) Maternal transfer of fluoride. *J. oral Ther. Pharmacol., 1*, 652-665

APPENDIX 1

APPENDIX I

APPENDIX 1

EPIDEMIOLOGICAL EVIDENCE RELATING TO THE POSSIBLE CARCINOGENIC EFFECTS OF HAIR DYES IN HAIRDRESSERS AND USERS OF HAIR DYES

The carcinogenic and mutagenic effects in experimental systems of several of the biologically active components of hair-dye formulations were evaluated previously (IARC, 1978). In the present volume, two additional compounds which have been patented for use as hair dyes (4-chloro-*ortho*- and 4-chloro-*meta*-phenylenediamine) have been evaluated, and the monograph on 2,4- diaminoanisole sulphate has been updated. The following compounds that may be found in hair dyes have thus been considered: (Those marked with an asterisk were also considered, but no monograph was prepared because adequate data on their carcinogenicity were not available.)

4-amino-2-nitrophenol (3-nitro-4-hydroxyaniline) (Vol. 16, p. 43); 2-amino-4- nitrophenol*; 2-amino-5-nitrophenol*; 2,4-diaminoanisole (3-amino-4-methoxyaniline; 4-methoxy-*meta*-phenylenediamine) (Vol. 16, p. 51; Vol. 27, p. 103); 1,2-diamino-4-nitrobenzene (2-amino-4-nitroaniline; 4-nitro-*ortho*-phenylenediamine) (Vol. 16, p. 63); 1,4-diamino-2-nitrobenzene (2-nitro-*para*- phenylenediamine; 2-nitro-4-aminoaniline) (Vol. 16, p. 73); 2,4- diaminotoluene (3-amino-4-methylaniline) (Vol. 16, p. 83); 2,5-diaminotoluene (2-methyl-4-aminoaniline) (Vol. 16, p. 97); *meta*-phenylenediamine (3- aminoaniline) (Vol. 16, p. 111); *ortho*-phenylenediamine (2-aminoaniline)*; *para*-phenylenediamine (4-aminoaniline) (Vol. 16, p. 125); *N*-phenyl-*para*-phenylenediamine*.

While data on chemical and physical properties, technical products and impurities, production and use, and analytical methods refer to the chemical itself, biological data come in many instances from experiments in which the chemical was administered as a constituent of a mixture, the exact composition of which was not known. In those cases, it was not possible to evaluate the carcinogenicity of the pure compound. The same limitation applies, and to a much greater extent, to evaluations of the carcinogenicity of these compounds on the basis of human data.

At best, epidemiological studies can document brand, type (e.g., whether temporary, semi-permanent or permanent) and extent of use of hair dyes. Rarely is it possible to document exposure to specific compounds; this is even more difficult in the case of people with occupational exposure to hair dyes, e.g., hairdressers. They may be exposed not only to many different hair dyes but also to other possible carcinogens, for example, aerosol propellants (vinyl chloride, fluorocarbons, etc.). Nonetheless, hair service occupation have been considered here as well as personal exposure to hair dyes. However, the problem of mixed exposure must be kept in mind when interpreting the results.

Hair-dye formulations and mode of use (Wall, 1972)

Hair dyes can be divided into three types, depending on whether they colour the hair temporarily, semi-permanently or permanently. The dyes, dye intermediates and other ingredients involved vary according to the type of hair product.

(a) Temporary dyes

These consist of a miscellaneous group of chemicals designed to effect a change in the shade of the hair and are generally applied in the form of shampoos; if they are not re-applied, they are removed by washing at the next shampoo. A commercial formulation may contain between 0.5 and 2.0% of the permitted colour, to give the appropriate shade, together with surface-active agents. These colour shampoos may also contain nitro compounds as colourants, together with anionic detergents, and compounds such as urea that increase solubility.

(b) Semi-permanent dyes

Such dyes remain on the hair for several washings and are only gradually washed out by successive shampoos. Many of the aromatic, nitro and amino compounds used in the formulation of permanent dyes (see following sub-section) may be used in semi-permanent dyes, without the addition of an oxidizing agent.

A typical semi-permanent formulation contains:
Monoethanolamine lauryl sulphate 20.0%
Ethylene glycol monostearate   5.0%
Diethanolamine fatty-acid salts   3.0%
Diamino-nitrobenzenes (dyes)   1.5%
Perfume }
Water   } to 100%

(c) Permanent dyes (oxidation dyes)

With this group of dyes, development of the colour shade depends on the addition of an oxidizing agent (usually hydrogen peroxide). These dyes can penetrate the hair shaft and, when oxidized, are irreversibly bound within it, allowing the colour to remain. A vast number of formulations are used as permanent dyes; 2-amino-4-nitrophenol, 2-amino-5-nitrophenol, 4-amino-2-nitrophenol, 2,4-diaminoanisole, 2,5-diaminoanisole, 1,2-diamino-4-nitrobenzene, 1,4-diamino-2-nitrobenzene, 2,4-diaminotoluene, 2,5-diaminotoluene, *meta*-phenylenediamine, *ortho*-phenylenediamine, *para*-phenylenediamine and *N*-phenyl-*para*-phenylenediamine have been used in such dye formulations, often as mixtures in order to give the desired shade. Typical formulations contain between 1 and 4% of different dye intermediates.

A typical permanent formulation contains:

Oleic acid	20.0%
Oleyl alcohol	15.0%
Solubilized lanolin	3.0%
Propylene glycol	12.0%
Isopropanol	10.0%
EDTA	0.5%
Sodium sulphite	0.5%
Ammonium hydroxide (28% w/v)	10.0%
Deionized water	26.0%
Dye intermediates	3.0%

The oxidizing agent rapidly oxidizes the aromatic group of the dye to form products which bind to the hair shaft, at the same time lowering absorption by the skin of the specific aromatic amines. However, it should be recognized that some of the products formed by the oxidation process, such as the three-ringed 'Bandrowski-base', may also be absorbed through the skin and may thus pose an additional health hazard.

Further information on the ingredients used in hair rinses and hair dyes is given by Gosselin *et al.* (1976).

Occupational exposure

The studies currently available deal mainly with possible risks of bladder cancer; there are as yet relatively few data concerning cancers at other sites. These studies also relate to exposures occurring at different times over the last 30 years or more, during which period there have been changes in both the types and quantities of hair dyes used (Wall, 1972).

The term 'hairdresser' has different meanings in different countries. In the US, for example, it applies to those involved in the care of hair, primarily women's hair, in beauty salons (also called beauticians); it excludes barbers. In the UK, the term is more inclusive; it is used to denote ladies' hairdressers, and it may also mean barbers.

(a) Bladder cancer

Wynder *et al.* (1963) studied the distribution of smoking habits and occupations in patients with carcinomas of the bladder. They combined interview data from two hospital-based series covering a total of 300 male patients in New York City during the period 1957-1961. A case-control design was used, and controls were matched by age and sex; 93% of cases and 82% of controls were smokers. Among patients, four had worked as hairdressers (all smokers); there were none among the controls. [No statistical analysis of these data was provided by the authors.]

Dunham et al. (1968) compared the most recent occupations of 265 white, male patients with bladder cancer with those of comparable controls in New Orleans, Louisiana, during 1958-1964. They found four barbers in this group, while 1.45 were expected (relative risk, 2.76). During the last three-and-a-quarter years of this study, 132 cases were compared with 136 controls for history of 'use of tonics, lotions and other preparations for the hair and scalp'. The percentage of controls who used such preparations (36%) was slightly higher than that of cases (32%).

Anthony & Thomas (1970) studied a series of 1030 patients (812 men and 218 women) with bladder tumours in Leeds, UK, interviewed in an eight-year series (1959-1967); 43% had known smoking histories. Occupations were defined according to the Registrar General's Classification of Occupations (General Register Office, 1966), looking separately at predominant occupation, occupations ever undertaken and occupations pursued for 20 or more years. Of several alternative analyses presented by the authors, the most reliable appears to be based upon a comparison of the male cases with a control group who had benign surgical disease and were matched to the cases by sex, age, place of residence and amount smoked. Among the cases, there were four hairdressers, of whom three had worked more than 20 years in the occupation; one control was a hairdresser. The relative risks of bladder tumour for hairdressers were, therefore, 4.1 (predominant occupation), 4 (occupation ever undertaken) and 3 (20 or more years in the occupation). Among the 218 female cases, there were no hairdressers, although 0.6 were expected. None of these differences was significant at the $P = 0.05$ level.

Cole et al. (1971, 1972) studied occupation in a systematic sample of patients (356 men, 105 women) aged 20-89 with transitional- or squamous-cell carcinoma of the lower urinary tract in a defined population in Boston, Massachusetts, during an 18-month period in 1967-1968. Controls were comparable with respect to age and sex. Risk was assessed for each of 42 selected occupational titles derived from a modification of the standard US Bureau of the Census classification. Among men, there was no increased risk for barbers (4 observed cases, 7.2 expected), and among women, no increased risk for hairdressers (1 observed case, 0.9 expected).

Howe et al. (1980) reported a study of 480 men and 152 women with bladder cancer and individually matched controls (one control per case). Among men, three cases but no controls were barbers; and among women, two cases but no controls were hairdressers. In addition, eight male cases (including two of the barbers), but no male controls, gave a history of personal use of hair dyes ($P = 0.004$, one-tailed test); only one of them had used hair dyes for more than six years before diagnosis of bladder cancer. There was no evidence in women of an increased risk of bladder cancer associated with personal use of hair dyes (relative risk, 0.7; 95% confidence interval, 0.3-1.4 for ever *versus* never use).

(b) Other cancers

Cancer mortality among men and single women in England and Wales was examined for the period 1949-1953 (Registrar General, 1958). For male barbers and hairdressers, the only indication of excess mortality was that from cancer of the lung and bronchus (114 observed, 99 expected). For single female hairdressers and manicurists, the numbers of observed deaths exceeded those expected from cancers at all sites combined (43:37) and from cancers of the lung and bronchus (4:2), breast (13:9) and cervix uteri (4:1). There was no excess of cancer at any other site.

Similar data from the Registrar General (1971) for the period 1959-1963 showed no excess mortality from cancer at any site in male hairdressers and barbers. Among single female hairdressers and manicurists, there were 21 observed deaths from breast cancer *versus* 12 expected (P <0.05); for cancers of the cervix uteri and other parts of the uterus the ratios of observed to expected deaths were 3:2 and 4:2, respectively. A further report from England and Wales (Registrar General, 1978), in which occupational mortality for the period 1970-1972 was analysed, did not indicate a significant excess of any cancer in men in those occupations. Single female hairdressers and manicurists had a Standardized Mortality Ratio (SMR) of 86 for all causes of deaths; for cancer deaths, no Proportional Mortality Ratio (PMR) was reported to reach statistical significance at the 0.01 level. [It should be noted that the excess of mortality from breast cancer reported for 1959-1963 (Registrar General, 1971), if still present, would not necessarily be mentioned in the most recent presentation of the data.]

Milham (1976) reported the results of a mortality study by occupation carried out in Washington state. The study was based on a proportional mortality analysis of statements of occupation on the death certificates of all white male residents over the age of 20 who had died in the state during the period 1950-1971. For barbers, all causes of death were examined; proportional mortality was stated to be significantly increased (P < 0.05) for oesophageal cancer and multiple myeloma.

Viadana *et al.* (1976) reported an increase in the incidence of laryngeal cancer among barbers in a hospital-based case-control study conducted in Buffalo, New York, between 1956 and 1965. A total of 11 591 white men (76% with cancer - the cases, and 24% without cancer following investigations for cancer - the controls) were studied with respect to both occupations ever held and occupations held for five or more years. Each occupation was considered in relation to a particular cancer site only if there were five or more men with that occupation among all men with the cancer. Results were standardized separately for age and smoking. Barbers had an age-adjusted relative risk of 2.83 (P <0.05) for laryngeal cancer (ten cases); this excess persisted in those who had been barbers for five or more years (2.49, P >0.05) (eight cases). When standardized for smoking, but not age, the relative risk for laryngeal cancer in barbers was 3.39, or 3.19 in those who had been barbers for five or more years (P <0.05 for both). There was no significant excess of bladder cancer among the barbers.

Garfinkel et al. (1977) studied the risk of any cancer in beauticians in Alameda County, California, during the period 1958-1962 by analysis of death certificates that gave cancer as the underlying cause of death. Of 25 beauticians who had died from cancer, 24 were women; they had died of cancers at the following sites: lung (6), breast (5), cervix (4), ovary (3), brain (1), bladder (1), stomach (1), synovium (1) and unspecified (2). The one man died of liver cancer. On the basis of 1000 controls who had no mention of cancer on their death certificates, an expected number of 21.8 beauticians was calculated for the 3460 adult women who died of cancer ($P = 0.43$ for the difference from the 24 observed). The hypothesis that female beauticians were at increased risk specifically of lung cancer was tested, using a matched-pairs design. Adult women who died from causes other than cancer were matched to those who died from lung cancer by age, race, date of death and county of residence. Of 176 lung cancer cases, six were in beauticians, compared with one among the 176 controls. The relative risk was 6.0, with one-tailed $P = 0.06$.

Menck et al. (1977) used data on 15 230 men and 22 792 women, aged 20-64, white and with non-Spanish names, from the Los Angeles County Cancer Surveillance Program to investigate the association of cancers with the latest occupation as reported by the patient or next-of-kin at hospital admission. Twenty of the cases of cancer found in 135 female beauticians were in the lung. Proportional incidence ratios (PIR) and standardized incidence ratios (SIR) were computed, the latter being derived from the sex-specific populations at risk by occupation as ascertained in a 2% sample census of Los Angeles County. Both PIRs and SIRs for lung cancer among beauticians were statistically significantly increased (about two-fold; $P < 0.05$). No data on smoking habits were available. The authors mentioned parenthetically, without giving details, that a case-control study of 199 lung cancer cases and 187 controls had shown a relative risk of 0.94 for beauticians.

Clemmesen (1977) studied the incidence of malignant neoplasms among hairdressers in Denmark during five-year periods from 1943-1972. Using as denominators the numbers of men and women described as hairdressers in successive censuses, he calculated expected numbers of cases of cancer among them using the age- and sex-specific cancer registration rates. For male hairdressers, the expected total number of malignant neoplasms during 1943-1972 was 517.4 and the observed number was 447. There was no excess in any of 13 groupings of cancer sites. In women, there was a large overall excess, with 872 malignant neoplasms observed and only 475.4 expected. This excess appeared in each five-year period and for each of 14 site groupings. [The consistent excess of cancers at all sites in women in this study suggests the possibility of a bias in the reporting of occupation.]

Alderson (1980) studied a sample of about 2000 male hairdressers identified in the 1961 census of England and Wales to test the hypothesis that hairdressers are at risk of certain cancers. The cohort was followed from April 1961 to December 1978, and the number of deaths observed was compared with the number expected on the basis of mortality data for all men in England and Wales. As of December 1978, 504 men had died and 168 men had been lost to follow-up. Mortality from all cancers was similar to that expected (134 observed, 126.1 expected). In addition, none of the SMRs for cancer at specific sites was significantly elevated: cancer of the oesophagus, 5 observed *versus* 3.4 expected; cancer of the larynx, 1 observed *versus* 1.3 expected; cancer of the lung, 52 observed *versus* 50.8 expected; cancer of the bladder, 7 observed *versus* 5.6 expected; and leukaemia, 3 observed *versus* 2.7 expected.

Non-occupational exposure

Shafer & Shafer (1976) asked 100 consecutive breast cancer patients in a clinical practice for a history of hair-dye use, obtaining such history when necessary from the next-of-kin of deceased patients. Only women who had used permanent hair-colouring formulations for five or more years were considered to have had such a history. Of 100 patients, 87% had been regular users of permanent dyes, compared with 26% of age-comparable controls. [No other information was provided concerning the number, nature or selection of the controls or their relationship to cases with respect to potential confounding variables.]

Kinlen *et al.* (1977) reported a study of 191 breast cancer patients interviewed in hospital in 1975 and 1976 in Oxford, UK, and 561 controls without cancer, matched to the patients by age (within three years), marital status and social class. Seventy-three cases and 213 controls had used permanent or semi-permanent hair dyes, giving a relative risk of 1.01. There was no evidence of an increasing risk of breast cancer with increasing duration of use of hair dyes or with use beginning more than four or more than nine years before diagnosis. Stratification by age at first pregnancy showed a deficit of cases in which hair-dye use was reported in those whose first pregnancy occurred at ages 15-19 years (33.8% of cases used hair dyes, compared with 64.7% of controls), and an excess of cases with use of hair dyes by those whose first pregnancy was at 30 years of age or older (38.3% of cases and 25.5% of controls). There were two hairdressers among cases (1.0%) and 10 among controls (1.8%).

Hennekens *et al.* (1979) carried out a cross-sectional mail questionnaire survey in 1976 on 172 413 married female nurses, aged 30-55, in 11 US states, whose names appeared in the 1972 register of the American Nurses' Association. Of the 120 557 responders, 38 459 reported that at some time they had used permanent hair dyes, while 773 reported that they had been hospitalized for cancer or had been diagnosed as having a cancer. The risk ratio for the association of cancers at all sites with hair-dye use (at any time) was 1.10 ($P = 0.02$). When 16 cancer sites were examined separately, significant associations with permanent hair-dye use were found for cervix uteri (relative risk, 1.44; $P < 0.001$) and for vagina and vulva (relative risk, 2.58; $P = 0.02$). These associations were reduced but remained statistically significant after adjustment for smoking habits. There was no consistent trend of cancer risk with increasing interval from first use of hair dyes, although women who had used permanent dyes 21 years or more before the onset of cancer had a significant increase in risk for cancer at all sites combined (relative risk, 1.38; $P = 0.02$), largely because of an excess of breast cancers [relative risk, 1.48]. The authors mentioned that similar analyses of cases of cancer that had occurred only after 1972 (the year the study population was defined from the nurses' register) yielded essentially the same results, thus indicating that early retirement, and loss from the professional register, was not a source of bias in the study. [Other biases cannot, however, be excluded, notably in view of the 30% proportion of non-responders.]

Shore *et al.* (1979) compared the hair-dye use of 129 breast cancer patients and 193 control subjects aged 25 and over, taken from the records of a multiphasic screening clinic. Information was obtained from existing clinical questionnaires and from *ad hoc*, structured telephone interviews. For 23% of the breast cancer cases and 8% of controls, the respondent to the telephone interview was the spouse, sibling or other relative of the subject; 19% of cases and 2% of controls were dead at the time of the interview. Nineteen variables bearing on personal and family medical history, reproductive history, education and occupation, were taken into account (combined in a multivariate confounder score) when comparing dye use in cases and controls. Adjusted relative risks for use of permanent hair dyes 0, 5, 10 and 15 years prior to appearance of breast cancer (or an equivalent time for controls) were, respectively, 1.08, 1.31, 1.58 and 1.44 (none significantly different from 1). A statistically significant relationship was detected between a measure of cumulative hair-dye use (number of years times frequency per year) and breast cancer. This relationship also held if the analysis was limited to cases in which the patient herself had responded to the telephone interview. Among women who had used hair dyes 10 years before developing breast cancer, the relationship held only for women at 'low risk' (as assessed from the distribution of the multivariate confounder score) and for those 50-79 years old. [Use of a multivariate confounder score for the control of confounding can produce misleading results in some situations (Pike *et al.*, 1979).]

Stavraky *et al.* (1979) compared, with respect to hair-dye use, 50 breast cancer cases from a cancer treatment centre and 100 hospitalized controls in London, Ontario; 35 breast cancer cases and 70 neighbourhood controls in Toronto, Ontario; and 36 endometrial cancer cases and 72 neighbourhood controls also in Toronto. Cases and controls were individually matched for age within five years. Relative risks for breast cancer from use of permanent hair dyes (at any time) were 1.3 (95% confidence limits, 0.6-2.5) in London and 1.1 (0.5-2.4) in Toronto, while the relative risk for endometrial cancer was 1.1 (0.5-2.3). Further statistical analyses, allowing for smoking habits, family history of cancer and age at first birth, did not show any statistically significant relationship between hair-dye use and breast cancer; similarly negative results were obtained for endometrial cancer after allowing for age at menarche, age at interview and (separately) country of birth.

Jain *et al.* (1977) reported data on hair-dye use in a study of 107 patients with bladder cancer and an equal number of controls matched to the cases by age ($\pm$ five years) and sex. All male controls had benign prostatic hypertrophy, and all female controls had stress incontinence. The relative risk of bladder cancer in association with any exposure to hair dyes (based on 19 pairs discordant for use of hair dye) was 1.1, with a 95% confidence interval of 0.41-3.03.

Neutel *et al.* (1978) reported data on hair-dye use in a sub-set of 50 case-control pairs (matched by sex and 10-year age group) re-interviewed after a previous, larger case-control study of bladder cancer. Use of hair dyes was reported for 18/50 cases and 19/50 controls. Frequent use of hair dyes and hairdressing as an occupation, however, were said to show *protective* effects (the former being statistically significant, $P < 0.01$) against bladder cancer, although the numbers on which these statements were based are not given in the report.

Nasca *et al.* (1980) reported a study of 118 patients, aged 20 to 84 years, with breast cancer and 233 controls matched to the patients by age and county of residence (115 matched triplets and three matched pairs). For all cases and controls there was no statistically significant association between breast cancer and use of any type of hair dye (relative risk, 1.28; 90% confidence interval, 0.86-1.90) or use of permanent and semi-permanent dyes (relative risk, 1.11), nor was an increase in risk seen with increasing numbers of times hair dyes were used or increasing time since first use. Women who dyed their hair to change its colour (as distinct from those who dyed their hair to mask greyness), however, had a significantly increased risk of breast cancer (relative risk, 3.13; 95% confidence interval, 1.50-6.54). In this group there was a significant trend towards increasing risk with increasing numbers of exposures to hair dyes. Examination of the risk of breast cancer with use of hair dyes in subgroups of women defined by other risk factors for breast cancer showed: a relative risk of breast cancer of 4.5 (95% confidence interval, 1.20-16.78) with use of hair dyes in women who had a past history of benign breast disease; a relative risk of 1.75 ($P = 0.03$, one-tailed test) with use of hair dyes in women who had 12 or more years of schooling; and a relative risk of 3.33 (95% confidence interval, 1.10-10.85) with use of hair dyes in women aged 40-49 years. (The relative risk was near unity for all other age groups.) These effects appeared to be independent of one another and were not explained by confounding on past pregnancy, age at first pregnancy, history of artificial menopause, or age of menarche. The authors stated that the associations observed in the subgroups should be considered newly-generated hypotheses requiring further testing.

Summary of data reported

In four out of five relevant case-control studies of bladder cancer reviewed by the Working Group there was an excess, based on rather small numbers, of barbers or hairdressers (who may or may not have had occupational exposure to hair dyes) among patients with bladder cancer. In one of these studies, men (but not women) with bladder cancer were also more likely to have used hair dyes than were controls. Two additional case-control studies of bladder cancer did not show an increased risk of bladder cancer in hair-dye users. In the two 'positive' studies in which it was mentioned, confounding on cigarette smoking did not appear to explain the observed excess of bladder cancer in barbers or hairdressers.

The occupational mortality surveys of the Registrar General for England and Wales have not shown an increased mortality from bladder cancer in male hairdressers or barbers or in single female hairdressers or manicurists. They have, however, shown, in some periods but not others, increased mortality rates from cancers of the lung and bronchus in both male and female hairdressers and increased rates of cancers of the breast and cervix uteri in female hairdressers. There was no increased mortality from bladder cancer or from cancer at any other single site in a cohort of 2000 British male hairdressers followed for 17 years from 1961.

The risk of breast cancer in relation to exposure to hair dyes has been addressed in several studies in addition to those of the Registrar General for England and Wales. Two, one based on cancer registry data and one based on mortality data of a US county, did not show an increased mortality from breast cancer in female hairdressers but did show an increased risk of lung cancer. One incompletely reported case-control study showed an increased risk of breast cancer in relation to personal use of hair dyes. Three further case-control studies, although not showing a statistically significant increase in risk of breast cancer with use of hair dyes in all subjects, showed significantly increased risks in variously defined subgroups of those studied. Two other well-conducted case-control studies were 'negative'.

In addition to the increased rates of cancers of the lung and cervix mentioned above, increases have been reported at other sites: Male hairdressers were reported to have increased rates of cancers of the larynx and oesophagus and of myeloma; and users of permanent hairdyes had increased rates of cancers of the cervix and of the vulva and vagina. It was not possible in most of these studies to control for the confounding effects of smoking and sexual activity (where relevant).

Evaluation

The evidence relating bladder cancer or any other cancer to the occupation of hairdressing or to personal use of hair dyes is inconclusive. In view of the possible hazard suggested by experimental studies of some components of hair dyes, the subject merits further study, perhaps by a cohort study of hairdressers with documented exposure to hair dyes.

References

Alderson, M. (1980) Cancer mortality in male hairdressers. *J. Epidemiol. Community Health, 34*, 182-185

Anthony, H.M. & Thomas, G.M. (1970) Tumors of the urinary bladder: an analysis of the occupations of 1,030 patients in Leeds, England. *J. natl Cancer Inst., 45*, 879-895

Clemmesen, J. (1977) Statistical studies in the aetiology of malignant neoplasms. V. Trends and risks. Denmark 1943-72. *Acta pathol. microbiol. scand. (Suppl.), 261*, 1-286

Cole, P., Monson, R.R., Haning, H. & Friedell, G.H. (1971) Smoking and cancer of the lower urinary tract. *New Engl. J. Med., 284*, 129-134

Cole, P., Hoover, R. & Friedell, G.H. (1972) Occupation and cancer of the lower urinary tract. *Cancer, 29*, 1250-1260

Dunham, L.J., Rabson, A.S., Stewart, H.L., Frank, A.S. & Young, J.L., Jr (1968) Rates, interview, and pathology study of cancer of the urinary bladder in New Orleans, Louisiana. *J. natl Cancer Inst., 41*, 683-709

Garfinkel, J., Selvin, S. & Brown, S.M. (1977) Brief communication: possible increased risk of lung cancer among beauticians. *J. natl Cancer Inst., 58*, 141-143

General Register Office (1966) *Classification of Occupations*, London, Her Majesty's Stationery Office

Gosselin, R.E., Hodge, H.C., Smith, R.P. & Gleason, M.N. (1976) *Clinical Toxicology of Commercial Products*, 4th ed., Baltimore, Williams & Wilkins, Section VI, pp. 116-119

Hennekens, C.H., Speizer, F.E., Rosner, B., Bain, C.J., Belanger, C. & Peto, R. (1979) Use of permanent hair dyes and cancer among registered nurses. *Lancet, i*, 1390-1393

Howe, G.R., Burch, J.D., Miller, A.B., Cook, G.M., Esteve, J., Morrison, B., Gordon, P., Chambers, L.W., Fodor, G. & Winsor, G.M. (1980) Tobacco use, occupation, coffee, various nutrients, and bladder cancer. *J. natl Cancer Inst., 64*, 701-713

IARC (1978) *IARC Monographs on the Evaluation of the Carcinogenic Risk of Chemicals to Man, Vol. 16, Some Aromatic Amines and Related Nitro Compunds - Hair Dyes, Colouring Agents and Miscellaneous Industrial Chemicals*, Lyon

Jain, M., Morgan, R.W. & Elinson, L. (1977) Hair dyes and bladder cancer. *Can. med. Assoc. J., 117*, 1131-1132

Kinlen, L.J., Harris, R., Garrod, A. & Rodriguez, K. (1977) Use of hair dyes by patients with breast cancer: a case-control study. *Br. med. J., ii*, 366-368

Menck, H.R., Pike, M.C., Henderson, B.E. & Jing, J.S. (1977) Lung cancer risk among beauticians and other female workers: brief communication. *J. natl Cancer Inst., 59*, 1423-1425

Milham, S., Jr (1976) *Occupational Mortality in Washington State, 1950-1971*, Vol. 1, Cincinnati, OH, US Department of Health, Education, & Welfare, National Institute for Occupational Safety & Health, p. 36

Nasca, P.C., Lawrence, C.E., Greenwald, P., Chorost, S., Arbuckle, J.T. & Paulson, A. (1980) Relationship of hair dye use, benign breast disease, and breast cancer. *J. natl Cancer Inst., 64*, 23-28

Neutel, C.I., Nair, R.C. & Last, J.M. (1978) Are hair dyes associated with bladder cancer? *Can. med. Assoc. J., 119*, 307-308

Pike, M.C., Anderson, J. & Day, N. (1979) Some insights into Miettinen's multivariate confounder score approach to case-control study analysis. *J. Epidemiol. Community Health, 33,* 104-106

Registrar General (1958) *The Registrar General's Decennial Supplement, England and Wales, 1951, Occupational Mortality,* Part 2, Vol. 2, Tables, London, Her Majesty's Stationery Office, pp. 154, 291

Registrar General (1971) *The Registrar General's Decennial Supplement, England and Wales, 1961, Occupational Mortality Tables,* London, Her Majesty's Stationery Office, p. 185

Registrar General (1978) *The Registrar General's Decennial Supplement, England and Wales, 1970-1972, Occupational Mortality,* London, Her Majesty's Stationery Office, p. 214

Shafer, N. & Shafer, R.W. (1976) Potential of carcinogenic effects of hair dyes. *N.Y. State J. Med., 76,* 394-396

Shore, R.E., Pasternack, B.S., Thiessen, E.V., Sadow, M., Forbes, R. & Albert, R.E. (1979) A case-control study of hair dye use and breast cancer. *J. natl Cancer Inst., 62,* 277-283

Stavraky, K.M., Clarke, E.A. & Donner, A. (1979) Case-control study of hair dye use by patients with breast cancer and endometrial cancer. *J. natl Cancer Inst., 63,* 941-945

Viadana, E., Bross, I.D.J. & Houten, L. (1976) Cancer experience of men exposed to inhalation of chemicals or to combustion products. *J. occup. Med., 18,* 787-792

Wall, F.E. (1972) *Bleaches, hair colorings and dye removers.* In: Balsam, M.S. & Sagarin, E., eds, *Cosmetics Science and Technology,* 2nd ed., Vol., 2, New York, Wiley-Interscience, pp. 279-343

Wynder, E.L., Onderdonk, J. & Mantel, N,. (1963) An epidemiological investigation of cancer of the bladder. *Cancer, 16,* 1388-1407

SUPPLEMENTARY CORRIGENDA TO VOLUMES 1-26

SUPPLEMENTARY CORRIGENDA TO VOLUMES 1-26

Corrigenda covering Volumes 1-6 appeared in Volume 7, others appeared in Volumes 8, 10-13, 15-26.

Volume 11

p. 49	para 3, lines 3-5	*delete* Cadmium sulphide ... (Harada, 1973).
p. 67	Harada (1973)	*delete*

Supplement 2

p. 372	penultimate line	*replace* 192 *by* 191

CUMULATIVE INDEX TO MONOGRAPHS

CUMULATIVE INDEX TO IARC MONOGRAPHS ON THE EVALUATION OF THE CARCINOGENIC RISK OF CHEMICALS TO HUMANS

Numbers in bold indicate volume, and other numbers indicate page. References to corrigenda are given in parentheses. Compounds marked with an asterisk (*) were considered by the working groups, but monographs were not prepared because adequate data on carcinogenicity were not available.

A

Acetamide	**7**, 197	
Acetylsalicyclic acid*		
Acridine orange	**16**, 145	
Acriflavinium chloride	**13**, 31	
Acrolein	**19**, 479	
Acrylic acid	**19**, 47	
Acrylic fibres	**19**, 86	
Acrylonitrile	**19**, 73	
Acrylonitrile-butadiene-styrene copolymers	**19**, 91	
Actinomycin C*		
Actinomycins	**10**, 29	
Adipic acid*		
Adriamycin	**10**, 43	
Aflatoxins	**1**, 145	(corr. **7**, 319)
		(corr. **8**, 349)
	10, 51	
Aldrin	**5**, 25	
Amaranth	**8**, 41	
5-Aminoacenaphthene	**16**, 243	
2-Aminoanthraquinone	**27**, 191	
para-Aminoazobenzene	**8**, 53	
ortho-Aminoazotoluene	**8**, 61	(corr. **11**, 295)
para-Amonobenzoic acid	**16**, 249	
4-Aminobiphenyl	**1**, 74	(corr. **10**, 343)
1-Amino-2-methylanthraquinone	**27**, 199	
2-Amino-5-(5-nitro-2-furyl)-1,3,4-thiadiazole	**7**, 143	
4-Amino-2-nitrophenol	**16**, 43	
2-Amino-4-nitrophenol*		
2-Amino-5-nitrophenol*		
6-Aminopenicillanic acid*		
Amitrole	**7**, 31	
Amobarbital sodium*		
Anaesthetics, volatile	**11**, 285	

Aniline	**4,** 27	(corr. **7,** 320)
	27, 39	
Aniline hydrochloride	**27,** 40	
ortho-Anisidine and and its hydrochloride	**27,** 63	
para-Anisidine and its hydrochloride	**27,** 65	
Anthranilic acid	**16,** 265	
Apholate	**9,** 31	
Aramite ®	**5,** 39	
Arsenic and arsenic compounds	**1,** 41	
	2, 48	
	23, 39	

    Arsanilic acid
    Arsenic pentoxide
    Arsenic sulphide
    Arsenic trioxide
    Arsine
    Calcium arsenate
    Dimethylarsinic acid
    Lead arsenate
    Methanearsonic acid, disodium salt
    Methanearsonic acid, monosodium salt
    Potassium arsenate
    Potassium arsenite
    Sodium arsenate
    Sodium arsenite
    Sodium cacodylate

Asbestos	**2,** 17	(corr. **7,** 319)
	14	(corr. **15,** 341)
		(corr. **17,** 351)

    Actinolite
    Amosite
    Anthophyllite
    Chrysotile
    Crocidolite
    Tremolite

Asiaticoside*		
Auramine	**1,** 69	(corr.**7,** 319)
Aurothioglucose	**13,** 39	
5-Azacytidine	**26,** 37	
Azaserine	**10,** 73	(corr. **12,** 271)
Azathioprine	**26,** 47	
Aziridine	**9,** 37	
2-(1-Aziridinyl)ethanol	**9,** 47	
Aziridyl benzoquinone	**9,** 51	

Azobenzene **8**, 75

B

Benz[c]acridine	**3**, 241	
Benz[a]anthracene	**3**, 45	
Benzene	**7**, 203	(corr. **11**, 295)
Benzidine	**1**, 80	
Benzo[b]fluoranthene	**3**, 69	
Benzo[j]fluoranthene	**3**, 82	
Benzo[a]pyrene	**3**, 91	
Benzo[e]pyrene	**3**, 137	
Benzyl chloride	**11**, 217	(corr. **13**, 243)
Benzyl violet 4B	**16**, 153	
Beryllium and beryllium compounds	**1**, 17	
	23, 143	(corr. **25**, 391)

    Bertrandite
    Beryllium acetate
    Beryllium acetate, basic
    Beryllium-aluminium alloy
    Beryllium carbonate
    Beryllium chloride
    Beryllium-copper alloy
    Beryllium-copper-cobalt alloy
    Beryllium fluoride
    Beryllium hydroxide
    Beryllium-nickel alloy
    Beryllium oxide
    Beryllium phosphate
    Beryllium silicate
    Beryllium sulphate and its tetrahydrate
    Beryl ore
    Zinc beryllium silicate

Bis(1-aziridinyl)morpholinophosphine sulphide	**9**, 55	
Bis(2-chloroethyl)ether	**9**, 117	
N,N-Bis(2-chloroethyl)-2-naphthylamine	**4**, 119	
Bischloroethyl nitrosourea (BCNU)	**26**, 79	
Bis(2-chloroisopropyl)ether*		
1,2-Bis(chloromethoxy)ethane	**15**, 31	
1,4-Bis(chloromethoxymethyl)benzene	**15**, 37	
Bis(chloromethyl)ether	**4**, 231	(corr. **13**, 243)
Bleomycins	**26**, 97	
Blue VRS	**16**, 163	

Boot and shoe manufacture and repair	**25,** 249	
Brilliant blue FCF diammonium and disodium salts	**16,** 171	
1,4-Butanediol dimethanesulphonate (Myleran)	**4,** 247	
Butyl-*cis*-9,10-epoxystearate*		
β-Butyrolactone	**11,** 225	
γ-Butyrolactone	**11,** 231	

C

Cadmium and cadmium compounds	**2,** 74	
	11, 39	(corr. **27,** 321)
Cadmium acetate		
Cadmium chloride		
Cadmium oxide		
Cadmium sulphate		
Cadmium sulphide		
Calcium cyclamate	**22,** 58	(corr. **25,** 391)
Calcium saccharin	**22,** 120	(corr. **25,** 391)
Cantharidin	**10,** 79	
Caprolactam	**19,** 115	
Carbaryl	**12,** 37	
Carbon tetrachloride	**1,** 53	
	20, 371	
Carmoisine	**8,** 83	
Carpentry and joinery	**25,** 139	
Catechol	**15,** 155	
Chlorambucil	**9,** 125	
	26, 115	
Chloramphenicol	**10,** 85	
Chlordane	**20,** 45	(corr. **25,** 391)
Chlordecone (Kepone)	**20,** 67	
Chlorinated dibenzodioxins	**15,** 41	
Chlormadinone acetate	**6,** 149	
	21, 365	
Chlorobenzilate	**5,** 75	
1-(2-Chloroethyl)-3-cyclohexyl-1-nitrosourea (CCNU)	**26,** 137	
Chloroform	**1,** 61	
	20, 401	
Chloromethyl methyl ether	**4,** 239	
4-Chloro-*ortho*-phenylenediamine	**27,** 81	
4-Chloro-*meta*-phenylenediamine	**27,** 82	
Chloroprene	**19,** 131	
Chloropropham	**12,** 55	

Chloroquine	**13,** 47	
para-Chloro-ortho-toluidine and its hydrochloride	**16,** 277	
5-Chloro-ortho-toluidine*		
Chlorotrianisene	**21,** 139	
Chlorpromazine*		
Cholesterol	**10,** 99	
Chromium and chromium compounds	**2,** 100	
	23, 205	
Barium chromate		
Basic chromic sulphate		
Calcium chromate		
Chromic acetate		
Chromic chloride		
Chromic oxide		
Chromic phosphate		
Chromite ore		
Chromium carbonyl		
Chromium potassium sulphate		
Chromium sulphate		
Chromium trioxide		
Cobalt-chromium alloy		
Ferrochromium		
Lead chromate		
Lead chromate oxide		
Potassium chromate		
Potassium dichromate		
Sodium chromate		
Sodium dichromate		
Strontium chromate		
Zinc chromate		
Zinc chromate hydroxide		
Zinc potassium chromate		
Zinc yellow		
Chrysene	**3,** 159	
Chrysoidine	**8,** 91	
C.I. Disperse Yellow 3	**8,** 97	
Cinnamyl anthranilate	**16,** 287	
Cisplatin	**26,** 151	
Citrus Red No. 2	**8,** 101	(corr. **19,** 495)
Clofibrate	**24,** 39	
Clomiphene and its citrate	**21,** 551	
Conjugated oestrogens	**21,** 147	
Copper 8-hydroxyquinoline	**15,** 103	
Coumarin	**10,** 113	

meta-Cresidine	**27,** 91	
para-Cresidine	**27,** 92	
Cycasin	**1,** 157	(corr. **7,** 319)
	10, 121	
Cyclamic acid	**22,** 55	(corr. **25,** 391)
Cyclochlorotine	**10,** 139	
Cyclohexylamine	**22,** 59	(corr. **25,** 391)
Cyclophosphamide	**9,** 135	
	26, 165	

D

Dacarbazine	**26,** 203
2,4-D and esters	**15,** 111
D & C Red No. 9	**8,** 107
Dapsone	**24,** 59
Daunomycin	**10,** 145
DDT and associated substances	**5,** 83 (corr. **7,** 320)
DDD (TDE)	
DDE	
Diacetylaminoazotoluene	**8,** 113
N,N'-Diacetylbenzidine	**16,** 293
Diallate	**12,** 69
2,4-Diaminoanisole and its sulphate	**16,** 51
	27, 103
2,5-Diaminoanisole*	
4,4'-Diaminodiphenyl ether	**16,** 301
1,2-Diamino-4-nitrobenzene	**16,** 63
1,4-Diamino-2-nitrobenzene	**16,** 73
2,6-Diamino-3-(phenylazo)pyridine and its hydrochloride	**8,** 117
2,4-Diaminotoluene	**16,** 83
2,5-Diaminotoluene and its sulphate	**16,** 97
Diazepam	**13,** 57
Diazomethane	**7,** 223
Dibenz[a,h]acridine	**3,** 247
Dibenz[a,j]acridine	**3,** 254
Dibenz[a,h]anthracene	**3,** 178
7H-Dibenzo[c,g]carbazole	**3,** 260
Dibenzo[h,rst]pentaphene	**3,** 197
Dibenzo[a,e]pyrene	**3,** 201
Dibenzo[a,h,]pyrene	**3,** 207
Dibenzo[a,i]pyrene	**3,** 215
Dibenzo[a,l]pyrene	**3,** 224

1,2-Dibromo-3-chloropropane	**15,** 139	
	20, 83	
ortho-Dichlorobenzene	**7,** 231	
para-Dichlorobenzene	**7,** 231	
3,3'-Dichlorobenzidine	**4,** 49	
trans-1,4-Dichlorobutene	**15,** 149	
3,3'-Dichloro-4,4'-diaminodiphenyl ether	**16,** 309	
1,2-Dichloroethane	**20,** 429	
Dichloromethane	**20,** 449	
Dichlorvos	**20,** 97	
Dicyclohexylamine	**22,** 60	(corr. **25,** 391)
Dieldrin	**5,** 125	
Dienoestrol	**21,** 161	
Diepoxybutane	**11,** 115	(corr. **12,** 271)
1,2-Diethylhydrazine	**4,** 153	
Diethylstilboestrol	**6,** 55	
	21, 173	(corr. **23,** 417)
Diethylstilboestrol dipropionate	**21,** 175	
Diethyl sulphate	**4,** 277	
Diglycidyl resorcinol ether	**11,** 125	
Dihydrosafrole	**1,** 170	
	10, 233	
Dihydroxybenzenes	**15,** 155	
Dihydroxymethylfuratrizine	**24,** 77	
Dimethisterone	**6,** 167	
	21, 377	
Dimethoate*		
Dimethoxane	**15,** 177	
3,3'-Dimethoxybenzidine (*ortho*-Dianisidine)	**4,** 41	
para-Dimethylaminoazobenzene	**8,** 125	
para-Dimethylaminobenzenediazo sodium sulphonate	**8,** 147	
trans-2[(Dimethylamino)methylimino]-5-[2-(5-nitro-2-furyl)vinyl]-1,3,4-oxadiazole	**7,** 147	
3,3'-Dimethylbenzidine (*ortho*-Tolidine)	**1,** 87	
Dimethylcarbamoyl chloride	**12,** 77	
1,1-Dimethylhydrazine	**4,** 137	
1,2-Dimethylhydrazine	**4,** 145	(corr. **7,** 320)
Dimethyl sulphate	**4,** 271	
Dimethylterephthalate*		
Dinitrosopentamethylenetetramine	**11,** 241	
1,4-Dioxane	**11,** 247	
2,4'-Diphenyldiamine	**16,** 313	
Diphenylthiohydantoin*		
Disulfiram	**12,** 85	

Dithranol	**13,** 75	
Dulcin	**12,** 97	

E

Endrin	**5,** 157	
Enflurane*		
Eosin and its disodium salt	**15,** 183	
Epichlorohydrin	**11,** 131	(corr. **18,** 125)
		(corr. **26,** 387)
1-Epoxyethyl-3,4-epoxycyclohexane	**11,** 141	
3,4-Epoxy-6-methylcyclohexylmethyl-3,4-epoxy-6-methylcyclohexane carboxylate	**11,** 147	
cis-9,10-Epoxystearic acid	**11,** 153	
Ethinyloestradiol	**6,** 77	
	21, 233	
Ethionamide	**13,** 83	
Ethyl acrylate	**19,** 57	
Ethylene	**19,** 157	
Ethylene dibromide	**15,** 195	
Ethylene oxide	**11,** 157	
Ethylene sulphide	**11,** 257	
Ethylenethiourea	**7,** 45	
Ethyl methanesulphonate	**7,** 245	
Ethyl selenac	**12,** 107	
Ethyl tellurac	**12,** 115	
Ethynodiol diacetate	**6,** 173	
	21, 387	
Evans blue	**8,** 151	

F

Fast green FCF	**16,** 187	
Ferbam	**12,** 121	(corr. **13,** 243)
Fluorescein and its disodium salt*		
Fluorides (inorganic, used in drinking-water and dental preparations)	**27,** 237	
Fluorspar		
Fluosilicic acid		
Sodium fluoride		
Sodium monofluorophosphate		
Sodium silicofluoride		
Stannous fluoride		

5-Fluorouracil	**26,** 217	
Formaldehyde*		
2-(2-Formylhydrazino)-4-(5-nitro-2-furyl)-thiazole	**7,** 151	(corr. **11,** 295)
The furniture and cabinet-making industry	**25,** 99	
Fusarenon-X	**11,** 169	

G

Glycidaldehyde	**11,** 175
Glycidyl oleate	**11,** 183
Glycidyl stearate	**11,** 187
Griseofulvin	**10,** 153
Guinea green B	**16,** 199

H

Haematite	**1,** 29
Haematoxylin*	
Hair dyes, epidemiology of	**16,** 29
	27, 307
Halothane*	
Heptachlor and its epoxide	**5,** 173
	20, 129
Hexachlorobenzene	**20,** 155
Hexachlorobutadiene	**20,** 179
Hexachlorocyclohexane (α-, β-, δ-, ε-, technical HCH and lindane)	**5,** 47
	20, 195
Hexachloroethane	**20,** 467
Hexachlorophene	**20,** 241
Hexamethylenediamine*	
Hexamethylphosphoramide	**15,** 211
Hycanthone and its mesylate	**13,** 91
Hydralazine and its hydrochloride	**24,** 85
Hydrazine	**4,** 127
Hydroquinone	**15,** 155
4-Hydroxyazobenzene	**8,** 157
17α-Hydroxyprogesterone caproate	**21,** 399
8-Hydroxyquinoline	**13,** 101
Hydroxysenkirkine	**10,** 265

I

Indenol[1,2,3-*cd*]pyrene	**3,** 229
Iron-dextran complex	**2,** 161
Iron-dextrin complex	**2,** 161 (corr. **7,** 319)
Iron oxide	**1,** 29
Iron sorbitol-citric acid complex	**2,** 161
Isatidine	**10,** 269
Isoflurane*	
Isonicotinic acid hydrazide	**4,** 159
Isophosphamide	**26,** 237
Isoprene*	
Isopropyl alcohol	**15,** 223
Isopropyl oils	**15,** 223
Isosafrole	**1,** 169
	10, 232

J

Jacobine	**10,** 275

L

Lasiocarpine	**10,** 281
Lead and lead compounds	**1,** 40 (corr. **7,** 319)
	2, 52 (corr. **8,** 349)
	2, 150
	23, 39, 205, 325
Lead acetate and its trihydrate	
Lead carbonate	
Lead chloride	
Lead naphthenate	
Lead nitrate	
Lead oxide	
Lead phosphate	
Lead subacetate	
Lead tetroxide	
Tetraethyllead	
Tetramethyllead	
The leather goods manufacturing industry (other than boot and shoe manufacture and tanning)	**25,** 279
The leather tanning and processing industries	**25,** 201

Ledate	**12**, 131	
Light green SF	**16**, 209	
Lindane	**5**, 47	
	20, 196	
The lumber and sawmill industries (including logging)	**25**, 49	
Luteoskyrin	**10**, 163	
Lynoestrenol	**21**, 407	
Lysergide*		

M

Magenta	**4**, 57	(corr. **7**, 320)
Maleic hydrazide	**4**, 173	(corr. **18**, 125)
Maneb	**12**, 137	
Mannomustine and its dihydrochloride	**9**, 157	
Medphalan	**9**, 168	
Medroxyprogesterone acetate	**6**, 157	
	21, 417	(corr. **25**, 391)
Megestrol acetate	**21**, 431	
Melphalan	**9**, 167	
6-Mercaptopurine	**26**, 249	
Merphalan	**9**, 169	
Mestranol	**6**, 87	
	21, 257	(corr. **25**, 391)
Methacrylic acid*		
Methallenoestril*		
Methotrexate	**26**, 267	
Methoxsalen	**24**, 101	
Methoxychlor	**5**, 193	
	20, 259	
Methoxyflurane*		
Methyl acrylate	**19**, 52	
2-Methylaziridine	**9**, 61	
Methylazoxymethanol	**10**, 121	
Methylazoxymethanol acetate	**1**, 164	
	10, 131	
Methyl bromide*		
Methyl carbamate	**12**, 151	
N-Methyl-N,4-dinitrosoaniline	**1**, 141	
4,4'-Methylene bis(2-chloroaniline)	**4**, 65	(corr. **7**, 320)
4,4'-Methylene bis(N,N-dimethyl)benzenamine	**27**, 119	
4,4'-Methylene bis(2-methylaniline)	**4**, 73	
4,4'-Methylenedianiline	**4**, 79	(corr. **7**, 320)

4,4'-Methylenediphenyl diisocyanate	**19,** 314	
Methyl iodide	**15,** 245	
Methyl methacrylate	**19,** 187	
Methyl methanesulphonate	**7,** 253	
2-Methyl-1-nitroanthraquinone	**27,** 205	
N-Methyl-N'-nitro-N-nitrosoguanidine	**4,** 183	
Methyl protoanemonin*		
Methyl red	**8,** 161	
Methyl selenac	**12,** 161	
Methylthiouracil	**7,** 53	
Metronidazole	**13,** 113	
Mirex	**5,** 203	
	20, 283	
Mitomycin C	**10,** 171	
Modacrylic fibres	**19,** 86	
Monocrotaline	**10,** 291	
Monuron	**12,** 167	
5-(Morpholinomethyl)-3-[(5-nitrofurfurylidene)amino]-2-oxazolidinone	**7,** 161	
Mustard gas	**9,** 181	(corr. **13,** 243)

N

Nafenopin	**24,** 125	
1,5-Naphthalenediamine	**27,** 127	
1,5-Naphthalene diisocyanate	**19,** 311	
1-Naphthylamine	**4,** 87	(corr. **8,** 349)
		(corr. **22,** 187)
2-Naphthylamine	**4,** 97	
Native carrageenans	**10,** 181	(corr. **11,** 295)
Nickel and nickel compounds	**2,** 126	(corr. **7,** 319)
	11, 75	
Nickel acetate and its tetrahydrate		
Nickel ammonium sulphate		
Nickel carbonate		
Nickel carbonyl		
Nickel chloride		
Nickel-gallium alloy		
Nickel hydroxide		
Nickelocene		
Nickel oxide		
Nckel subsulphide		
Nickel sulphate		

Niridazole	**13,** 123	
5-Nitroacenaphthene	**16,** 319	
5-Nitro-*ortho*-anisidine	**27,** 133	
4-Nitrobiphenyl	**4,** 113	
5-Nitro-2-furaldehyde semicarbazone	**7,** 171	
1[(5-Nitrofurfurylidene)amino]-2-imidazolidinone	**7,** 181	
N-[4-(5-Nitro-2-furyl)-2-thiazolyl]acetamide	**1,** 181	
	7, 185	
Nitrogen mustard and its hydrochloride	**9,** 193	
Nitrogen mustard N-oxide and its hydrochloride	**9,** 209	
Nitrosatable drugs	**24,** 297	
N-Nitrosodi-n-butylamine	**4,** 197	
	17, 51	
N-Nitrosodiethanolamine	**17,** 77	
N-Nitrosodiethylamine	**1,** 107	(corr. **11,** 295)
	17, 83	(corr. **23,** 419)
N-Nitrosodimethylamine	**1,** 95	
	17, 125	(corr. **25,** 391)
N-Nitrosodiphenylamine	**27,** 213	
para-Nitrosodiphenylamine	**27,** 227	
N-Nitrosodi-n-propylamine	**17,** 177	
N-Nitroso-N-ethylurea	**1,** 135	
	17, 191	
N-Nitrosofolic acid	**17,** 217	
N-Nitrosohydroxyproline	**17,** 304	
N-Nitrosomethylethylamine	**17,** 221	
N-Nitroso-N-methylurea	**1,** 125	
	17, 227	
N-Nitroso-N-methylurethane	**4,** 211	
N-Nitrosomethylvinylamine	**17,** 257	
N-Nitrosomorpholine	**17,** 263	
N'-Nitrosonornicotine	**17,** 281	
N-Nitrosopiperidine	**17,** 287	
N-Nitrosoproline	**17,** 303	
N-Nitrosopyrrolidine	**17,** 313	
N-Nitrososarcosine	**17,** 327	
N-Nitrosarcosine ethyl ester*		
Nitroxoline*		
Nivalenol*		
Norethisterone and its acetate	**6,** 179	
	21, 441	
Norethynodrel	**6,** 191	
	21, 461	(corr. **25,** 391)

Norgestrel	**6**, 201
	21, 479
Nylon 6	**19**, 120
Nylon 6/6*	

O

Ochratoxin A	**10**, 191
Oestradiol-17β	**6**, 99
	21, 279
Oestradiol 3-benzoate	**21**, 281
Oestradiol dipropionate	**21**, 283
Oestradiol mustard	**9**, 217
Oestradiol-17β-valerate	**21**, 284
Oestriol	**6**, 117
	21, 327
Oestrone	**6**, 123
	21, 343 (corr. **25**, 391)
Oestrone benzoate	**21**, 345
Oil Orange SS	**8**, 165
Orange I	**8**, 173
Orange G	**8**, 181
Oxazepam	**13**, 58
Oxymetholone	**13**, 131
Oxyphenbutazone	**13**, 185

P

Panfuran S (Dihydroxymethylfuratrizine)	**24**, 77
Parasorbic acid	**10**, 199 (corr. **12**, 271)
Patulin	**10**, 205
Penicilic acid	**10**, 211
Pentachlorophenol	**20**, 303
Pentobarbital sodium*	
Phenacetin	**13**, 141
	24, 135
Phenazopyridine and its hydrochloride	**24**, 163
Phenelzine and its sulphate	**24**, 175
Phenicarbazide	**12**, 177
Phenobarbital and its sodium salt	**13**, 157
Phenoxybenzamine and its hydrochloride	**9**, 223
	24, 185

Phenylbutazone	**13**, 183	
ortho-Phenylenediamine*		
meta-Phenylenediamine and its hydrochloride	**16**, 111	
para-Phenylenediamine and its hydrochloride	**16**, 125	
N-Phenyl-2-naphthylamine	**16**, 325	(corr. **25**, 391)
N-Phenyl-*para*-phenylenediamine*		
Phenytoin and its sodium salt	**13**, 201	
Piperazine oestrone sulphate	**21**, 148	
Polyacrylic acid	**19**, 62	
Polybrominated biphenyls	**18**, 107	
Polychlorinated biphenyls	**7**, 261	
	18, 43	
Polychloroprene	**19**, 141	
Polyethylene (low-density and high-density)	**19**, 164	
Polyethylene terephthalate*		
Polyisoprene*		
Polymethylene polyphenyl isocyanate	**19**, 314	
Polymethyl methacrylate	**19**, 195	
Polyoestradiol phosphate	**21**, 286	
Polypropylene	**19**, 218	
Polystyrene	**19**, 245	
Polytetrafluoroethylene	**19**, 288	
Polyurethane foams (flexible and rigid)	**19**, 320	
Polyvinyl acetate	**19**, 346	
Polyvinyl alcohol	**19**, 351	
Polyvinyl chloride	**7**, 306	
	19, 402	
Polyvinylidene fluoride*		
Polyvinyl pyrrolidone	**19**, 463	
Ponceau MX	**8**, 189	
Ponceau 3R	**8**, 199	
Ponceau SX	**8**, 207	
Potassium bis(2-hydroxyethyl)dithiocarbamate	**12**, 183	
Prednisone	**26**, 293	
Procarbazine hydrochloride	**26**, 311	
Proflavine and its salts	**24**, 195	
Progesterone	**6**, 135	
	21, 491	
Pronetalol hydrochloride	**13**, 227	(corr. **16**, 387)
1,3-Propane sultone	**4**, 253	(corr. **13**, 243)
		(corr. **20**, 591)
Propham	**12**, 189	
β-Propiolactone	**4**, 259	(corr. **15**, 341)
n-Propyl carbamate	**12**, 201	

Propylene	**19**, 213
Propylene oxide	**11**, 191
Propylthiouracil	**7**, 67
The pulp and paper industry	**25**, 157
Pyrazinamide*	
Pyrimethamine	**13**, 233
Pyrrolizidine alkaloids	**10**, 333

Q

Quinoestradol*	
Quinoestrol*	
para-Quinone	**15**, 255
Quintozene (pentachloronitrobenzene)	**5**, 211

R

Reserpine	**10**, 217
	24, 211 (corr. **26**, 387)
Resorcinol	**15**, 155
Retrorsine	**10**, 303
Rhodamine B	**16**, 221
Rhodamine 6G	**16**, 233
Riddelliine	**10**, 313
Rifampicin	**24**, 243

S

Saccharated iron oxide	**2**, 161
Saccharin	**22**, 111 (corr. **25**, 391)
Safrole	**1**, 169
	10, 231
Scarlet red	**8**, 217
Selenium and selenium compounds	**9**, 245 (corr. **12**, 271)
Semicarbazide hydrochloride	**12**, 209 (corr. **16**, 387)
Seneciphylline	**10**, 319
Senkirkine	**10**, 327

Sodium cyclamate	**22**, 56	(corr. **25**, 391)
Sodium diethyldithiocarbamate	**12**, 217	
Sodium equilin sulphate	**21**, 148	
Sodium oestrone sulphate	**21**, 147	
Sodium saccharin	**22**, 113	(corr. **25**, 391)
Soot, tars and mineral oils	**3**, 22	
Spironolactone	**24**, 259	
Sterigmatocystin	**1**, 175	
	10, 245	
Streptozotocin	**4**, 221	
	17, 337	
Styrene	**19**, 231	
Styrene-acrylonitrile copolymers	**19**, 97	
Styrene-butadiene copolymers	**19**, 252	
Styrene oxide	**11**, 201	
	19, 275	
Succinic anhydride	**15**, 265	
Sudan I	**8**, 225	
Sudan II	**8**, 233	
Sudan III	**8**, 241	
Sudan brown RR	**8**, 249	
Sudan red 7B	**8**, 253	
Sulfafurazole (sulphisoxazole)	**24**, 275	
Sulfamethoxazole	**24**, 285	
Sunset yellow FCF	**8**, 257	

T

2,4,5-T and esters	**15**, 273	
Tannic acid	**10**, 253	(corr. **16**, 387)
Tannins	**10**, 254	
Terephthalic acid*		
Terpene polychlorinates (Strobane®)	**5**, 219	
Testosterone	**6**, 209	
	21, 519	
Testosterone oenanthate	**21**, 521	
Testosterone propionate	**21**, 522	
2,2',5,5'-Tetrachlorobenzidine	**27**, 141	
1,1,2,2-Tetrachloroethane	**20**, 477	
Tetrachloroethylene	**20**, 491	
Tetrafluoroethylene	**19**, 285	
Thioacetamide	**7**, 77	

4,4'-Thiodianiline	**16,** 343
	27, 147
Thiouracil	**7,** 85
Thiourea	**7,** 95
Thiram	**12,** 225
2,4-Toluene diisocyanate	**19,** 303
2,6-Toluene diisocyanate	**19,** 303
ortho-Toluenesulphonamide	**22,** 121
ortho-Toluidine and its hydrochloride	**16,** 349
	27, 155
Toxaphene (polychlorinated camphenes)	**20,** 327
Treosulphan	**26,** 341
1,1,1-Trichloroethane	**20,** 515
1,1,2-Trichloroethane	**20,** 533
Trichloroethylene	**11,** 263
	20, 545
2,4,5- and 2,4,6-Trichlorophenols	**20,** 349
Trichlorotriethylamine hydrochloride	**9,** 229
Trichlorphon*	
Triethylene glycol diglycidyl ether	**11,** 209
2,4,5-Trimethylaniline and its hydrochloride	**27,** 177
2,4,6-Trimethylaniline and its hydrochloride	**27,** 178
Tris(aziridinyl)-para-benzoquinone (Triaziquone)	**9,** 67
Tris(1-aziridinyl)phosphine oxide	**9,** 75
Tris(1-aziridinyl)phosphine sulphide (Thiotepa)	**9,** 85
2,4,6-Tris(1-aziridinyl)-s-triazine	**9,** 95
1,2,3-Tris(chloromethoxy)propane	**15,** 301
Tris(2,3-dibromopropyl) phosphate	**20,** 575
Tris(2-methyl-1-aziridinyl)phosphine oxide	**9,** 107
Trypan blue	**8,** 267

U

Uracil mustard	**9,** 235
Urethane	**7,** 111

V

Vinblastine sulphate	**26,** 349
Vincristine sulphate	**26,** 365
Vinyl acetate	**19,** 341

THE LIBRARY
UNIVERSITY OF CALIFORNIA
San Francisco
666-2334

THIS BOOK IS DUE ON THE LAST DATE STAMPER BELOW
Books not returned on time are subject to fines according to the Library Lending Code. A renewal may be made on certain materials. For details consult Lending Code.

14 DAY SEP 9 1982 RETURNED AUG 2 6 1982 14 DAY OCT 2 8 1982 RETURNED OCT 2 1 1982 14 DAY NOV 1 2 1982 RETURNED NOV 1 0 1982	14 DAY MAY 2 5 1992 RETURNED MAY 2 6 1992 14 DAY DEC - 2 1992 renewed by phone due 12/30/92 RETURNED DEC 2 9 1992	Series 4128